Power Systems-On-Chip

Power Systems-On-Chip

Practical Aspects of Design

Edited by

Bruno Allard

iSTE

WILEY

First published 2016 in Great Britain and the United States by ISTE Ltd and John Wiley & Sons, Inc.

ISTE Ltd
27-37 St George's Road
London SW19 4EU
UK

www.iste.co.uk

John Wiley & Sons, Inc.
111 River Street
Hoboken, NJ 07030
USA

www.wiley.com

Library of Congress Control Number: 2016950149

British Library Cataloguing-in-Publication Data
A CIP record for this book is available from the British Library
ISBN 978-1-78630-081-2

Contents

Preface

The Internet of things (IoT), as well as any portable system, brings to the fore the issue of powering the system. Cutting energy consumption is mandatory to limit the weight of batteries and/or extend the lifetime of the device before recharging. It is well known that electricity needs to be adequately converted to the precise requirements of the consumer. Power management units are required. Unfortunately, their number is expected to increase as well as their variety. A change of paradigm calls for integration to support dissemination. Ultimate integration of a power converter means either embedding all components in silicon or stacking them closely in a 3D manner. High switching frequency is also mandatory for reducing the values of passive devices, hence the need to make the power system management smaller. Silicon integration places a preference on switched-capacitor converters with respect to inductive converters because of the magnetic devices, especially when a magnetic material is involved, which is the case for high power density. The latter components may be fabricated in an integrated approach but the fabrication process is not compatible in essence and cost with standard silicon process. A heterogeneous integration must be considered, which is a cost-effective solution.

Such solutions experimenting, demonstrating and reporting about many technological issues, control issues or design issues have to be addressed jointly. Obviously, the success of a power system-on-chip solution depends on a difficult trade-off. The focus of this book is to pragmatically address all the facets of the design trade-off toward a successful fabrication of non-isolated high switching frequency DC/DC converters. Readers will find a clear message about many issues necessary to understand advanced solutions. It is not a textbook but references to textbooks will be given. It is not a collection of research articles: industrial reality is mostly considered. It is not a detailed analysis of one best solution but a comprehensive presentation of typical solutions with insights on the reason of some success of given solutions and the necessary background to understand them. It is not a CMOS design of this or that book as CMOS is reduced to a minimum to give room essentially to trade-offs, design practice and industrial analysis of results.

The *Introduction* talks about a paradigm in power management inside advanced and IoT products. From a case shown in the introduction, conclusions about candidate paradigms and their limitations can be drawn. One paradigm is proposed according to development in use in industry. This book highlights this paradigm.

Chapter 1, "*Control Strategies and CAD Approach*", by P. Alou, J. A. Oliver, J. A. Cobos, B. Labbe, B. Allard, A. Radic and A. Prodic, explains a system-level approach for analysis, simulation and optimization of a converter with the necessary models and tools. Practical cases are illustrated to demonstrate the superiority of a systematic approach over a lack of convenient approach, as compared to a succession of local optimization or a suite of truncated analyses.

Chapter 2, "*Magnetic Components for Increased Power Density*", by C. O Mathuna, S. Kulkarni and C. Martin, addresses inductors and transformers necessary for inductive DC/DC converters. An engineer must face a cost-to-performance challenge, which is even more difficult to meet if availability of values is considered. The offer in magnetic devices is quite large and ever evolving. The chapter wishes to recall the quantities that prevail in dimensioning a magnetic device and choosing a technology. Now, mature integrated solutions are discussed as nothing else can be envisaged in an IoT device, supposed to be as small as possible. Many device details are given as well as perspectives.

Chapter 3, "*Dielectric Components for Increased Power Density*" by F. Voiron shows that capacitors through their banal appearance are devices of great technologies. An IoT device does not offer many possibilities for selecting capacitor technologies. The limitations of existing, small foot-print, low form factor technologies are recalled just to exhibit the specifications of a dream solution. A near highly satisfying technology is then detailed along with the method to practically implement such fabricated capacitors. Many device details are given as well as perspectives.

Chapter 4, "*On-board Power Management DC/DC Inductive Converter*", by B. Labbé, covers the case of local, stand-alone PMUs. These are placed between the battery (or input source of energy) and the consumer integrated circuits and peripherals inside the IoT device. This front-end DC/DC converter corresponds to specifications where efficiency is high as well as the transient capabilities. The power density is a consequence but solutions exist to guarantee the former targets with satisfying results on the latter. A full DC/DC example is detailed with simulation and experimental results.

Chapter 5, "*On-Chip Power Management DC/DC Switched-Capacitor Converter*", by G. Pillonnet, T. Souvignet and B. Allard, revisits a first kind of DC/DC converter when high switching frequency is considered. Several books exist on the switched-capacitor converter but the converter is analyzed as stand-alone. It is necessary to revisit the converter in the frame of on-chip integration. A full DC/DC example is detailed with simulation and experimental results.

Chapter 6, "*High-Switching Frequency Inductive DC/DC Converters*", by A. Prodic, Z. Lukic, C. Martin F. Neveu and B. Allard, demonstrates that pushing the operation frequency very high offers an interesting degree of freedom. However the impact is detrimental on the efficiency of the converter unless particular schemes are considered. The chapter extends the practicality of previous ones to cover more advanced materials. The possible schemes to limit losses while taking advantage of higher operating frequency are listed and the ones considered in more mature alternative solutions in industry are discussed. Full DC/DC examples are detailed with simulation and experimental results.

Chapter 7, "*Hybrid and Multi-level converter topologies for on-chip implementation of reduced voltage-swing converters*", by A. Prodic, S.M. Ahsanuzzaman, B. Mahdavikhah, and T. McRae, covers a new kind of converter that are not only switched-capacitor based but also inductor based. To save power density, solutions taken from large and medium power applications are revisited. A mix of both kinds of converters as previously detailed is possible. Hybrid DC/DC conversion should be considered with a renewed state of mind. A full DC/DC example is detailed with simulation and experimental results.

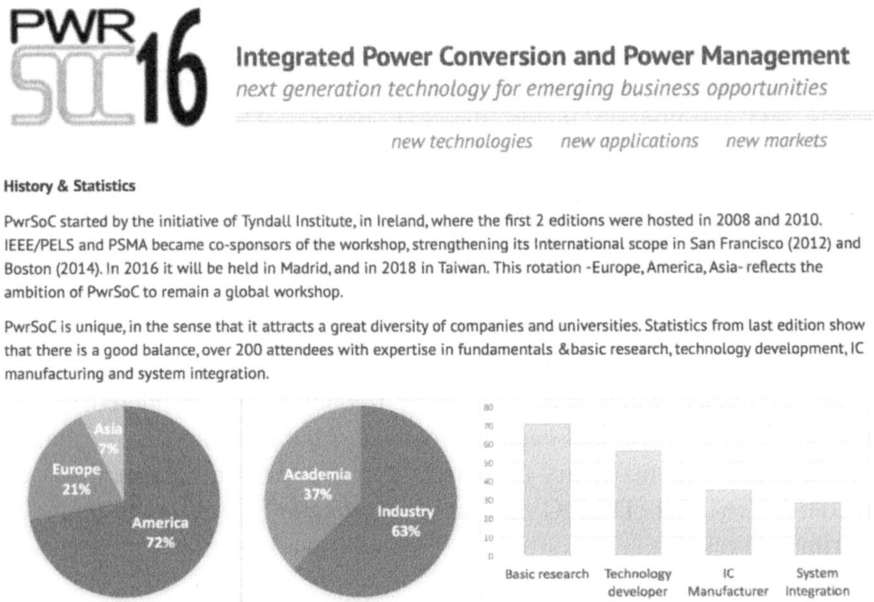

Figure 1. *PowerSoC–SiP dedicated website*

The content of this book and many references have been discussed during sessions of the dedicated workshop for power supply solution integration in system-on-chip or in system-in-package: IEEE PowerSoC-SiP International Workshop. A unique website (http://pwrsocevents.com) (Figure 1) introduces an exciting community of industrial and academic actors in the field.

Bruno ALLARD
September 2016

Introduction

Many electronic products are designed for compactness and each function or block is designed accordingly. Unfortunately, power supplies occupy a significant part of the board and passive components are the major contributors.

Figure I.1 shows two examples of printed circuit boards (PCB) of two mobile devices. The active components of multiple converters (power switches and controllers) are often integrated inside a single or multiple power management integrated circuit (or power module IC) (PMIC), while the reactive components of those converters (inductors and capacitors) are usually discrete. These discrete components often take a very large portion of the overall device volume and/or PCB area and in some cases are among the main contributors to the overall device volume.

Depending on the application, the passives often take between 12% and 80% of the overall PCB area [CHE 05]. In the space-limited applications, the reactive components, especially inductors, are the main obstacles to further minimization of electronics devices and, in some cases, prevent the introduction of novel functional blocks that could potentially add new features to those devices. In the targeted applications, the volume of relatively large inductors compared with that of the capacitors, which is often reflected through their height, is related to a much lower energy storing capacity of the inductors. Comparisons of the energy capacity between the inductors and

---

Introduction written by Bruno ALLARD.

capacitors used in the targeted low-power applications show that, inside the same volume, the capacitors can store between 100 and 1000 times more energy [SEE 10].

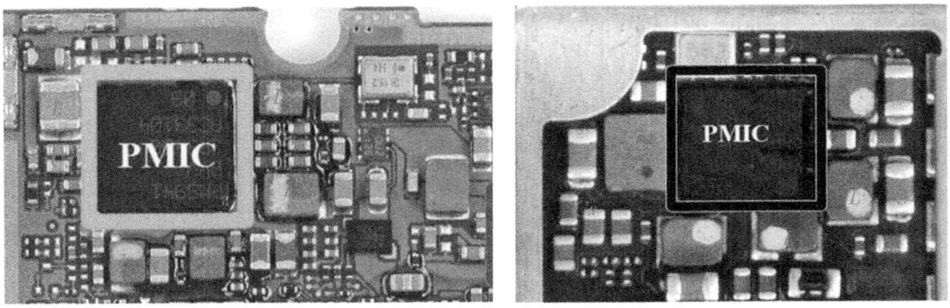

Figure I.1. *Printed circuit boards of iPad air tablet (left) and Nexus 5 cell phone, where the inductors and capacitors are labeled*

It is a common knowledge that an industrial trend is to dramatically improve the power density of the power supply. Innovations cover the following fields: converter architectures, passive devices, control, design and packaging [WAL 13].

Recent literature has dealt with many demonstrations of switched-capacitor converters either on-chip or in-package. Switched-capacitor (SC) converters perform power processing without inductive components and therefore allow for much more compact implementation than their conventional counterparts [SEE 10, SAN 13]. Beyond the impressive results [PAS 15], switched-capacitor converters face a limitation due to the available density of capacitors.

The efficiency of SC converters drastically degrades when required to operate away from the fixed (and optimal) conversion ratio. Also, compared with conventional solutions, the SC suffers from inferior dynamic performance. As such, SC converters are not always the most suitable solution when variable conversion ratio and/or fast transient response are required. This also translates into a limitation in power density [VIL 13] where inductive converters offer a better trade-off at a similar level of efficiency.

This book focuses on high power density inductive buck converters for embedded products: this is called power-system-on-chip [WAL 13], [MAT 14]. A review of high-switching frequency buck converters is available in [NEV 14], [NEV 16]. The following statements may be derived:

– The natural corner frequency of the output filter of the converter correlates with the switching frequency: the high switching frequency is used to enhance the converter transient performances but this is not a driving design indicator so far. Higher switching frequency is intended to reduce the values of passive components.

– More interestingly, the efficiency drops with increasing frequency. The figure must be related to the output-to-input voltage conversion ratio. Selecting the literature results for given ratios, the different efficiencies drop in a similar manner. One result seems to be above the general trend (ratio of 0.5, [BUR 14], [KUR 15]).

– Efficiency profits from thin silicon technologies irrespective of the switching frequency.

– A penalizing effect of thin technologies is the limitation in input voltage range.

There is therefore a real challenge to handle a standard input voltage range (3.3 V) with a thin technology (CMOS 40 nm, 1.2 V). Efficiency in the vicinity of 90 % is reported in the literature, but this is for a high ratio of output-to-input voltage (larger than 0.85) [SON 14]. A challenge is evidently to achieve similar efficiency but for a ratio below 0.3.

Various strategies are reported in the literature to fight power losses inside a non-isolated DC/DC converter, namely a buck converter as considered here.

– Soft switching, either zero-voltage or zero-current schemes, has been applied to integrated buck converters [ABE 07]. Variability and parasitic devices are limitations to the optimization of the architecture. Moreover, a shift in the actual switching frequency complexifies the filtering of noise.

– Multiphase architecture: the current is shared upon several phases and the Joule losses are reduced [ABE 07]. Independent inductors are considered. Moreover, a phase shedding scheme makes it possible to operate the exact amount of silicon necessary at one moment with respect to the output current.

– Multiple inductors may be coupled [WIB 08]. One magnetic device will then be affected by a current waveform at twice the switching frequency. At similar induct values, the ripple is reduced. At similar ripple and induct values, the switching frequency can be lowered, i.e. the switching losses in the active devices are reduced.

– Resonant gate drivers have been experimented [BAT 12a]. The gain on losses is rather limited and does not compensate the penalty on silicon area to accommodate the air-core inductance needed for each single driver.

– Transistor width segmentation is an ultimate solution to adapt, on the fly, the silicon area to the actual output current [BAT 12a]. This necessitates an evaluation of the output current. The scheme superimposes a modulation to the pulse-width modulation generally considered at high switching frequency. A risk exists for the stability of the converter.

– High-voltage MOSFETs create more switching and conduction losses than low-voltage counterparts. Cascode association of low-voltage MOSFETs has been experimented to recreate a high-voltage MOSFET but with lower losses [WEN 08]. For a thin CMOS technology, a cascode power stage is a solid candidate to stand high input voltage [OST 14], [BUR 14].

Figure I.2. *Schematic of a heterogeneous power stage*

Table I.1 lists a set of general specifications. If given to an engineer used to MHz DC/DC converter, he will probably find a reason to select hard switching operation and simple voltage mode control as in [BAT 12b]. Of course, other control methods offer better performances. Regarding the input voltage, it is straightforward to select a convenient standard MOSFET with respect to the input or output voltage. A cascode power stage can also be considered, and in some conditions, it appears a better choice. Using coupled inductors inside a multi-phase architecture makes it possible to reduce the values of passive components for a targeted output current ripple. Obviously, adding the possible solutions is not straightforward. [BAT 12b] reported a 78% peak efficiency at the output power considered in Table I.1 with similar voltage conditions at input and output. The trade-off is not optimal.

Name	Min	Typ.	Max	Unit
V_{IN}	3.0	3.3	3.6	V
V_{OUT}	–	1.2	–	V
I_{LOAD}	50	280	500	mA
I_{SLEEP}	–	15	–	mA
$I_{STANDBY}$	–	150	–	uA
voltage accuracy	–2.5	–	+2.5	%
Dynamic voltage accuracy	–6	–	+6	%
Efficiency at $I_{LOAD,TYP}$	90	–	–	%
Efficiency at $I_{LOAD,MIN}$	80	–	–	%
Efficiency at I_{SLEEP}	70	–	–	%
F_{SW}	100	–	–	MHz
Ambient temperature	–40	55	125	Celsius
Package height	–	1.2	4	mm
Total area	–	–	10	mm^2
Technology platform	CMOS			

Table I.1. *Specifications of the target converter*

1

Control Strategies and CAD Approach

This chapter recalls the fundamentals of the switched-mode power supply control strategy on the one hand, and some general issues on the computer-aided design (CAD) approach on the other. Section 1.2 introduces the fundamentals on the buck, boost and buck-boost non-isolated converters. Relevant issues on MOSFET switching behavior are summarized in section 1.3.1 with emphasis on parameter identification for the system-level analysis of converters with respect to the control strategy. Optimization of the power stage with respect to specifications is presented in section 1.3.3 for the reader's convenience. So far, it is considered that the reader has sufficient background knowledge to understand the operation of non-isolated converters.

The focus is then turned to the transient performances of a given converter. Fast response to the line transient or load transient is a key issue for power density at the highest possible efficiency [COR 15a]. The control strategy is critical with regard to transient performances. A fast converter can limit the output capacitor, for example. Various control strategies may be envisaged, as shown in Figure 1.1. The control strategies are listed in section 1.4 [COR 15d]. Figure 1.1 shows that an extension of the so-called $V^2\mathrm{I}_C$ control generalizes ripple contributions to build a control quantity. The so-called ripple-based control is finding adoption as analogue implementations [CHE 16].

Chapter written by Pedro ALOU, José A. COBOS, Jesus A. OLIVER, Bruno ALLARD, Benôit LABBE, Aleksandar PRODIC and Aleksandar RADIC.

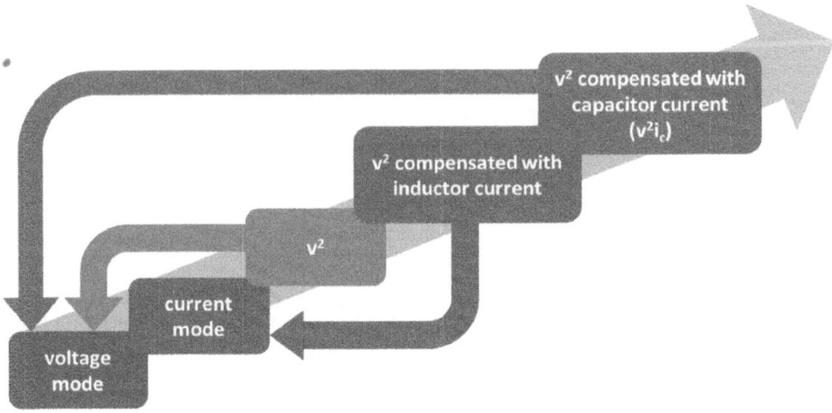

Figure 1.1. *Chart of a classical control strategy for non-isolated DC/DC converters. For the color version of this figure, see www.iste.co.uk/allard/systems.zip*

Analogue versus digital implementation provides insights into the control strategy. A rich literature exists on this topic [GUO 09, LI 12a]. A non-classical control strategy in digital form is presented in section 1.5 as a specific example from this perspective; however, the example covers the important issue of the minimum voltage deviation strategy.

Finally, section 1.6 introduces the fundamentals for a system-level optimization of a given converter architecture. Necessary models as well as optimization concerns are recalled. Examples are provided for the exploration of converter capabilities offered by a CAD design approach.

1.1. Objectives

Power converters that supply microprocessors, digital signal processors (DSP), field programmable gate arrays (FPGA) and similar digital loads must meet very demanding specifications:

– Steady-state specifications: accurate regulation of the supply voltage, very low-voltage ripple ($<1\%$ of V_{OUT} amplitude), high efficiency;

– Fast dynamic response under load steps (Figures 1.2 and 1.3): the converter must meet a tight voltage regulation under aggressive load current steps from a full load to an idle state; it is an effective way of saving energy in the microprocessor, but it is very demanding for the power converter;

– Adaptive voltage positioning (Figure 1.4): the supply voltage depends on the current demanded by the load;

– Fast voltage tracking (Figure 1.5): the supply voltage must follow a dynamic reference. This technique is also used to save energy from the load side, but it becomes very demanding for the power converter. For example, dynamic voltage scaling to supply microprocessors or similar loads [LU 15], [SOT 07], [SU 08a] and [BUR 00] or envelope tracking use in RF amplifiers to minimize losses in the amplifier [CHE 14b].

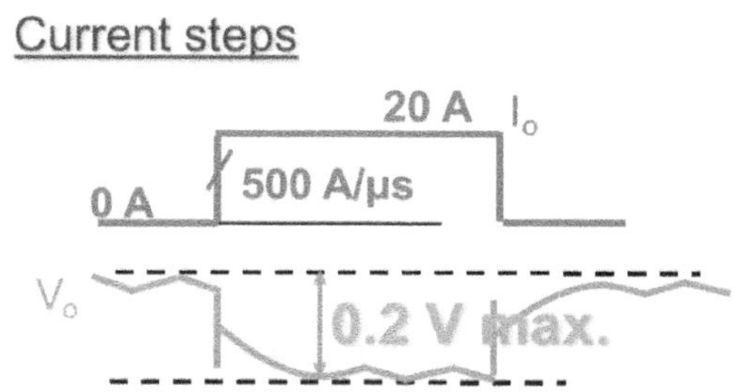

Figure 1.2. *Load-step transient. For the color version of this figure, see www.iste.co.uk/allard/systems.zip*

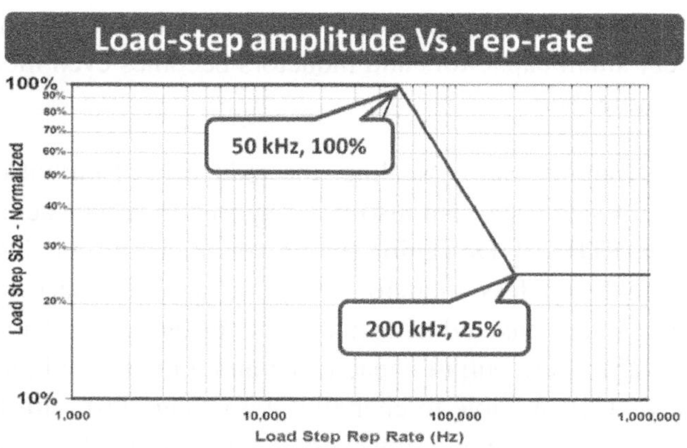

Figure 1.3. *Frequency of the load step for an INTEL microprocessor*

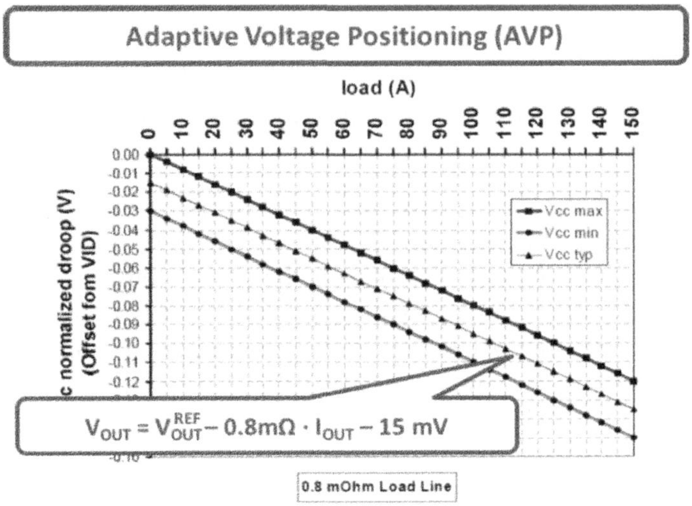

Figure 1.4. *Adaptative voltage positioning. For the color version of this figure, see www.iste.co.uk/allard/systems.zip*

The traditional design of the voltage regulation module (VRM) has been developed with a huge effort to reduce the size and cost of the output capacitor by including a very fast converter with a very low inductance and a very fast control. A slow converter would require huge output capacitors. To achieve this goal, it is required to operate at very high switching frequency and to use very fast control techniques (even nonlinear techniques) to optimize the capabilities of the power stage. Designing integrated power converters, very small capacitors and inductors becomes even more critical to be able to integrate them.

Regarding dynamic voltage scaling, the reduction of output capacitors is not only a cost or size issue, but also it is required to meet tracking time and tracking energy [STR 99]. Designing for a given tracking time, less charge (or discharge) current and less inductor slew rate are necessary if the output capacitance is small, thus reducing the tracking energy and increasing the efficiency. Clearly, a low output capacitance and a relatively high inductance design would be very suitable to changing the voltage fast with low tracking energy and high efficiency. However, there would be two problems encountered with this solution: (1) poor regulation under load current steps and (2) large size of the inductor.

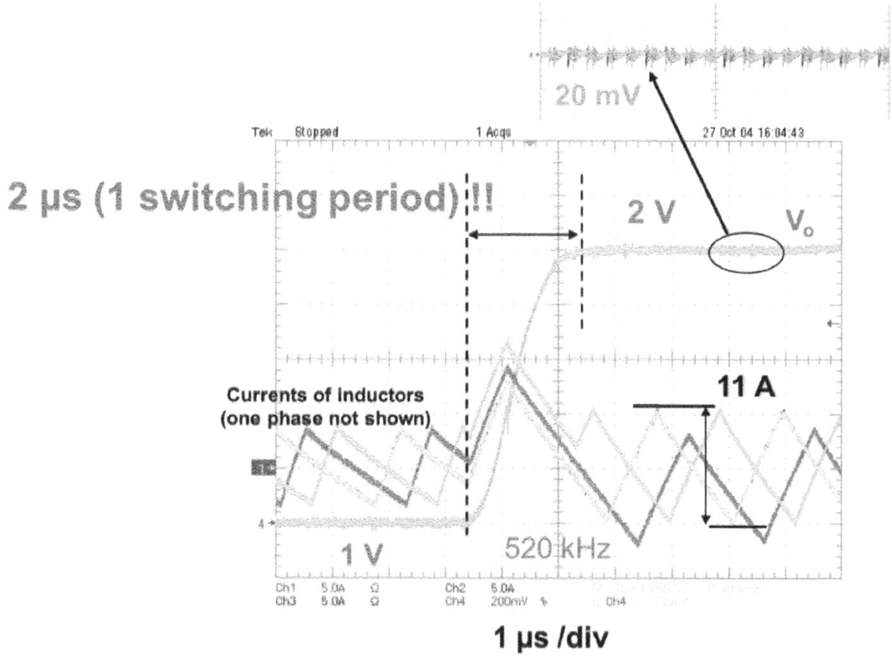

Figure 1.5. *Example of a dynamic voltage scaling. For the color version of this figure, see www.iste.co.uk/allard/systems.zip*

High switching frequency is very convenient from the perspective of the dynamic response and size, but from the point of view of the efficiency, it becomes a burden. The multiphase concept, several converters in parallel and time-shifted, can help to meet the dynamic response specs without an excessive increment of the switching frequency. A new degree of freedom appears to find the appropriate trade-off between switching frequency, efficiency and number of phases.

Many questions arise with respect to the design: which output filter (inductance and capacitance) must be used, which switching frequency, how many phases, which is the most appropriate control strategy?

The specifications are very demanding, which results in a very complex task to optimize the design that meets all the required specifications. To optimize this power converter, CAD tools become mandatory. These CAD tools must account for both the power stage and the control stage.

It is important to highlight that even with an ideal control, the LC filter represents a limit to the fastest response that can provide the power converter. The minimum time for a state transition in the buck converter is obtained by applying the maximum principle or Pontryagin's principle [MAS 16]. This principle gives the necessary conditions for the input to a system if a defined state has to be reached within a minimum time. In [SOT 07], this principle is applied to a buck converter, for obtaining the minimum time control law: there must be at most one transition of the voltage applied to the filter, from 0 to V_{in} or vice versa to achieve minimum time (Figure 1.6).

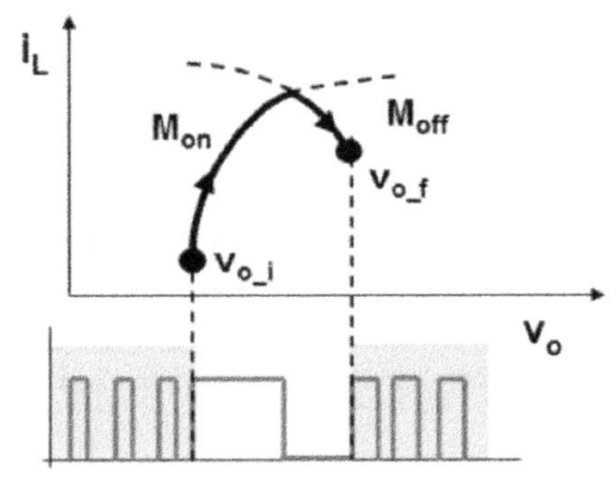

Figure 1.6. *Up-step transition in the buck converter: phase-plane trajectories (top) and minimum time control law (bottom)*

Figure 1.7 shows the minimum time transition when the control law is applied to a buck converter to move from one state to another. Figure 1.8 shows the comparison of the minimum time transition with the duty cycle step transition in the same converter. Clearly, the minimum time transition is much better. There is no overshoot and the stable state is reached within the minimum time achievable by the converter.

The minimum time concept provides the maximum dynamic response that a given LC filter can achieve. This is really valuable information for designing a converter. This concept can be applied to either voltage steps or current steps, which are used in the following section to determine the LC filter design space that can meet the defined specifications.

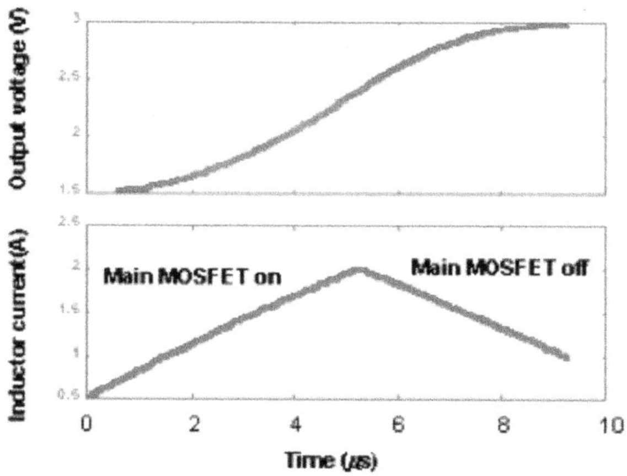

Figure 1.7. *Minimum time transition of the voltage step*

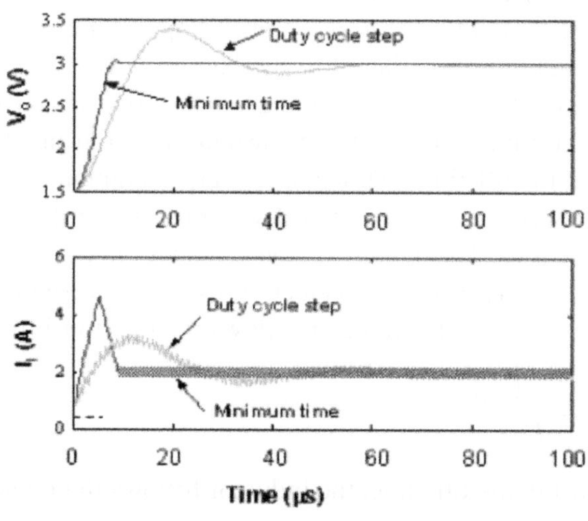

Figure 1.8. *Comparison of the minimum time transition with the response to a duty cycle step. For the color version of this figure, see www.iste.co.uk/allard/systems.zip*

1.2. Operation principle of three non-isolated converters

Converters for Internet of things (IoT) and mobile devices are mostly non-isolated converters. Galvanic isolation is required for line-connected equipment, mainly for safety purposes. However, a typical mobile device operates on a battery with reduced voltage levels; thus, isolation or large voltage transformation ratio is not required. This section briefly describes three non-isolated power converters that can step-down, step-up or follow the input voltage.

1.2.1. *Buck converter operation*

A buck or step-down regulator is shown in Figure 1.9 (top). An inductor and a capacitor form the output filter that is switched to the input voltage or to the ground by the high-side switch (sw1) or the low-side switch (sw2), respectively. Most low-voltage buck converters use the low-side switch as a freewheeling diode. However, this switch can be replaced by a diode at the expense of higher conduction losses. The output voltage is generated by alternatively switching the inductor to the battery voltage during an on-time, as shown in Figure 1.9 (middle), or the ground during an off-time, as shown in Figure 1.9 (bottom). The inductor current rises during the on-time and decreases during the off-time, while the capacitor maintains almost a constant output voltage. A conduction cycle is defined as one on-time, T_{on}, followed by one off-time, T_{off}. The length of the conduction cycle is the switching period, T_{sw}, and the rate of the conduction cycle is the switching frequency, F_{sw}. During the on-time, the inductor follows the relationship:

$$\frac{dI_L}{dt} = \frac{V_{in} - V_{out}}{L} \qquad [1.1]$$

Similarly, during the off-time, the inductor follows the relationship:

$$\frac{dI_L}{dt} = -\frac{V_{out}}{L} \qquad [1.2]$$

During a steady-state operation, the inductor current at the end of the cycle reaches the same value as at the beginning of the cycle. Therefore, the overall current variation through the inductor is zero:

$$\Delta I_L = I_L\left(T_{sw}\right) - I_L\left(0\right) = 0 \qquad [1.3]$$

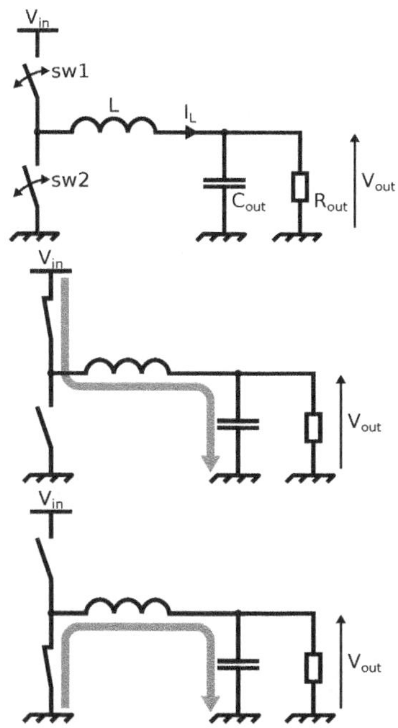

Figure 1.9. *Synchronous buck converter and its conduction states*

Using equations 1.1 and 1.2 in 1.3, we obtain:

$$\Delta I_L = 0 = \frac{V_{in} - V_{out}}{L} T_{on} - \frac{V_{out}}{L} T_{off} \qquad [1.4]$$

Rearranging the terms, we obtain:

$$V_{out} = \frac{T_{on}}{T_{on} + T_{off}} V_{in} = \frac{T_{on}}{T_{sw}} V_{in} = DV_{in} \qquad [1.5]$$

Thus, the output voltage of the buck converter depends on the input voltage and the ratio of the on-time over the switching period, which is also called the duty cycle, D. This simple equation assumes perfect elements that are not available to the designer; however, non-idealities will only introduce small variations. The output voltage is theoretically limited between the battery and the ground. Practical limitations such as a limited duty cycle and the resistance of various elements will reduce that range.

Input and output voltages are part of the specifications of the system. The designer focuses on the time-domain variables by selecting a switching frequency and a control mode that will adjust the on-time or off-time, or both. This also outlines the fact that time-domain simulations are extensively used to design a switched-mode converter since the mode of operation of the converter is based on time sequences.

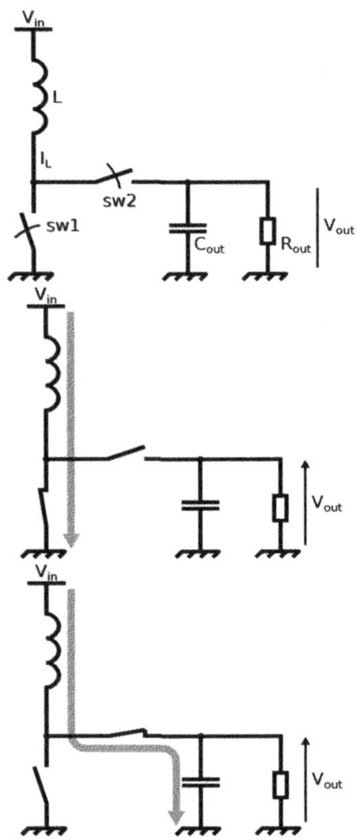

Figure 1.10. *Synchronous boost converter and its conduction states*

1.2.2. *Boost converter operation*

A boost or step-up converter is presented in Figure 1.10 (top). During the on-time, the switch sw1 connects the switching end of the inductor to the ground, as shown in Figure 1.10 (middle). The inductor current rises according to the relationship:

$$\frac{dI_L}{dt} = \frac{V_{in}}{L} \qquad [1.6]$$

During the off-time, the switch sw2 connects the switching end of the inductor to the output, as shown in Figure 1.10 (bottom). A diode can be used instead of a switch at the expense of higher conduction losses. The inductor current decreases according to the relationship:

$$\frac{dI_L}{dt} = \frac{V_{in} - V_{out}}{L} \qquad [1.7]$$

For the buck converter, charge-balance principle during a steady-state operation can be used to determine the DC transfer ratio:

$$V_{out} = \frac{T_{on} + T_{off}}{T_{off}} V_{in} = \frac{1}{1 - D} V_{in} \qquad [1.8]$$

The output voltage of the boost converter depends on time-related ratios as well as the input voltage. Second-order elements come into play, but they do not change the fundamental results of the time-domain analysis.

The output voltage of the boost converter can only be controlled to be above its input voltage. The output voltage can theoretically reach an infinitely high value. However, practical limitations are induced by parasitic resistances and maximum duty cycles that can be generated by the control circuit.

1.2.3. *Buck-boost converter operation*

A four-switch buck-boost converter is shown in Figure 1.11 (top). It can be used as a buck converter by turning on switch sw4, turning off switch sw3 and using switches sw1 and sw2 with the states shown in Figures 1.11 (middle-down) and 1.11 (bottom). Similarly, it can be used as a boost converter by turning on switch sw1, turning off switch sw2 and using switches sw3 and sw4 with the states shown in Figures 1.11 (middle-up) and 1.11 (middle-down). A third operation mode can be achieved by switching on switches sw1 and sw3 during the on-time and turning on switches sw2 and sw4 during the off-time using the states shown in Figures 1.11 (middle-up) and 1.11 (bottom). During the on-time, the inductor current rises according to the relationship:

$$\frac{dI_L}{dt} = \frac{V_{in}}{L} \tag{1.9}$$

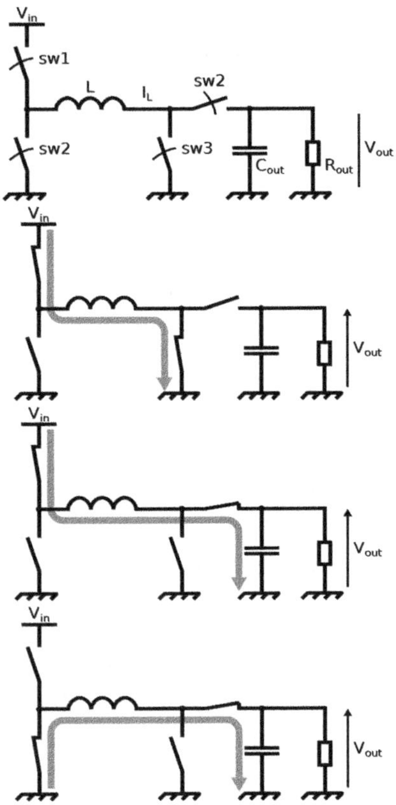

Figure 1.11. *Synchronous buck-boost converter and its conduction states*

During the off-time, the inductor current decreases according to the relationship:

$$\frac{dI_L}{dt} = \frac{-V_{out}}{L} \tag{1.10}$$

Using the charge-balance equation, we can determine the DC transfer ratio of the buck-boost converter as follows:

$$V_{out} = \frac{Ton}{T_{off}} V_{in} = \frac{D}{1-D} V_{in} \qquad [1.11]$$

Depending on the duty cycle, the output voltage can be either lower or higher than the input voltage of the converter. This flexibility has a cost as the four switches are constantly switching and the RMS current through the inductor is higher than in the boost-only or buck-only operation, hence a degraded efficiency. Therefore, we can use the buck-boost converter as a boost or buck when the input voltage is lower or higher than the output voltage, respectively. When the output voltage is close to the input voltage, the input voltage can also follow a four-switch buck-boost. We still benefit from the line rejection and the accuracy of the converter when the converter works in the buck-boost mode.

1.3. Power stage

The power stage of the converter includes the switches and their associated drivers as well as power passive components. For the three converters considered (buck, boost and buck-boost), it can be two or four switches and drivers, two capacitors and an inductor. We may include other subcircuits such as various current detectors and current sensors. The design of these elements as such is not detailed in this chapter.

1.3.1. *MOSFET switching an inductive load*

Converters that are taken into consideration are hard-switching converters, i.e. the on-time switch switches a non-zero voltage V_{drain} with an inductive load carrying a non-zero current I_{ind}. A simplified model of a MOSFET transistor is shown in Figure 1.12 (top). During the off-state, the transistor is modeled as three capacitors between each terminal and a diode, as shown in Figure 1.12 (middle-up). During the switching transition, the transistor behaves as a voltage-controlled current source, as shown in Figure 1.12 (middle-down). When in the on-state, the transistor can be modeled as a resistor, as shown in Figure 1.12 (bottom). The parasitic capacitors are not constant capacitors, i.e. their actual value changes when the voltage between their two terminals changes. The on-state resistance is not constant either, which mainly depends on the driving voltage.

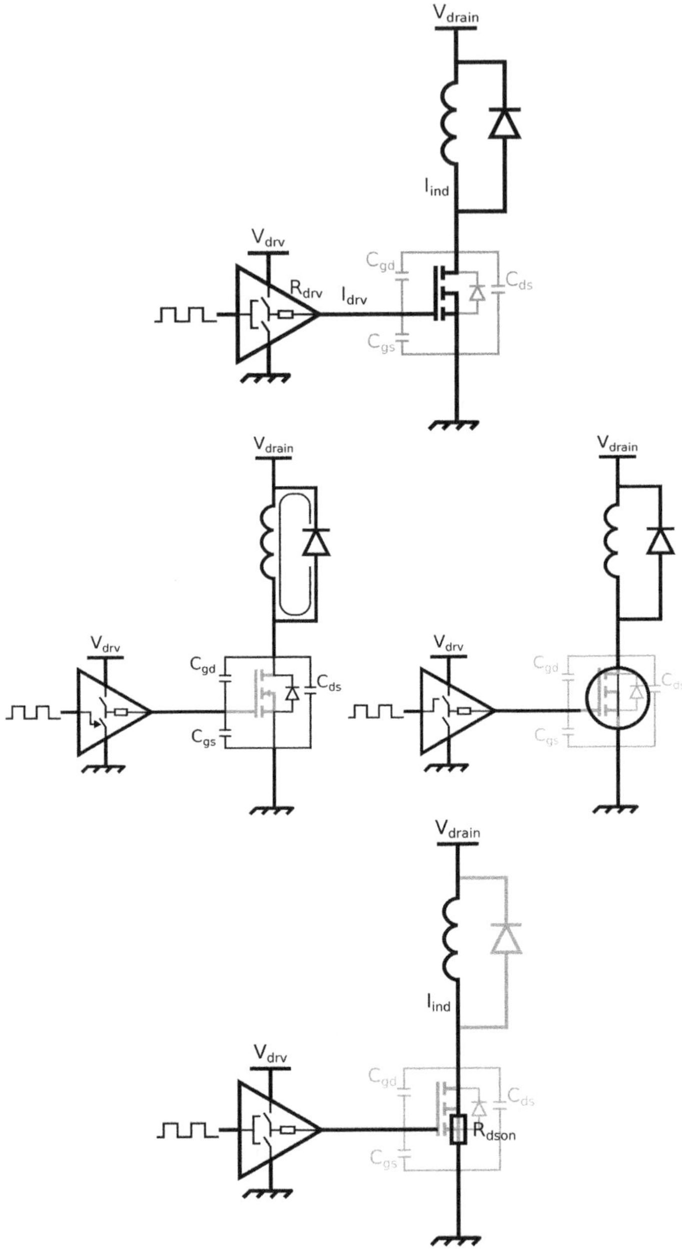

Figure 1.12. *Model of a transistor switching an inductive load*

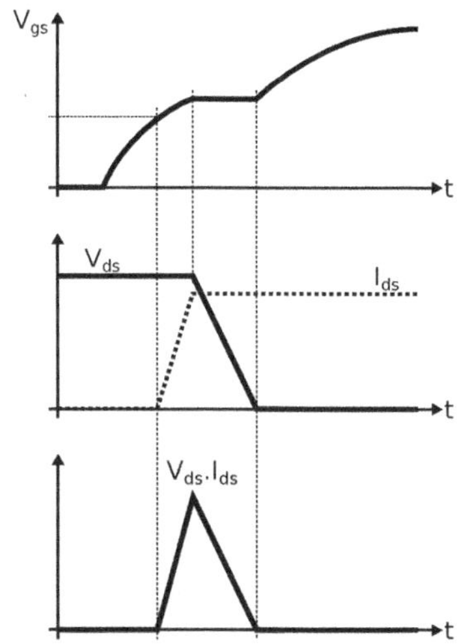

Figure 1.13. *Turn-on sequence of a MOS transistor switching an inductive load*

The turn-on sequence of an N-type MOSFET transistor switching an inductive load is shown in Figure 1.13. The inductor current is assumed to be constant over the transition time; therefore, the inductive load is modeled as a constant current source and an ideal diode. The gate driver is modeled as an ideal logic buffer with a constant output resistance R_{drv} and a supply voltage V_{drain}. When the transistor is initially in the off-state, the parasitic capacitor charges are:

$$Q_{gs} = 0$$

$$Q_{gd} = -C_{gd}V_{drain}$$

$$Q_{ds} = C_{ds}V_{drain}$$

When the transistor is in the on-state, the parasitic capacitor charges are:

$$Q_{gs} = C_{gs}V_{drv}$$

$$Q_{gd} = C_{gd}V_{drain}$$

$$Q_{ds} = 0$$

At this point, we can note that the charge and discharge of the parasitic capacitors dissipate a finite amount of energy. This energy is transferred from the power source to the transistor and discharged through the driver at each conduction cycle. Then, the average power dissipated by the driver and the transistor to charge and discharge the parasitic capacitors of the transistor can be approximated as:

$$P_{sw} = 0.5\left(C_{gs}V_{drv}^2 + C_{gd}\left(V_{drv} + V_{drain}\right)V_{drv} + C_{ds}V_{drain}^2\right)f_{sw} \qquad [1.12]$$

where f_{sw} is the switching frequency of the converter. Equation 1.12 describes the losses related to the charge and discharge of the parasitic capacitors of the transistor. Switching the transistor also generates crossover losses. These losses are related to the turn-on speed of the transistor.

Figure 1.13 shows the turn-on signals of a MOSFET. When the control signal rises, the driver charges the gate-source and gate-drain capacitors of the transistor. When the gate-source voltage reaches the threshold voltage of the transistor, the transistor becomes a voltage-controlled voltage source. The gate-source voltage increases until the drain-source current of the transistor reaches the inductor current value and the diode turns off. When the freewheeling diode turns off, the drain voltage decreases as well as the voltage of the gate-drain capacitor of the transistor. The decrease rate mainly depends on the gate-drain capacitor of the transistor and the driver strength. The transistor remains in the voltage-controlled current source until the drain-source voltage becomes equal to the gate-source voltage minus the threshold voltage of the transistor. Then, the nonlinear resistor shown in Figure 1.12 (bottom) applies and the driver fully charges the gate of the transistor.

During the turn-on of the transistor, the drain-source voltage and the drain-source current reaches a peak value of $V_{drain} \cdot I_{ind}$. The instantaneous power dissipated through the transistor is shown in Figure 1.13. These crossover losses depend on the drain voltage, the inductor current, the gate-to-drain capacitor of the transistor, the transistor transconductance and the driver strength. An estimate of these losses is proposed in [MAN 12]; however, the nonlinear behaviors of the transconductance and the driver strength make this estimate difficult. Furthermore, for low-voltage systems switching battery-level voltages, the crossover time is short and the switching speed is for the most part not limited by the device itself but by the packaging. Any impedance between the source of the transistor and the ground plane acts as degeneration impedance and reduces the switching speed. After a first initial sizing, it is often quicker to extensively simulate the system when accurate electrical models are available.

The turn-off of the transistor is similar to the turn-on sequence. The driver discharges the gate-source and gate-drain capacitor of the transistor until the transistor leaves the resistive model region defined as:

$$V_{ds} < V_{gs} - V_t \quad [1.13]$$

where V_t is the threshold voltage of the transistor. Then, the transistor becomes a voltage-controlled current source whose current is almost equal to the inductor current. Therefore, the inductor charges the drain-source and gate-drain parasitic capacitors of the transistor. For the turn-on of the switch, the rate of change of the drain voltage is highly dependent on the driver strength and the gate-to-drain capacitor of the transistor. When the drain voltage of the transistor reaches V_{drain}, the drain-source current through the transistor decreases significantly until the gate-source voltage crosses the threshold voltage of the transistor.

It is worth mentioning that the switching speed depends not only on the switch transistor but also on the driver. Nonetheless, the loop impedance comes into play very quickly as the switching speed increases. Packaging, decoupling and PCB impedance are of some importance. As a general guideline, we will try to minimize the parasitic inductances between the switch and its driver and optimize the decoupling of the driver supply voltage. Power modules including a pair of transistors and their associated drivers are commercially available. These modules minimize the parasitic elements by a tight integration of a different element in a single package as well as the integration of active elements on a single silicon die, or the use of an integrated power converter.

Equation 1.12 outlines the role of the driver supply voltage in the switching losses and assumes a 100% efficient driver supply. This is the case when the driver is supplied directly by the input power source, for instance. However, this is not the case when the maximum gate-source voltage is lower than the input voltage, or in a boost case or when a linear regulator supplies the driver. We need to emphasize the switching losses of the device related to the driver voltage with a proper driver efficiency for such a case. For instance, using an ideal linear regulator to supply the driver with a voltage V_{drv} from the power input of the driver whose voltage is V_{drain}, we obtain:

$$P_{sw} = 0.5\big(C_{gs}V_{drv}^2\frac{V_{drain}}{V_{drv}} + C_{gd}(V_{drv} + V_{drain})V_{drain} + C_{ds}V_{drain}^2\big)f_{sw} \quad [1.14]$$

1.3.2. *Extracting the parasitic capacitance values using simulations*

Parasitic capacitances of the transistor play a key role in the overall efficiency of the converter. However, these elements are often partially documented or not documented at all. As the capacitance values are voltage-dependent, equivalent values are extracted following the charge equation of an ideal capacitor:

$$C_i = \frac{Q_i}{V_i} \quad [1.15]$$

where Q_i is the amount of charge needed to charge the capacitor at the voltage V_i. Schematic representations of the extraction of the parasitic elements for an N-MOSFET are shown in Figure 1.14. This method requires accurate models for the parasitic capacitances of the transistor. The scheme shown in Figure 1.12 (top) is used to extract the output capacitance C_{oss} that combines the drain-to-source and part of the gate-to-drain capacitances. A voltage source increases the drain-source voltage of the transistor from 0 V up to the drain voltage of operation. Integrating the current delivered by the voltage source gives the amount of charge stored in the equivalent capacitors. Figure 1.12 (bottom) shows the extraction of the input capacitance of the transistor C_{iss}. This capacitance is a compound of the gate-to-source, gate-to-bulk and gate-to-drain capacitance. Again, integrating the current delivered by the voltage source gives the amount of charge stored in the equivalent capacitors. It is worth mentioning that the amount of charge needed to invert the voltage across the gate-drain parasitic capacitor, C_{gs}, depends on the drain voltage. Therefore, the equivalent capacitance seen by the driver increases when the off-state drain voltage of the transistor increases.

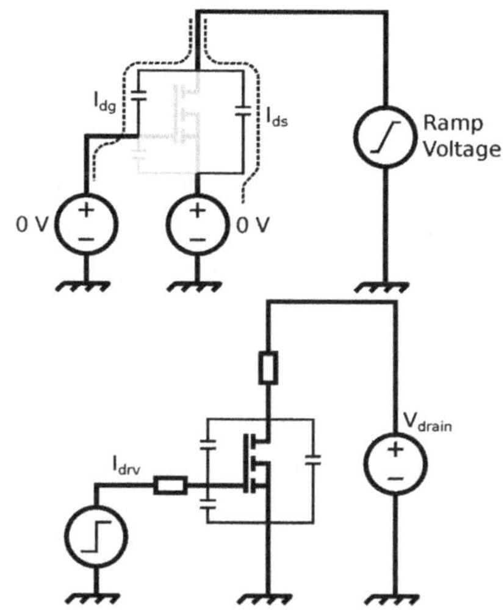

Figure 1.14. *Simulation setup for the extraction of the parasitic capacitances*

1.3.3. *Power-stage design issues*

The usual process to optimize a power stage is to select a switching frequency and an inductor, use a transistor or design a transistor so that its losses are shared equally between the switching losses and the conduction losses, and repeat this process for each transistor in the circuit and design the best possible drivers. This simplistic way of describing the process hides the reality of a much more iterative and nonlinear process. This section discusses how these few design choices interact together and with the requirements.

The inductor of the converter is a critical component that often defines a significant portion of the solution cost, volume and performances. Its choice results from a multivariable trade-off. We can define the, "Current ripple ratio", r, as:

$$r = \frac{I_{pkpk}}{I_{av}} \qquad [1.16]$$

where I_{av} and I_{pkpk} are the average current and the peak-to-peak current through the inductor, respectively. As a rule of thumb, a current ripple ratio close to $0.3 \approx 0.4$ ends up being a good trade-off for most applications. This statement can be used to discuss three features:

– What is the best average current to take into account? Is it at peak output current or at typical output current? The answer to this question leads to the specification of the load current profile. On the one hand, extreme current peaks that occur rarely probably need not be optimized. On the other hand, if the average load current is 10 times less than the maximum output current, a good trade-off is certainly in between the average and the peak value.

– The peak-to-peak current depends not only on the inductor value but also on the switching frequency, which is still not firmly defined. The higher the switching frequency, the smaller the size and value of the inductor. High switching frequencies tend to increase the power density of the converter. However, switching losses can become a significant dissipation factor as the switched voltages are defined by the application, and the parasitic capacitors of the devices cannot be decreased without an increase in the R_{dson} of the devices.

– The ripple factor remains a rule of thumb and there are good reasons to avoid its use. For instance, some low-noise applications tend to require less current ripple even if self-resonance frequencies and skin effects through the inductor can create switching frequency-dependent problems in the noise spectrum. The higher the inductance, the lower the current slew rate delivered by the converter. The best-in-class transient response converters tend to keep the inductance value low in order to maintain di/dt as high as possible. For instance in [LAB 13], the falling load current slew rate at the on-board converter output reaches values up to 6.55×10^6 A.s^{-1}, while the current slew rate through the inductor can be up to 2.5 times lower (2.55×10^6 A.s^{-1}). Even an ideal transient controller would require a minimum amount of decoupling capacitor to absorb the residual energy that is left in the inductor. Otherwise the excess of residual energy could lead to exceeding the maximum load voltage.

Generally, finding an optimum solution requires several iterations. A good starting point is to use as much space as possible given the project constraints. This allows low switching frequencies and low switching losses. Conduction losses for the inductor can be maintained under control, thanks to a large inductor package and a high value inductor. Transient performances will be discussed later, but adding decoupling output capacitors can mitigate the large inductance value and the poor current slew rate. However, non-mechanical requirements can come into play. Among others, start time of a variation of the DC output voltage of the load for different operation points restricts the amount of decoupling capacitor that can be used. This is the case for the average power tracking for a radio-frequency power amplifier. Between each plateau, the converter must charge and discharge the output decoupling capacitors as fast as required by the application.

The input power source of an Internet of things (IoT) device is likely to vary as well as the output voltage of the converter. The former varies because of the discharge of the battery and the latter varies because of the various power states of the load. A good knowledge of the load distribution helps us to find the best optimization point where switching losses of the devices equal conduction losses. We may choose two slightly different points for the switch and the rectifier of the converter in order to flatten the efficiency curve. A trial and error method and a good understanding of the loss mechanism often give satisfying results.

1.3.4. *Segmented power stage and multiphase operation*

For highly variable operation conditions, a single set of transistors may not be sufficient to keep the efficiency high. A technique is to connect several transistors in parallel and to switch only some of them when the current is low, as shown in Figure 1.15. For instance, two transistors can be connected in parallel with each independent driver. At high output current, both transistors are switched. The switching losses are the sum of the two transistor switching losses and the overall on-resistance is equivalent to the two R_{dson} in parallel. At low output current, only one transistor is switched while the other is turned off. The R_{dson} of the active transistor generates the conduction losses. The switching transistor generates switching losses as well as the output capacitance, C_{oss}, of the passive switch that is charged and discharged at the switching rate. Hence, the efficiency is degraded with respect to a single device at low current, but at high current, the efficiency is higher than that of a single transistor. Fine-grain automatic tuning of the sizing of the switches is presented in [MIC 14]. Despite a notable improvement at medium current, only a reduced number of segments are sufficient to extend the efficiency of the converter. Therefore, most commercial integrated products offer only a few segments.

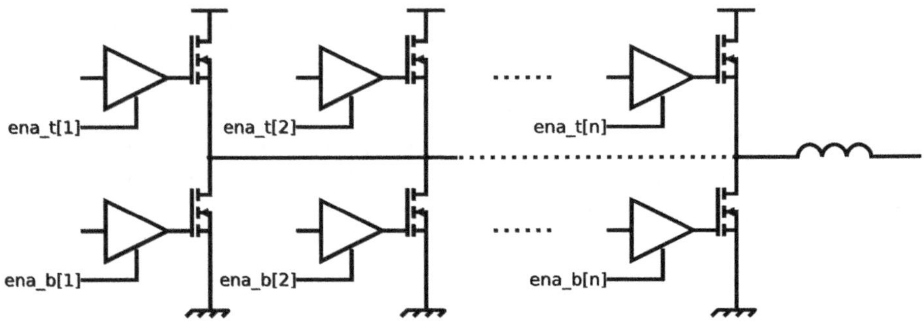

Figure 1.15. *Operation principle of a segmented power-stage*

We can take the segmentation principle further and use several inductors, as shown in Figure 1.16. Each pair of switches and inductors is called the phase, hence the name of the multiphase converter. At low output current, a single phase is used and the other ones are turned to high impedance mode. When the load current increases, the number of active phases also increases, maintaining each phase close to its optimum efficiency point. This concept can be used with two phases or with a large number of phases. It is worth mentioning that the current through each phase must be balanced. This requires an active current-balancing control circuit and some sort of current sensing through each phase.

1.3.5. *LC filter design space*

As explained above, the design and optimization of power converters is very complex since there are tight connections between dynamic and static specifications, LC filter design, switching frequency, number of phases, MOSFET design, inductor design and capacitor design. The LC filter design space provides physical insight, decoupling the design process in two steps:

1) LC filter design to meet dynamic and static specifications: determination of the constraints for the filter to meet tracking time and tight regulation under load current changes. Other requirements such as tracking energy and size of output capacitors and inductors can be considered at this point. Identification of the control speed constraints is done. The control speed of the converter limits the minimum output capacitance and hence the minimum tracking energy. If linear control is considered, the design process must account for the minimum switching frequency that the system bandwidth requires. However, if nonlinear

control is considered, it is possible to decouple the switching frequency and the dynamic response, thus making minimum output capacitance designs feasible at low switching frequency.

2) Power stage optimization: when the degrees of freedom are obtained in terms of dynamic response and control, in this step, the design of the power stage, the selection of the switching frequency, the number of phases, MOSFETs, inductances and output capacitors is done to achieve the efficiency and size requirements.

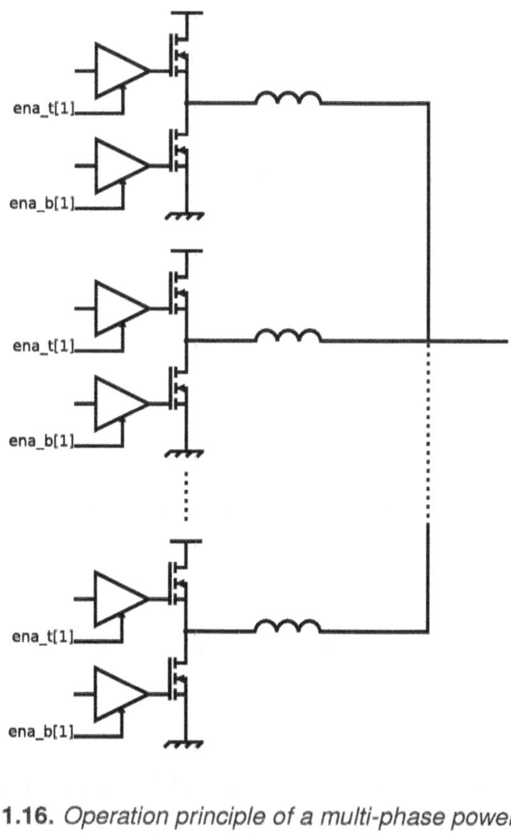

Figure 1.16. *Operation principle of a multi-phase power stage*

The key of this methodology is based on the one-phase equivalent concept (Figure 1.17). The one-phase equivalent has the same total silicon area for the MOSFETs than a generic N-phase solution. In the first case, the channel area per switch is N times higher than in the N phase solution to keep the same losses and the same size. The equivalent inductance L_{eq} of the one-phase equivalent is designed to have the same maximum current slew rate as a generic N-phase solution, $L_{eq} = L_{phase}/N$; therefore, both power stage solutions present the same maximum dynamic response.

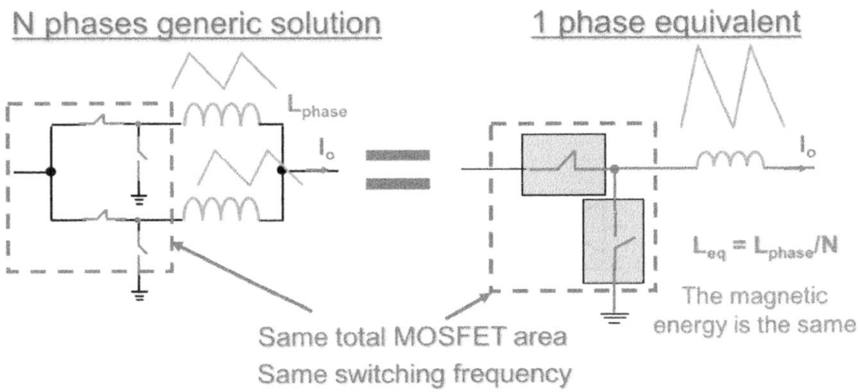

Figure 1.17. *One-phase equivalent: same behavior as N phases, same dynamic response, same efficiency and same size. For the color version of this figure, see www.iste.co.uk/allard/systems.zip*

The LC design space guarantees that the one-phase equivalent solution meets the dynamic and static specifications. Once the LC filter of the one-phase equivalent is selected, the design effort is focused on the power stage optimization: switching frequency, number of phases, MOSFET design, inductor design and capacitor design.

The LC design space is explained by means of a design example. The design specifications are:

– 5 V input voltage, 20 A load current;

– output voltage range from 1 V to 2 V and 2 λs tracking time;

– the tolerance band for dynamic regulation under current steps (20 A, 1000 A/λs) is 200 mV. The minimum output voltage at which the load step can occur is 1.5 V. The output voltage ripple should be less than 20 mV. The data refer to the values at the processor pins.

Figure 1.18 shows the LC design space to meet the dynamic specifications. The shaded area represents the design space that meets the dynamic specifications. The red border line is given by the voltage tracking time when the minimum time control law is applied. The blue border line corresponds to the load steps when the minimum time control law is again applied. The green border line corresponds to the maximum output capacitance that can be given by the maximum allowed tracking energy or by the bandwidth; if a linear control is used, the higher the bandwidth, the lower the output capacitance. All the LC values inside these borders (shaded area) meet the dynamic specifications. All the equations and details to determine these border lines are given in [SOT 05]

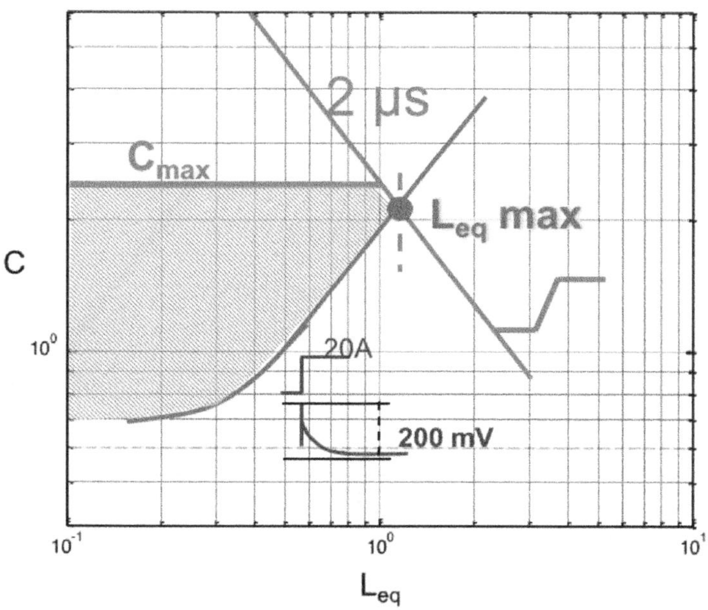

Figure 1.18. *LC design space to meet dynamic specifications: load steps, dynamic voltage scaling, energy tracking and bandwidth. For the color version of this figure, see www.iste.co.uk/allard/systems.zip*

The next step is to consider the output voltage ripple at high frequency. It defines an additional LC design space. Figure 1.19 shows the LC design space to meet the ripple specification, and the shaded area on the upper part corresponds to high values of L and C. The higher the switching frequency, the lower the required values for the inductance and the capacitance. Figure 1.20 shows both LC design spaces, corresponding to the dynamic and the static specifications. To achieve a LC design space that meets dynamic and static specifications, both shaded areas must overlap to have some common solutions. With a one-phase converter, the only way to find a solution is to increase the switching frequency. For the considered specifications, a one-phase converter would switch at 8 MHz to meet all the specifications.

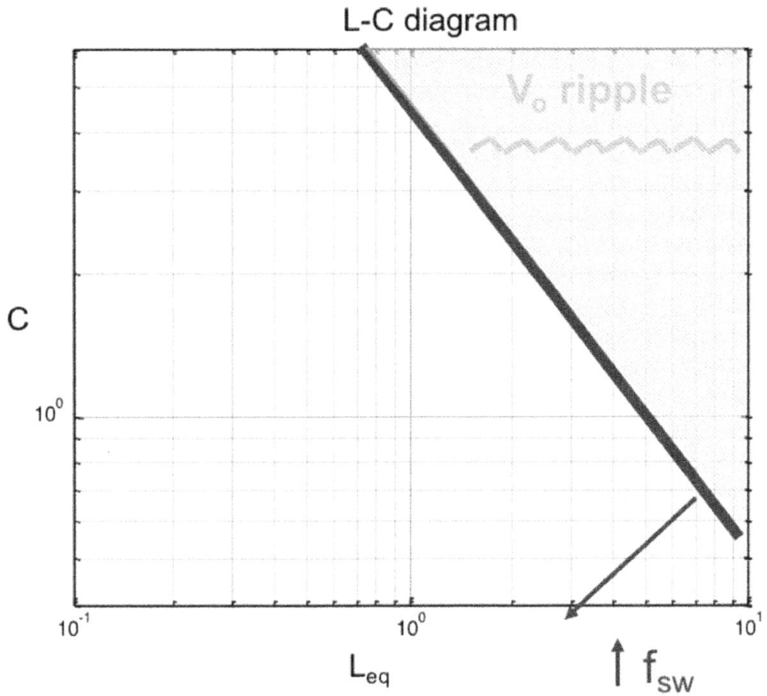

Figure 1.19. *LC design space to meet static specifications (output voltage ripple)*

However, the LC design space is valid not only for the one-phase equivalent but also for a generic N-phase solution. Therefore, a degree of freedom comes into play to optimize the design of the power topology, and the overlapping of the design spaces can be achieved either by increasing the switching frequency or by increasing the number of phases (Figure 1.21). For the considered specifications, a four-interleaved phase converter switching at 500kHz meets all the specifications (Figure 1.22). Only a one-phase converter running at 500 kHz presents a 300 mV ripple, a value much larger than the specified one (20 mV); however, by interleaving four phases, the ripple cancellation among phases allows us to meet the ripple specifications [SOT 05].

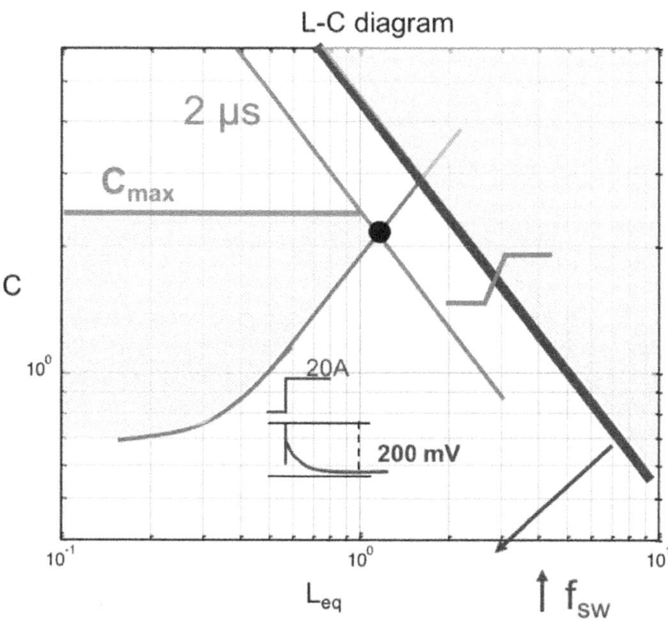

Figure 1.20. *LC design space to meet dynamic specifications (load steps, dynamic voltage scaling, energy tracking and bandwidth) and static specifications (output voltage ripple): increasing switching frequency to match both design spaces. For the color version of this figure, see www.iste.co.uk/allard/systems.zip*

In summary, the LC design space allows us to decouple the selection of the LC filter to meet the dynamic and static specifications from the power stage optimization. In the second stage of the design process, power stage design, switching frequency, number of phases, MOSFET design, capacitor design and inductor design are completed. A CAD tool could really help to optimize this complex design.

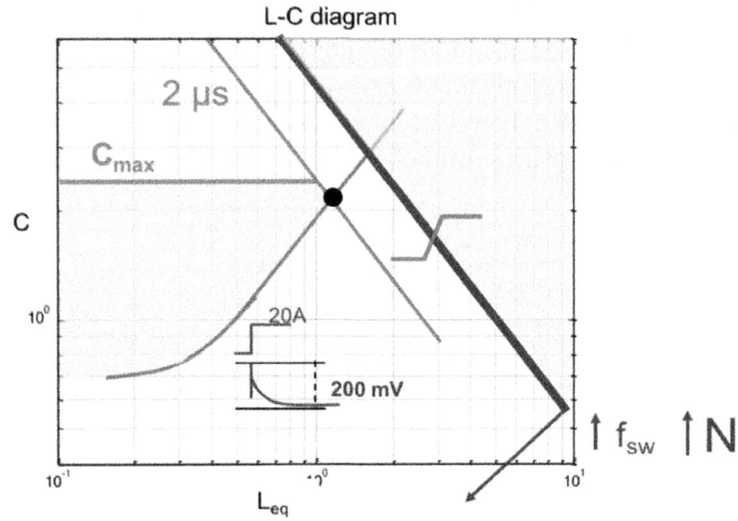

Figure 1.21. *LC design space to meet dynamic specifications (load steps, dynamic voltage scaling, energy tracking and bandwidth) and static specifications (output voltage ripple): increasing switching frequency or number of phases to match both design spaces. For the color version of this figure, see www.iste.co.uk/allard/systems.zip*

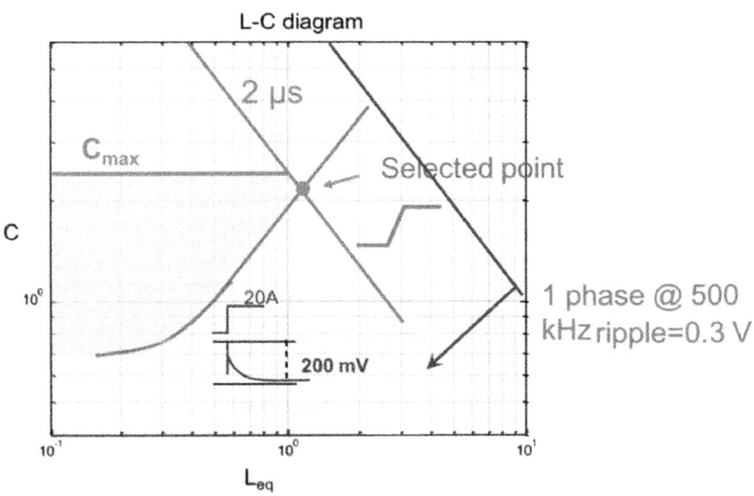

Figure 1.22. *LC design space: an example. For the color version of this figure, see www.iste.co.uk/allard/systems.zip*

1.4. Control stage

Numerous control strategies coexist to control the power stage. However, they are mainly derivatives of a few ones. First of all, the voltage-mode pulse width modulation (PWM), voltage mode control (VMC), is one of the most commonly used strategy. The current-mode control strategy is derived from the voltage-mode PWM strategy by the addition of an inner current loop. The control of two variables (the inductor current and the output voltage) provides additional advantages like load balance among converters operating in parallel, operation under overload or short-circuited conditions. From the perspective of dynamic response, current-mode control presents better line transient performance (audio susceptibility); however, under load steps, it presents worst performance since the output impedance is higher than the corresponding VMC. Since voltage-mode and current-mode PWM control use a clock-synchronized modulation, they are referred to as "hard-synchronized" methods, i.e. each cycle is synchronized with the reference clock. On the contrary, strategies where a conduction cycle can start out of phase with respect to a reference clock are considered as "soft-synchronized".

1.4.1. *Voltage-mode control of the buck converter*

A general PWM scheme of the voltage-mode control is shown in Figure 1.23. The control part is built around a PWM modulator whose input is a linear compensation function. The output of the modulation (duty cycle) is directly fed to the power stage.

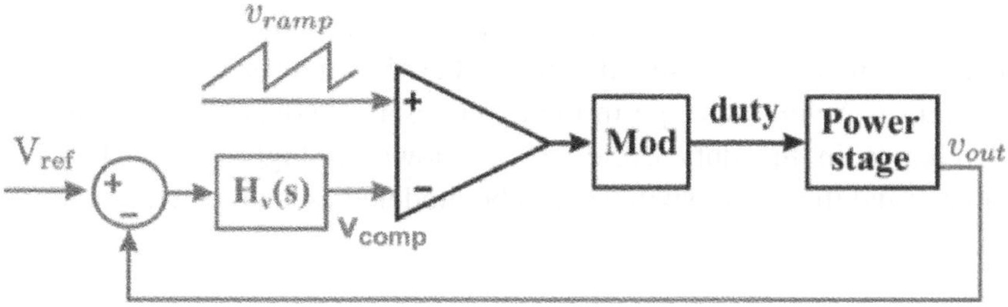

Figure 1.23. *General scheme of a voltage-mode PWM control. For the color version of this figure, see www.iste.co.uk/allard/systems.zip*

1.4.1.1. *Closing the loop of the voltage-mode-controlled buck converter*

A model of the converter has to be defined in order to determine the linear compensation function. The simplest methodology is the small-signal AC characterization associated with phase-margin and gain-margin criteria [ERI 01]. Although it is widely used in conventional design, some limitations have been outlined in [TRO 05]. However, this methodology offers a good first step and provides sufficiently accurate results to properly model the converter behaviors.

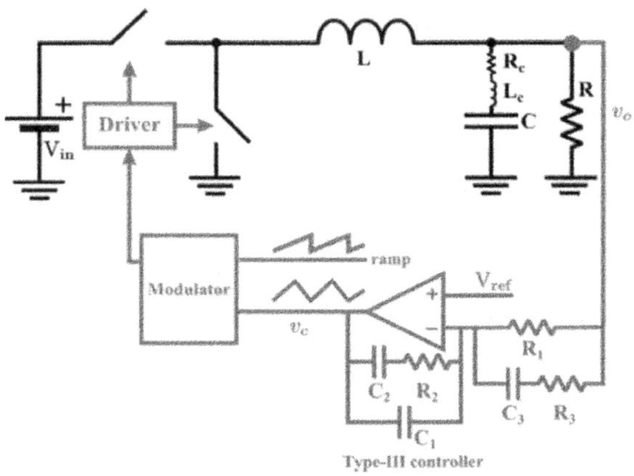

Figure 1.24. *Voltage-mode PWM control of a buck converter with type III regulator*

The scheme of buck converter with voltage mode PWM control is shown in Figure 1.24. PWM modulator signals are shown in Figure 1.25. The input signal is compared with a sawtooth and the output signal has a duty cycle proportional to the input signal. The higher the input signal, the wider the output pulse. A small change in the compensation voltage, v_{comp}, induces a small change in the duty cycle, *d*, as shown in Figure 1.26. The PWM modulator and the power switches can be modeled as a single gain.

$$G_{pwm} = \frac{V_{in}}{V_H} \quad [1.17]$$

where V_H is the height of the modulation ramp. Using feedforward action on the power input voltage, the sawtooth height V_H can be adjusted to V_{in}. In doing so, the PWM modulator and power switches become a unitary gain.

This technique is commonly used to reduce the line-to-output sensitivity and simplifies the compensation function design. Assuming the feedforward action on the PWM modulator rising slope so that $V_H = V_{in}$, the control to output transfer function of the buck converter becomes:

$$\frac{\hat{v}_{out}(s)}{\hat{v}_{comp}(s)} = \frac{1}{\frac{s^2}{\omega_0^2} + 2m\frac{s}{\omega_0} + 1} \quad [1.18]$$

where m is the damping factor of the output filter, which is load-dependent, and ω_0 is the time constant of the LC filter. The linear compensation function provides key characteristics:

– A high DC gain to diminish the static output error. This gain is provided by an integration function;

– Two zeros to compensate the filter double pole;

– A wide small-signal bandwidth to provide satisfying transient performances.

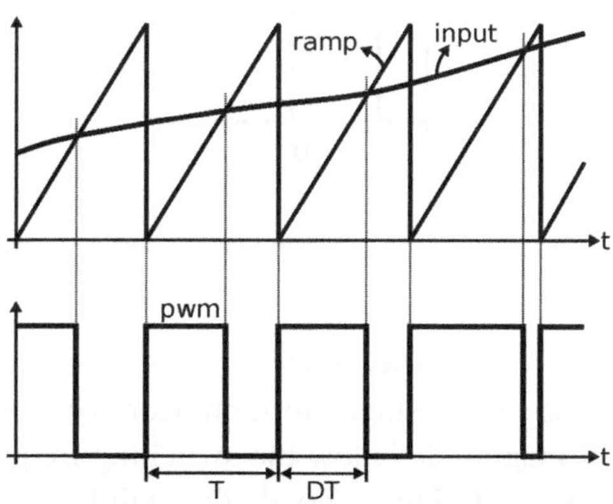

Figure 1.25. *Operation of a PWM modulator*

A type III provides these characteristics. A possible implementation is presented in 1.23. The two higher frequency poles are designed in order to set the cut-off frequency 5 to 10 times below the switching frequency. This widely used trade-off is set in order to average the switching effects while maximizing the regulation speed. A common design method is to set:

– The two zeroes at the LC double pole;

– A high-frequency pole at the capacitor ESR zero;

– The other high-frequency pole at the switching frequency or higher;

– The integration constant in order to provide an open-loop bandwidth 5 to 10 times lower than the switching frequency.

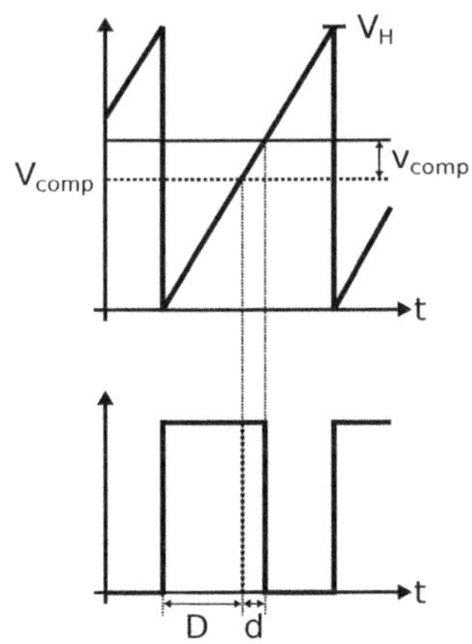

Figure 1.26. *Small signal modeling of a PWM modulator*

This methodology for stability suffers from approximations. First, the switching events of the power stage are nonlinear. The converter bandwidth has to be an order of magnitude below the switching frequency to ensure the validity of the stability criterion. Second, the stability analysis is valid for steady-state analysis only. It does not prevent large-scale instabilities that can result from a large output transient. Last but not least, it does not prevent period doubling or chaotic behaviors.

1.4.1.2. *Transient performances of the voltage-mode-controlled buck converter*

Load-transient performances of a converter can be studied in both the time domain and the frequency domain. Time-domain characterization consists of several measurements of overshoot and undershoots of the output voltage when applying various load current steps. Using load transient profiles that represent the load behaviors, it can be determined whether the transient response of the converter is good enough to meet the specifications. However, the transient response of a highly optimized, fixed-frequency PWM converter varies depending on the phase of the converter when a transient occurs. This phenomenon is shown in Figure 1.27, where 554 load current steps are measured and plotted with persistence. If a load transient occurs at the beginning of the conduction cycle, the converter reacts as fast as allowed by the feedback loop. On the other hand, if a load transient occurs at the beginning of the second subcycle, i.e. when the switching node is tied to the ground, the converter waits for the beginning of the next conduction cycle to react. This behavior leads to a non-negligible variability in the load transient response and should be carefully checked during the design phase of the converter.

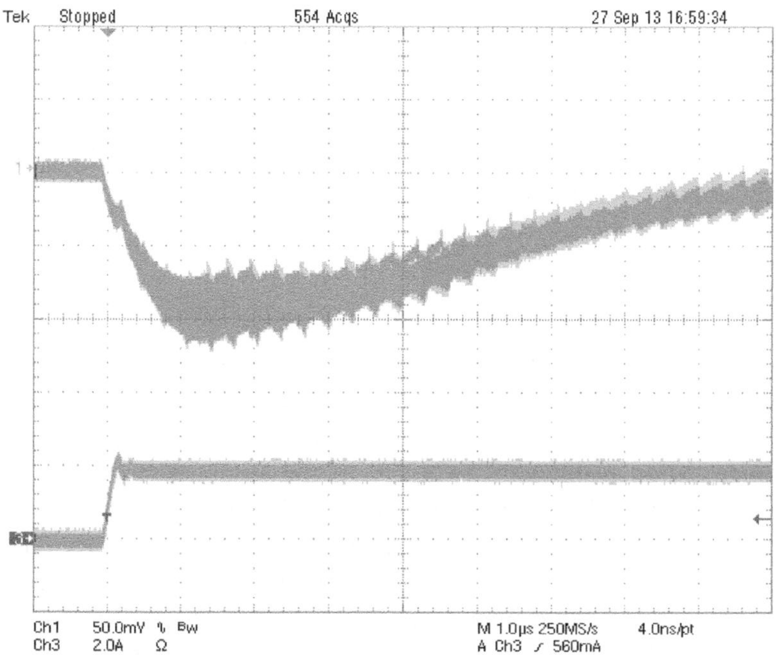

Figure 1.27. *Load step response of a general-purpose voltage-mode-controlled PWM converter, up: output voltage (AC coupled), down: load current*

Frequency-domain characterization is useful for simulating the converter behaviors while taking into account the power delivery network model (Power Delivery Network (PDN)). The latter can be extracted using 3D electromagnetic solvers, and unwanted resonances between the converter and the PDN can be avoided. The output impedance aims to characterize the converter behavior when a small and repetitive output transient occurs. A small change in the output current yields a small change in the output voltage. The resulting output impedance is defined as:

$$Z_o(s) = \left. \frac{\hat{v}_{out}(s)}{\hat{i}_o(s)} \right|_{\hat{v}_{ref}(s)=0, \hat{v}_{in}(s)=0} \qquad [1.19]$$

Without control, the buck open-loop output impedance is simply the parallel combination of the passive output network [MAN 12]:

$$Z_{O_{OL}}(s) = R_o \parallel \frac{1}{C_o s} \parallel (R_l + Ls) \qquad [1.20]$$

where R_o is the equivalent load impedance. The open-loop impedance exhibits a peak at the filter self-oscillation frequency that can degrade the load. A proper feedback control aims to divide the output impedance by its gain as:

$$Z_o(s) = \frac{Z_{O_{OL}}(s)}{1 + T_{comp}(s)} \qquad [1.21]$$

where $T_{comp}(s)$ is the feedback transfer function. Thus, a high DC gain feedback is best suited to compensate the uncompensated DC drop caused by the power-stage parasitic elements. The amount of high-frequency gain, i.e. the compensation bandwidth, to apply to the feedback controller is more difficult to evaluate. The output impedance in the open loop ($Z_{out}(s)$) and in the closed loop ($Z_{CLout}(s)$) is shown in Figure 1.28. It clearly shows the effect of closing the loop; the closed-loop output impedance is much lower than the open-loop output impedance up to the bandwidth. In this case, the bandwidth is 700 kHz for a 5 MHz switching buck converter.

We can consider that increasing the controller bandwidth even within the switching frequency limit is sufficient to diminish the output impedance of the converter. This technique hits the linearization limits, i.e. the current though the inductor is slew rate limited, and quantification effects due to the switching frequency create nonlinear behaviors. Simulation and experimental measurement of the output impedance can be performed using a controlled load that generates a sinusoidal load current at the measurement frequency. The output voltage and load currents spectrum are extracted using an FFT and the relative amplitudes are used to determine the output impedance at the frequency of interest. During the experiment, we must make sure that the system operates in the small-signal linear domain. This can be verified by checking the harmonic distortion of the current through the inductor. The latter increases rapidly when nonlinear effects come into play.

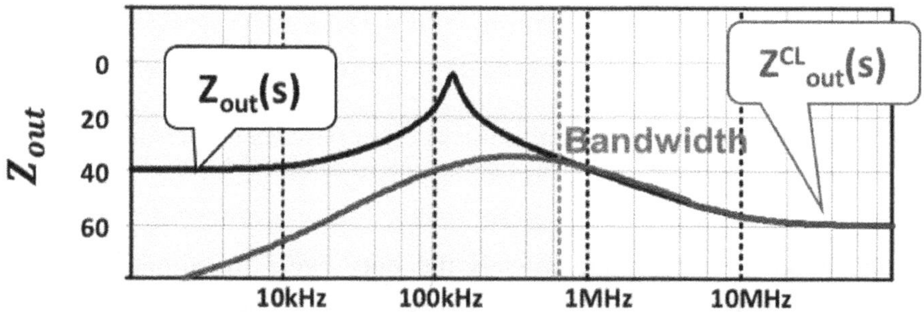

Figure 1.28. *Output impedance in the open loop and closed loop of a voltage-mode PWM buck converter with type III regulator. For the color version of this figure, see www.iste.co.uk/allard/systems.zip*

For a given power stage design, the voltage loop is responsible for the dynamic response of the converter. Figure 1.29 shows the response of a voltage-mode PWM buck converter under a fast load down step. The loop demands a very low duty cycle, v_{comp}, but it is lower than the minimum one, saturating the duty cycle and falling out of the small signal linear domain. It is called the large signal response of a converter, which must be taken into account for fast dynamic specifications.

1.4.2. *The RHP zero of the boost converter*

Several methods for determining a linear transfer function for a switched-mode converter have been proposed and discussed in the literature. Among them, the state-space averaging technique gives a consistent framework for the derivation of any transfer function of most common converters [RAS 10].

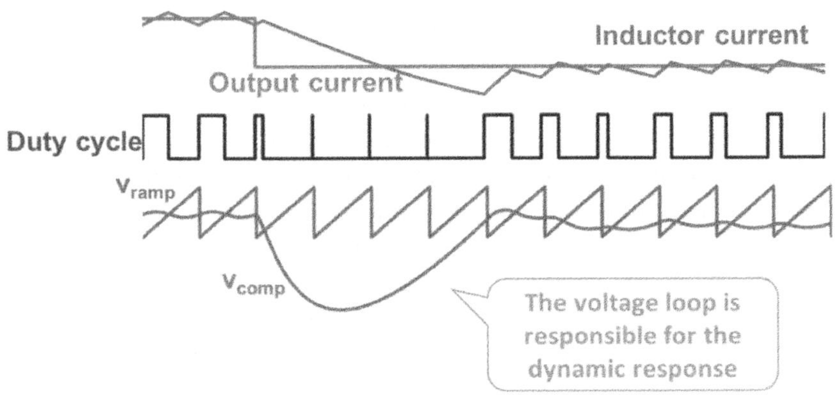

Figure 1.29. *Large signal response of a voltage-mode PWM control buck converter under a load down step. For the color version of this figure, see www.iste.co.uk/allard/systems.zip*

This modeling technique models the duty cycle to output the voltage transfer function, $\frac{\hat{v}_{out}}{\hat{d}}$, of a boost converter or a buck-boost converter in four-switch mode with a transfer function containing a positive zero, which is also called a right-half plane zero. For instance, the transfer function of a boost converter proposed in [ERI 01] is given by:

$$\frac{\hat{v}_{out}(s)}{\hat{d}(s)} = G_{d0}\frac{1-\frac{s}{\omega_z}}{1+\frac{s}{Q\omega_0}+\left(\frac{s}{\omega_0}\right)^2} \quad [1.22]$$

where

$$G_{d0} = V_{out}\frac{T_{on}+T_{off}}{T_{off}} \quad [1.23]$$

$$\omega_z = \left(\frac{T_{off}}{T_{on}+T_{off}}\right)^2 \frac{R_{load}}{L} \quad [1.24]$$

Both G_{d0} and ω_z are positive; therefore, the numerator of the transfer function contains the positive zero term. From a harmonic-domain perspective, the positive zero increases the gain of the transfer function but decreases the phase. The system becomes a non-minimum phase system and its closed-loop control is difficult.

From a time-domain perspective, this can be described as follows: when a load step occurs, the output voltage decreases. The inductor current must increase in order to compensate for this output voltage drop. However, increasing the inductor current means that the boost (or four-switch buck-boost) locally increases its on-time. During the on-time, the output voltage decreases as energy is not transferred to the output capacitor by the inductor. Therefore, to compensate for an output drop, the converter must allow the output voltage to decrease further in order to increase the average inductor current and to compensate for the drop later on. This non-intuitive control justifies the need for a current-mode control, i.e. detecting a change in the output voltage, the compensation function must control the inductor current to increase. The output capacitor converts this current change into a voltage change.

1.4.3. *Current-mode control*

The general scheme of the current-mode control (current-mode control (CMC)) of a converter is shown in Figure 1.30. An inner loop is used to regulate the inductor current, while an outer loop regulates the output voltage. The ripple of the inductor current modulates the duty cycle in the steady state (Figure 1.31).

A current-mode-controlled PWM buck regulator is shown in Figure 1.32. The current loop is a peak current-type loop, where the positive peak current of the inductor is regulated. On the contrary, valley current control regulates the negative peak current of the inductor. A peak current loop works as follows: when the clock cycle starts, the clock rising edge sets an RS-type or a D-type flop, with the switching node being tied to Vin, and the inductor current rises. An artificial slope is sometimes added to the measured current and compared with the loop input voltage, V_{comp}, to stabilize the converter when the duty cycle is greater than 50%. A larger control voltage implies a larger inductor peak current that results in a larger average inductor current.

The inductor current is regulated and the resulting current loop can be considered as a voltage-controlled current source. A simple approach is to model the inner loop as a voltage-to-current gain. This gain is only determined by the current-sensing circuit gain. Then, the output filter is now a first-order system with a single pole formed by the output capacitor, which is much easier to stabilize. The current loop gain does not depend on the input voltage. Therefore, the line to output characteristic is improved with respect to that of the voltage mode control. The control of the inductor current removes the LC filter resonance, making the design of the voltage loop simpler. However, this simple model is only valid for very slow dynamic specifications. As explained below, within this chapter, more advanced models are required to maximize the dynamic response of the CMC, and are mandatory to include the current ripple information inside the model.

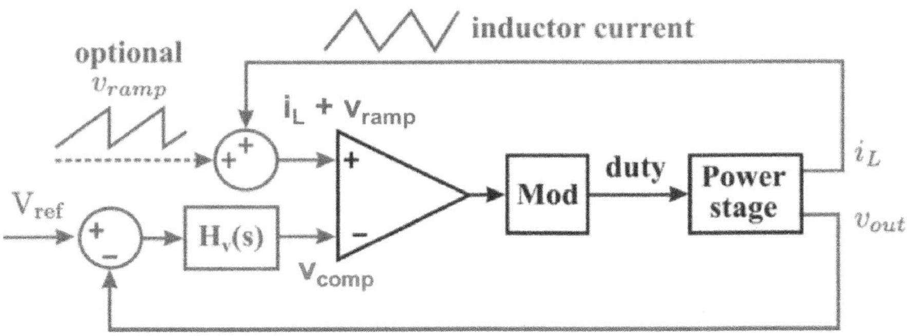

Figure 1.30. *General scheme of a current-mode control (CMC) of power converter. For the color version of this figure, see www.iste.co.uk/allard/systems.zip*

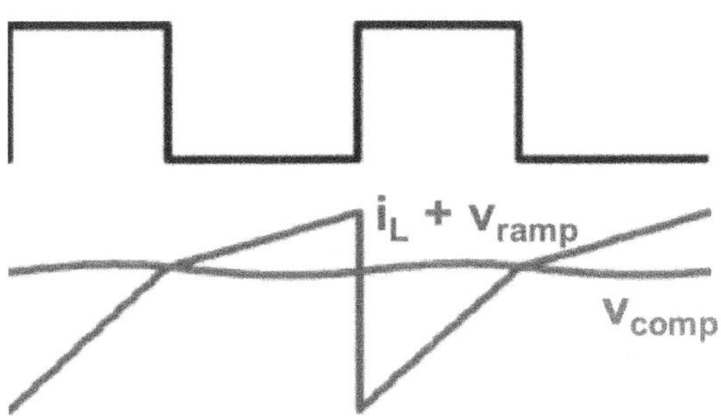

Figure 1.31. *Current-mode control (CMC): duty cycle modulation. For the color version of this figure, see www.iste.co.uk/allard/systems.zip*

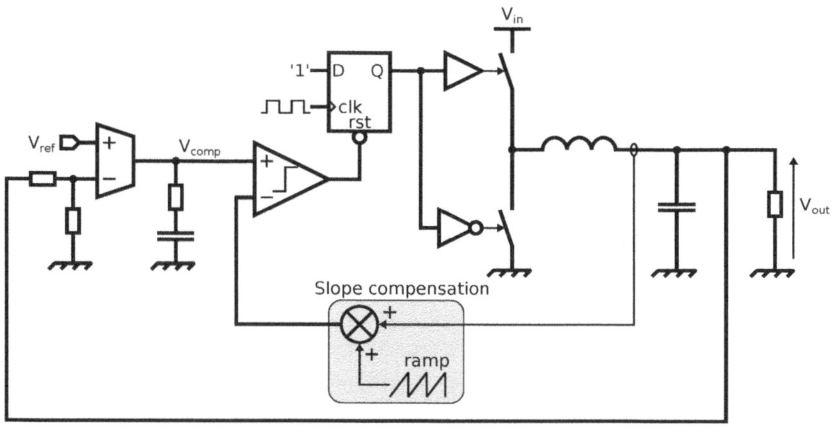

Figure 1.32. *Current-mode PWM converter: simplified diagram*

However, the peak current of the inductor does not reflect the average current through the inductor over a single cycle. A more accurate model of the current loop can be derived using a state-space averaging method or the similar approach proposed in [ERI 01]. It reintroduces a small line to output characteristic and a second pole to the control to output transfer function. Nevertheless, this second pole appears at high frequency and can usually be neglected, making the previous simplified assumption of a first-order current control to output transfer function valid.

Advanced models for peak current-mode control that take into account the subharmonic oscillations are reviewed in section 1.4.5.

Current-mode control alters the positive zero of the boost and the buck-boost converters so that it becomes negligible for most designs. The two-step approach for compensating a boost converter presented in the previous section is performed by design with an inner-current loop and an outer voltage loop. The voltage loop compensates for the output voltage changes while the current loop is in charge of controlling the required inductor current.

The main design issue when designing current-mode control is to sense the inductor current with a sufficient bandwidth and accuracy. This issue becomes even more difficult with the increase in the switching frequency and the output current dynamic. A common solution that preserves the converter efficiency is to sense the voltage drop across a switch. However, a high current-low voltage power stage presents a low voltage drop that requires a high current sense gain and accuracy. Furthermore, a few MHz of switching frequency require a sensor of much wider bandwidth than the switching frequency that is difficult to obtain. On the other hand, current-mode control eases the design of the compensation function, that can become a type-II compensation function with a high-DC gain, a low-frequency pole and a single zero. The gain-bandwidth product requirements of the amplifier are simplified and even a transconductance amplifier can be used.

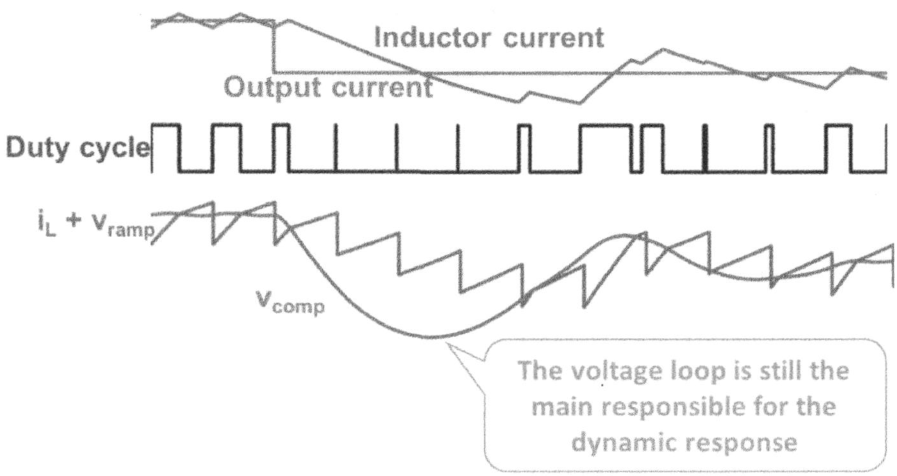

Figure 1.33. *Current-mode control (CMC): dynamic response under a load down step. For the color version of this figure, see www.iste.co.uk/allard/systems.zip*

The dynamic response of a buck converter with peak current mode (peak current-mode control (PCM)) control is shown in Figure 1.33. This figure clearly shows that the dynamic response of the converter still depends on the voltage loop rather than the current loop. The derivative of the current loop ($i_L + v_{ramp}$) is limited by the inductor current slope while the voltage loop (v_{comp}) is the variable that reacts faster to the load step.

1.4.4. *Hysteretic and sliding-mode control*

Limitations of linear-based control of a DC/DC converter make these solutions unsuitable for some high-performance mobile devices. The first limitation is the difficulty in reducing the output filter of a voltage mode-controlled PWM converter. The second limitation is the lack of fast and power-friendly solutions for current sensing in a current-mode-controlled PWM converter. To circumvent the limitations of linear-based analysis and design of DC/DC controllers, the simplest approach is to start from the other side of control theories, i.e. from the nonlinear approach.

1.4.4.1. *Hysteretic current loop and current-mode hysteretic converter*

A hysteretic current-mode loop is shown in Figure 1.34. As presented in section 1.2, the current through the inductor of the converter rises when the switch is turned on and decreases when the switch is turned off and the rectifier is active. Hence, the current can be maintained within a known interval using a comparator with a hysteresis cycle. When the inductor current is lower than the low boundary of the hysteresis cycle, the comparator is high and the converter is in on-time. Figure 1.35 shows this operation mode. The inductor current rises until it reaches the high bound of the hysteresis cycle. Then, the comparator switches low, and the converter starts the off-time. The inductor current decreases during the off-time until it reaches the low bound of the hysteresis cycle. Then, the comparator switches and starts a new switching sequence. This operation principle guarantees that the inductor current tracks the control input (i.e. the hysteresis cycle) as long as the current can reach the upper and lower bounds of the hysteresis cycle.

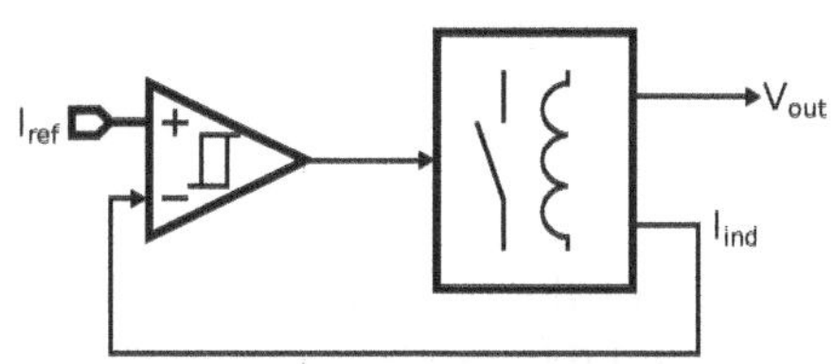

Figure 1.34. *Schematic of a hysteretic current loop. The power stage can be any of the three non-isolated converter presented in this chapter*

The hysteretic current-mode loop does not switch according to an externally provided reference clock. Its switching frequency depends on the ripple current and can be determined using the fundamental equations of the converters presented in section 1.2. For instance, the on-time of a buck converter with hysteretic current-mode control is:

$$T_{on} = \frac{L}{V_{in} - V_{out}} \Delta I_L \qquad [1.25]$$

And the off-time is:

$$T_{off} = \frac{L}{V_{out}} \Delta I_L \qquad [1.26]$$

where ΔI_L is the width of the hysteresis cycle. Hence, the switching frequency of the hysteretic current-mode loop for the buck converter is:

$$f_{sw} = \frac{V_{out}\left(V_{in} - V_{out}\right)}{V_{in} L \Delta I_L} \qquad [1.27]$$

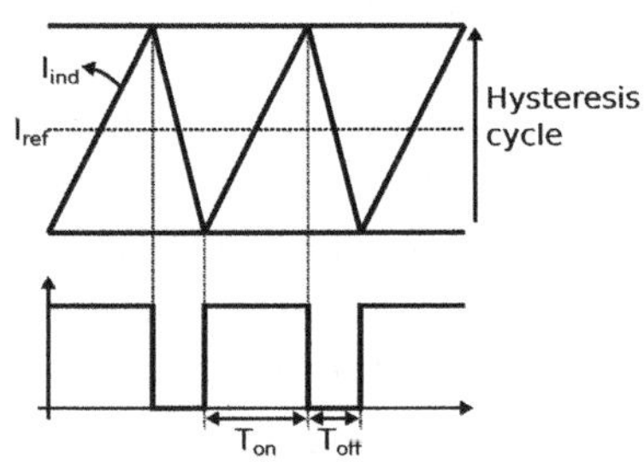

Figure 1.35. *Operation principle of a hysteretic current loop*

The switching frequency of the converter varies with respect to the input voltage, the output voltage and the width of the hysteresis cycle. We assume a fixed inductor value and neglect its variations so far. Variable switching frequency offers some benefits such as a simple control that does not require a slope compensation circuit, a nonlinear and fast transient response and its ability to reject high frequency phenomenon that are aliased by the sampling of a fixed-frequency modulation. This latter characteristic is useful in envelope tracking application, for instance. On the other hand, an uncontrolled switching frequency can be problematic. It makes the power-stage efficiency difficult to predict and optimize. Furthermore, the interferences generated by the converter are less predictable and more difficult to control. The application dictates its requirements, and when a wide input voltage range, a wide output voltage range and a precise switching frequency are required, the hysteretic current-mode loop itself is certainly not the best candidate.

At high switching frequency, the hysteresis cycle becomes small and the propagation delays through the comparator and the power stage come into play. Turning the hysteresis cycle down to zero, the switching frequency becomes:

$$f_{sw} = \left[\frac{V_{in}}{V_{in} - V_{out}} T_{don} + \frac{V_{in}}{V_{out}} T_{doff} \right]^{-1} \qquad [1.28]$$

where T_{don} and T_{doff} are the propagation delays. This is the so-called sliding-mode operation, as the current slides along the control surface as fast as possible.

So far, a nonlinear current loop that maintains the inductor current of the converter in the boundaries of a controlled value has been described. This current loop can then be used as an inner current loop for a current-mode-controlled converter. The design of the compensation function is similar to that of a peak current-mode control without the need for a slope compensation circuit.

1.4.4.2. *Current-mode sliding-mode converter*

This converter converges with the theoretical sliding-mode approach by the use of a sliding function that mixes the output voltage and the inductor current, as shown in Figure 1.36. This control structure is simply called the "hysteretic regulator" or the "hysteretic PWM regulator" [CAS 07, HAZ 05b, LI 11, RED 09] and is described as a way to increase the useful ripple. The added RC network integrates the voltage across the inductor and produces a scaled representation of the inductor current. Thus, "current-mode sliding-mode converter" or "hysteretic current-mode converter" are more explicit descriptions of this control scheme.

The AC part of the inductor current is equivalent to the output capacitor current since the output voltage is kept quasi-constant. Thus, another way to achieve the current-mode sliding-mode control is to measure or emulate the current through the output capacitor, C_o, as proposed in [HUE 13].

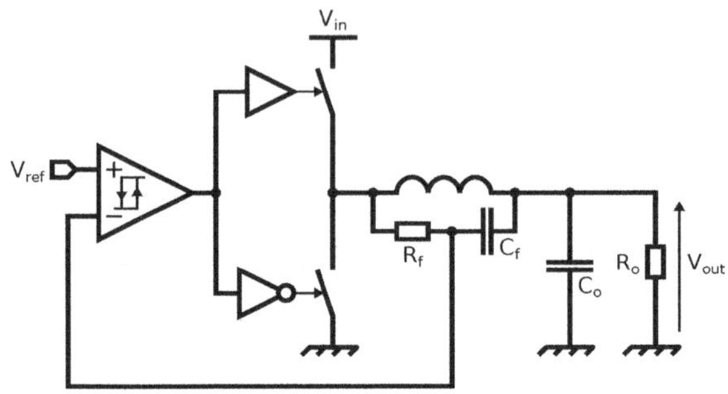

Figure 1.36. *Current-mode sliding-mode converter*

Considering the converter in Figure 1.36, the current-to-voltage transfer function of the current emulation circuit made up with R_f and C_f is [HUA 07]:

$$\frac{V_{C_f}(s)}{I_L(s)} = R_l \frac{1 + \frac{L}{R_L}s}{1 + R_f C_f s} = R_L \frac{1 + \tau_l s}{1 + \tau_c s} \qquad [1.29]$$

where R_L is the inductor equivalent series resistance, τ_l is the inductive branch time constant and τ_c is the capacitive branch time constant. This current measurement technique can be used in any current-mode control technique where the accuracy of the current measurement is not critical.

Once the sliding-mode existence condition is verified, the converter slides along a sliding surface S that can be expressed by making the comparator input error and input error dynamics equal to zero:

$$S(s) = V_{out}(s) + R_l \frac{1 + \tau_l s}{1 + \tau_c s} I_L(s) - V_{ref} = 0 \qquad [1.30]$$

When considering a single conduction cycle, the observer becomes a single gain K_{obs} (i.e. the slope gain) since the two time constants are larger than the switching period. Writing $i_L = I_L - I_o$, with i_l being the AC current component in the inductor, this assumption yields:

$$S(t) = V_{out}(t) + R_l I_o(t) + K_{obs} i_L(t) - V_{ref} = 0 \qquad [1.31]$$

Ideal sliding-mode operation implies that the converter switches with an infinite frequency; however, for the hysteretic current loop, the effective switching frequency is limited by delays in the control loop.

Neglecting the ripple in the inductor and averaging the converter behavior gives a simple model:

$$\overline{V}_{ref}(t) = \overline{V}_{out}(t) + R_l\overline{I}_o(t) \qquad [1.32]$$

where $\overline{X}$ is the average value of X.

It is obvious that the DC output impedance of the converter is non-negligible since the output voltage is load-dependent. A load-dependent drop is used to perform adaptive voltage positioning (AVP) as proposed in [CAS 07, LEE 09, MIY 14]. This method involves the reduction of the output voltage when the load current of a digital load increases. Voltage margin is not required to compensate for an output voltage undershoot when the load is operating at maximum current. Therefore, the output voltage of the converter can be reduced slightly without degradation of the performances of the digital circuit, and the power dissipation is slightly reduced.

1.4.5. *Ripple-based controls for fast dynamics*

Applications with highly demanding load steps and dynamic voltage scaling such as point-of-load converters and voltage regulator modules need very fast controls in order to comply with the dynamic requirement and still maintain an output capacitor as small as possible.

Ripple-based controls (Figure 1.37) are very appropriate to achieve a fast dynamic response. They are composed of a fast feedback path (FFB) and a slow feedback path (SFB). The FFB path is a rippled signal that provides information about the power stage, and it is responsible for the modulation of the duty cycle and the dynamic behavior of the control. The SFB path is an integrator designed to have a very low bandwidth, which is responsible for regulating the output voltage in the steady state. The peak current-mode control (Figure 1.30) is a ripple-based control, but the inductor current feedback cannot be considered a fast feedback since it is very limited by the maximum inductor current slope. However, if a feedforward of the output current is applied to the inductor current feedback, as shown in Figure 1.37, the corresponding loop becomes a fast feedback path. The duty cycle modulation in the steady state is defined by the inductor current ripple (Figure 1.38) as in the peak current-mode control. However, in the case of the ripple-based controls, the dynamic response is also defined by the ripple signal (Figure 1.39) while the voltage loop signal (v_{comp}) is almost constant during the load step.

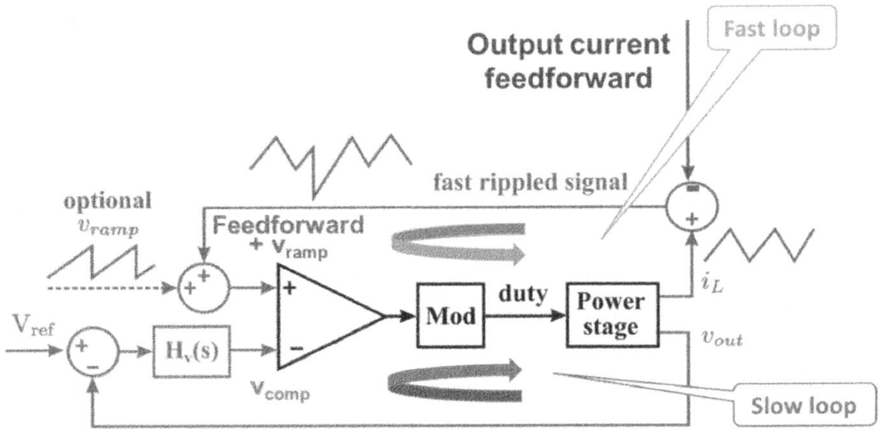

Figure 1.37. *Ripple-based control: general scheme dynamic. For the color version of this figure, see www.iste.co.uk/allard/systems.zip*

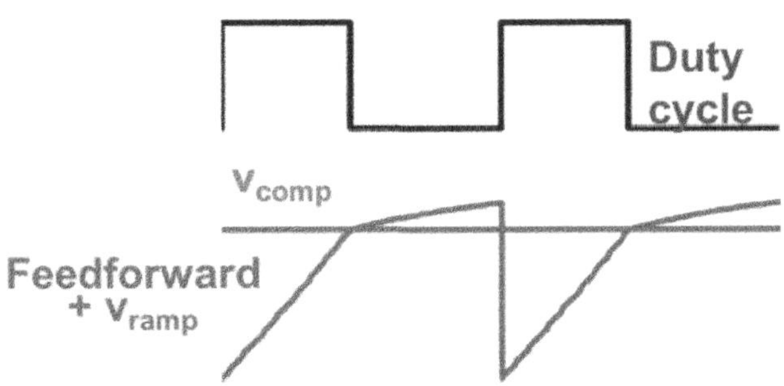

Figure 1.38. *Ripple-based control: duty cycle modulation. For the color version of this figure, see www.iste.co.uk/allard/systems.zip*

Analyzing the general scheme of the ripple-based control (Figure 1.37), it can be derived that the fast ripple signal, that is really fedback through the fast path, is the output capacitor current (Figure 1.40). This variable has all the needed information: 1) inductor current ripple for steady-state modulation and 2) feedforward of the output current for the load steps.

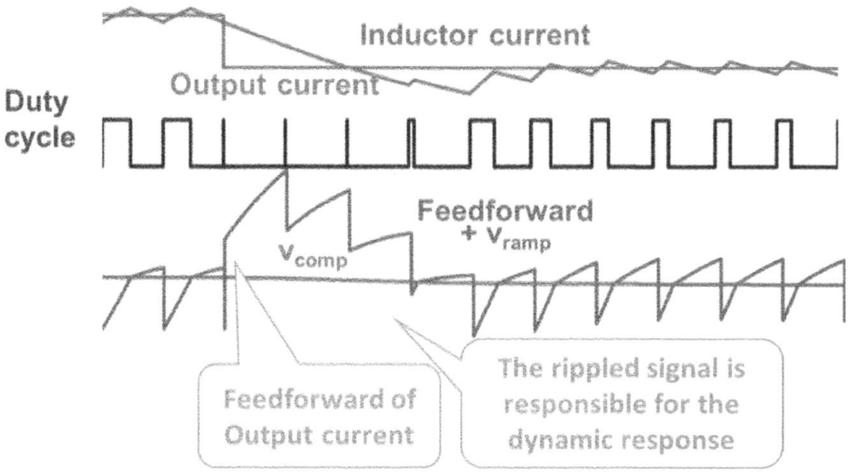

Figure 1.39. *Ripple-based control: response under a load down step*

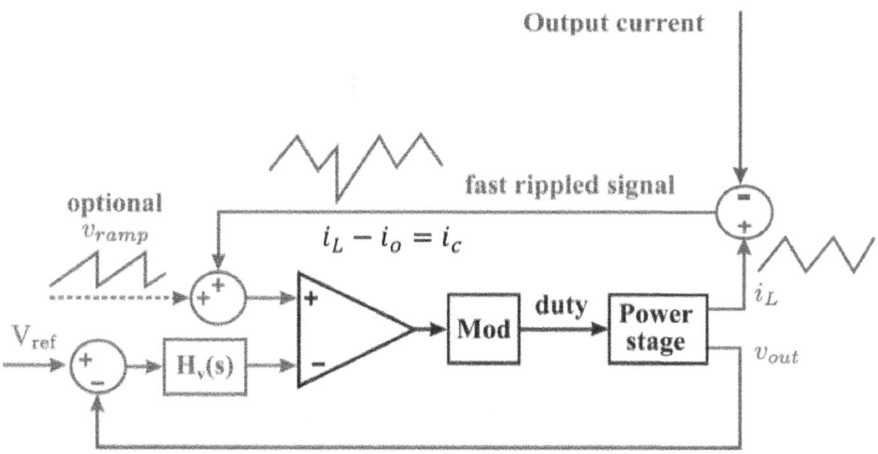

Figure 1.40. *Ripple-based control: output capacitor current feedback*

1.4.5.1. *V^2, quasi V^2 and V^2I_C converters*

The V^2 control is the most popular ripple-based control (Figure 1.41) [GOD 96], [ON 09]. It uses the output voltage in an accurate and slow loop in parallel with a fast loop that uses the output voltage ripple. In the V^2 control, the rippled signal of the FFB path is the output voltage (Figure 1.41a). In the case where the output capacitor has dominant equivalent series resistance (ESR), the output voltage ripple provides information about the capacitor current, which has the combined information of the inductor current and the output current. Consequently, the output voltage is shaped as the inductor current and can be used to modulate the duty cycle as in current-mode control, while exhibiting a fast dynamic response due to an inherent feedforward of the output current (Figure 1.41b).

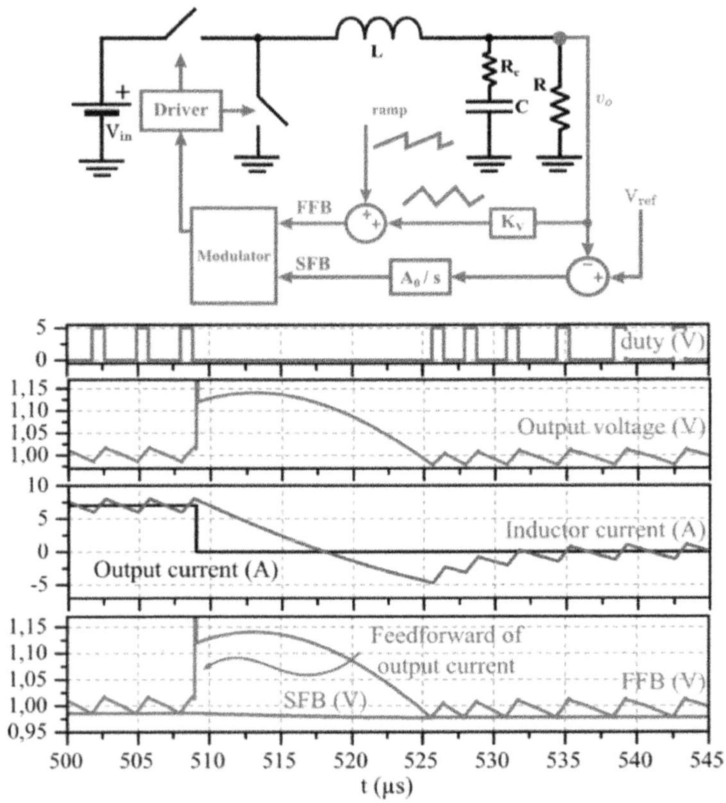

Figure 1.41. *V^2 control of a buck converter: (a) general scheme of a V^2 control on a buck converter; (b) load transient response for V^2 with constant on-time modulation. For the color version of this figure, see www.iste.co.uk/allard/systems.zip*

Note that the V^2 control only uses the output voltage but behaves like a current-mode control when the ESR of the output capacitor is large. For low ESR capacitors such as ceramic capacitors, the current information in the ripple of the output voltage is not dominant and, therefore, an additional current ripple has to be added to stabilize the converter. A popular way in the industry is to add the inductor current directly to the rippled signal (Figure 1.42), usually sensed with an R-C network. Then, the FFB path is composed of the sum of the output voltage and the inductor current (Figure 1.42a). This control is called enhanced-V^2 control [HUA 01], which is also named V^2 control with current injection [TEX 11], called V^2I_L for simplicity. As the information of the load current is limited, when low ESR capacitors are used, this control does not provide an optimum response under load transients since ceramic capacitors (low ESR) do not provide a good feedforward of the load current (Figure 1.42b).

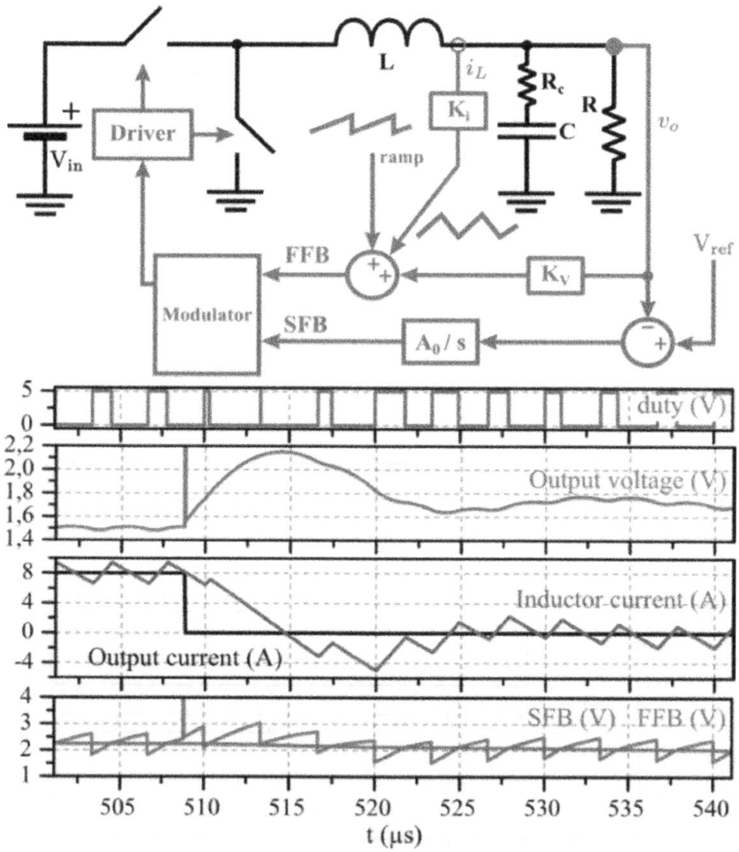

Figure 1.42. *IL control of a buck converter: (a) general scheme of a V^2I_L control of a buck converter; (b) load transient response for V^2I_L with constant frequency modulation. For the color version of this figure, see www.iste.co.uk/allard/systems.zip*

Instead of adding the inductor current, a better approach is to add the capacitor current (Figure 1.43). Then, the FFB path is composed of the sum of the output voltage and the capacitor current (Figure 1.43a). This control is named V^2I_C [VIE 11] and it was later proposed in [YAN 13] and [VOI 14] with different implementations but the same concept. As the capacitor current provides the control with information about the load current, V^2I_C reacts under load transient almost optimally even for low ESR capacitors [COR 14a] (Figure 1.43b). The dynamic response can be further improved by synchronizing the modulator with the load step [COR 15b].

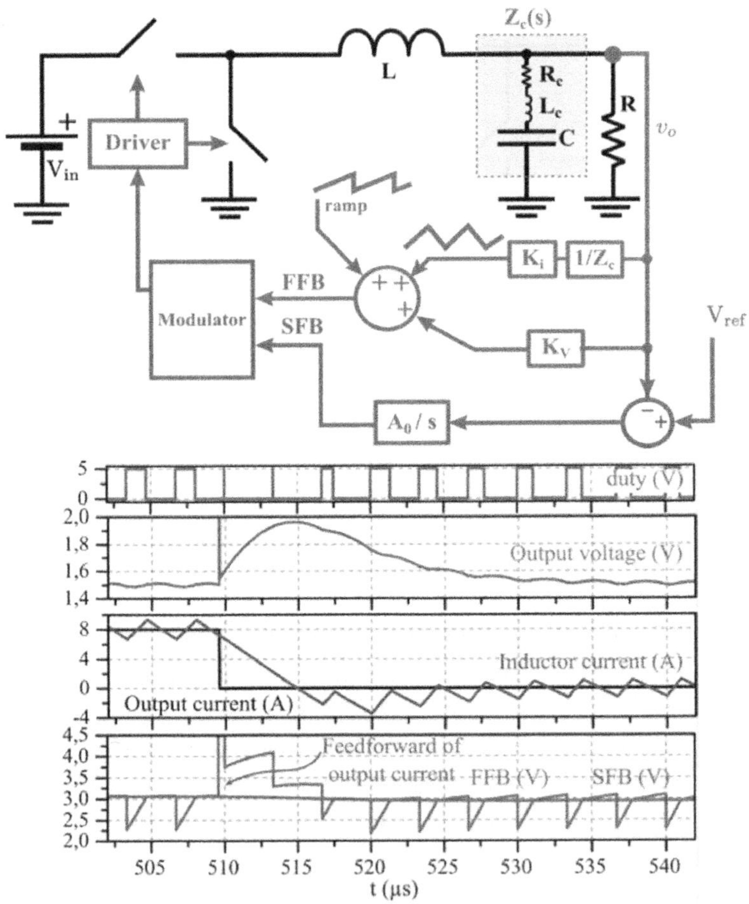

Figure 1.43. *V^2I_C control of a buck converter: (a) general scheme of a V^2I_C control of a buck converter; (b) load transient response for V^2I_C with constant frequency modulation. For the color version of this figure, see www.iste.co.uk/allard/systems.zip*

For the implementation of V^2I_C, the capacitor current can be estimated by sensing only the output voltage and designing a transimpedance amplifier with an impedance proportional to the impedance of the real output capacitor, including the ESR and the equivalent series inductance (ESL) [HUE 09a]. An alternative scheme of the implementation of V^2I_C is shown in Figure 1.44, where $K_i/Z_c(s)$ represents the sensor of the capacitor current. Note that, as with V^2 control with high-ESR capacitors, V^2I_C behaves like a current-mode control, but only senses the output voltage.

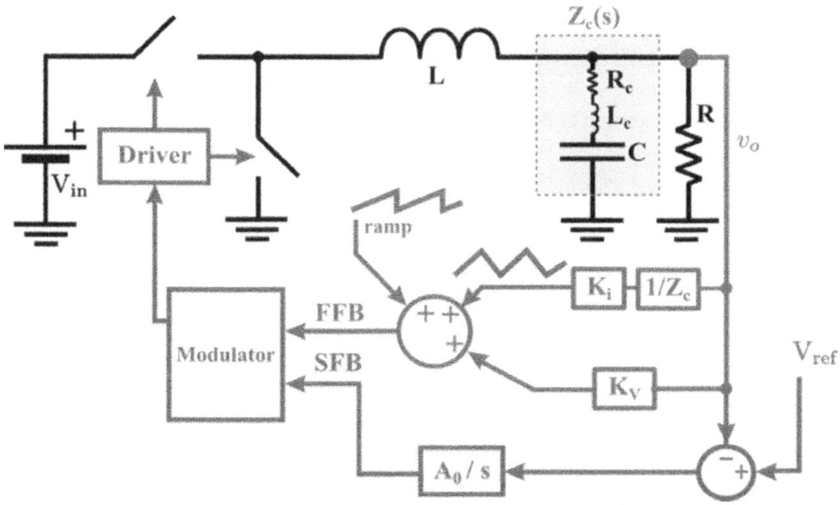

Figure 1.44. *V^2I_C scheme where capacitor current is estimated only by sensing the output voltage. This sensor of the capacitor current is explained in [HUE 09a]. For the color version of this figure, see www.iste.co.uk/allard/systems.zip*

V^2I_C control can be designed to provide an almost optimal response under load steps and under voltage reference steps for dynamic voltage scaling (DVS) applications or voltage tracking applications. Figure 1.45 shows the simulation results of design optimized for both dynamic specifications, load steps and voltage reference steps.

VMC, CMC, V^2, V^2I_L and V^2I_C are designed to be optimized for each control technique for the same specification and the same power stage. It is a 10 MHz buck converter, whose input voltage is 5V, output voltage is 1.2V and load steps range from 0 A up to 4 A and from 4 A down to 0 A. The worst transient is the down step since the steady-state duty cycle is very small. Figure 1.46 shows the results obtained with each control technique under the down step. V^2I_C presents the best results, with the output voltage deviation of 25% over the nominal output voltage. VMC, V^2 and V^2I_L present a very similar response with 33% output voltage deviation while CMC presents the worst response with 41% voltage deviation.

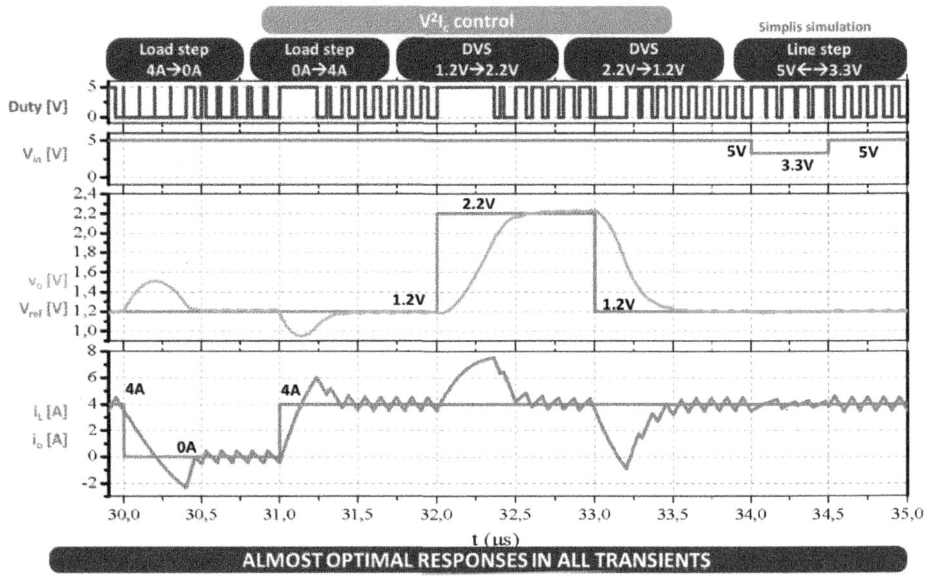

Figure 1.45. *V^2I_C simulations, designed optimized for both load steps and DVS. For the color version of this figure, see www.iste.co.uk/allard/systems.zip*

1.4.6. *V^1 concept: description and applicability*

Even if ripple-based controls only sense the output voltage (V^2 and V^2I_C), they are by nature current-mode controls. This is because the output voltage has inherent information of the capacitor current (from the ESR of the capacitor or by using a transimpedance amplifier). Also, the capacitor current itself has information from both the inductor current and the output current. This information about the output current is very important because it is what allows the V^2 and V^2I_C controls to behave almost time optimally under a load transient. Additionally, the inductor current has information from the input voltage during the on-time. This concept of only sensing once the output voltage and using its inherent information about the power stage is called V^1. Following the V^1 concept, the question then arises whether a traditional voltage-mode control can be designed in a way so that this intrinsic information is exploited and, consequently, it behaves like a current-mode control with a very fast dynamic response. This voltage-mode control could be modulated, as ripple-based and current-mode controls, with constant-frequency (peak or valley), constant on-time, constant off-time or hysteretic modulations. Figure 1.43 shows the structure of the V^2I_C control where the capacitor current is sensed with a transimpedance amplifier with an impedance proportional to the impedance of the output capacitor and implemented in two paths, the SFB and the FFB path. Now, as an alternative equivalent representation, both the sensing of the capacitor current and the output voltage of the FFB path can be deducted from the output of the integrator (Figure 1.47).

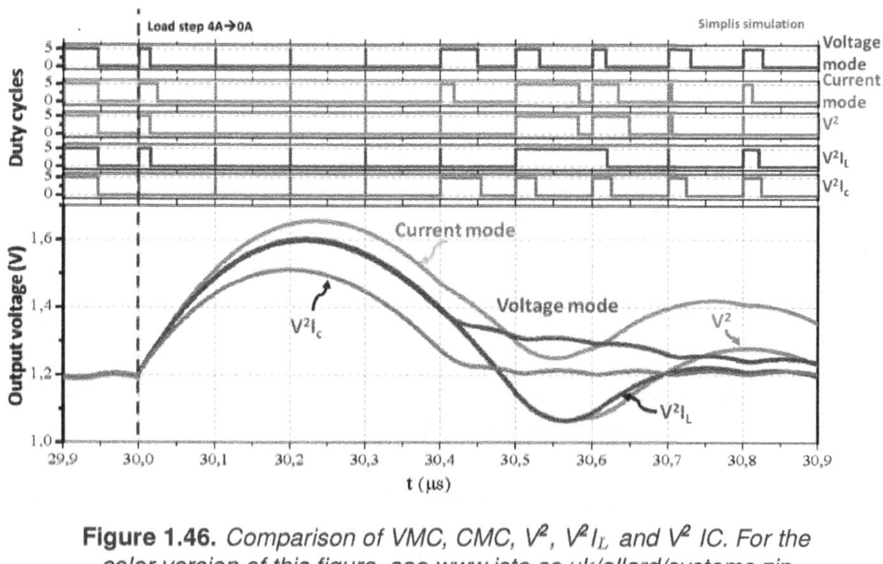

Figure 1.46. *Comparison of VMC, CMC, V^2, V^2I_L and V^2 IC. For the color version of this figure, see www.iste.co.uk/allard/systems.zip*

The equivalent regulator $H_t(s)$ is:

$$H_t(s) = \frac{A_0}{s} + K_v \qquad [1.33]$$

where Z_C(s) is the output capacitor impedance [1.34], with C being the capacitance value, RC the series resistance of the capacitor (ESR) and LC the series inductance of the capacitor (ESL):

$$Z_C(s) = (1 + sCR_C) \cdot (1 + s\frac{L_S}{R_C}) \qquad [1.34]$$

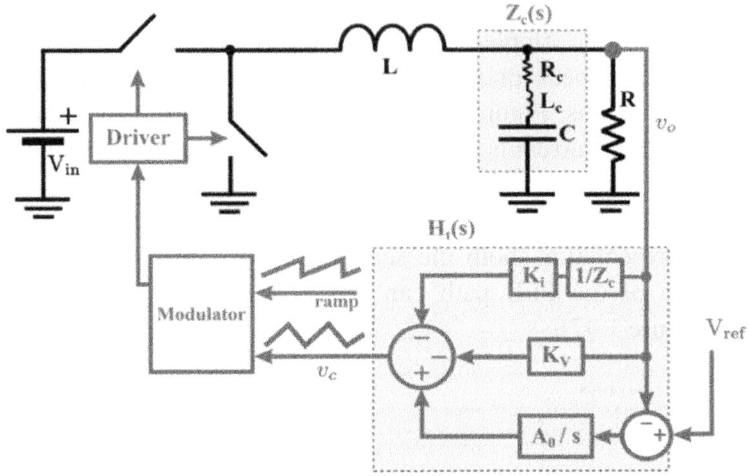

Figure 1.47. *Alternative representation of V^2I_C control. For the color version of this figure, see www.iste.co.uk/allard/systems.zip*

Figure 1.48 shows the Bode diagram of the controller H_t(s). This regulator has an integral action, A0/s, which closely regulates the output voltage, a proportional action Kv, that provides a zero to the regulator, boosting the phase, and a weighted estimator of the capacitor current, K_i/Z_C(s), that converts voltage information into current information. This current sensor is very important because it provides to the control a feedforward of the output current, needed to obtain a near time-optimal response under load transients.

In [COR 15d], it has been presented how this regulator can be designed with a classical type-III regulator of a VMC. It can be done for either low-Q capacitors or high-Q capacitors.

To obtain a voltage-mode control that behaves like a current mode with near time-optimal response under load transients, the regulator needs to mirror the impedance of the output capacitor at high frequencies. It is important to state that a perfect matching of the impedance of the output capacitor is not possible in an actual product. Tolerances of the output capacitor due to aging, temperature and DC bias will have to be considered when designing the controller. Its effect on the stability can be studied and the control can be optimized by means of the procedure proposed in [COR 15c] and [COR 14b], respectively.

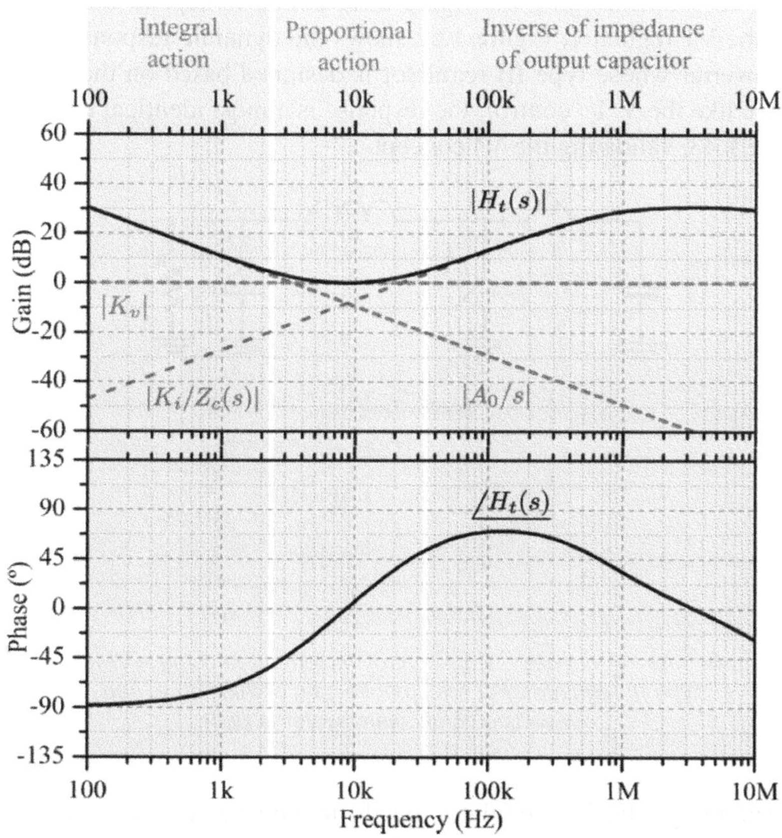

Figure 1.48. *Decomposition of the frequency response of the equivalent regulator of V^2I_C ($H_t(s)$). For the color version of this figure, see www.iste.co.uk/allard/systems.zip*

The key of the V^1 concept states the fact that a V^2I_C control can be implemented using a VMC with type III regulator (Figure 1.49). The block diagram shown in Figure 1.49 is the same as a traditional VMC with a type III regulator. The difference lies in the design of the regulator that must match equation [1.33]. Figure 1.50 shows two type III regulators; one designed as a traditional VMC and the other following the V^1 concept. The V^1 regulator follows the reverse of the output capacitor impedance at high frequencies, keeping the ripple information in the loop. On the contrary, a traditional VMC regulator filters the ripple information. Figure 1.51 shows the output of the regulator for both designs; the ripple information is kept in the loop in the case of the V1 regulator. Figure 1.52 shows the dynamic response of a VMC buck converter whose type III regulator is designed based on the V^1 concept to behave like the V^2I_C control; the response is almost identical to that shown in Figure 1.43, validating the V^1 concept.

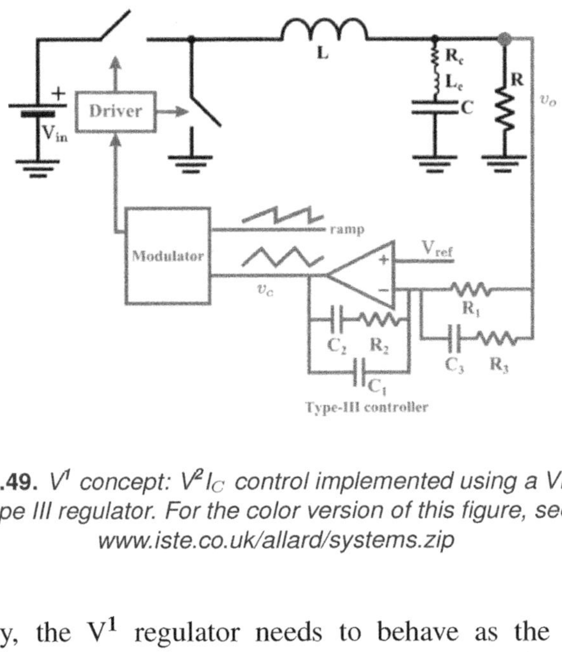

Figure 1.49. *V^1 concept: V^2I_C control implemented using a VMC with type III regulator. For the color version of this figure, see www.iste.co.uk/allard/systems.zip*

In summary, the V^1 regulator needs to behave as the inverse of the impedance of the output capacitor at high frequencies. In this way, the voltage information is converted to current information, and then the voltage loop has information about the capacitor current and therefore about the output current. This provides the control a kind of feedforward of the output current by only sensing the output voltage. As a result, the controller does not attenuate the side-band frequencies produced by the modulator. Thus, the control is prone to subharmonic oscillations as in ripple-based controllers.

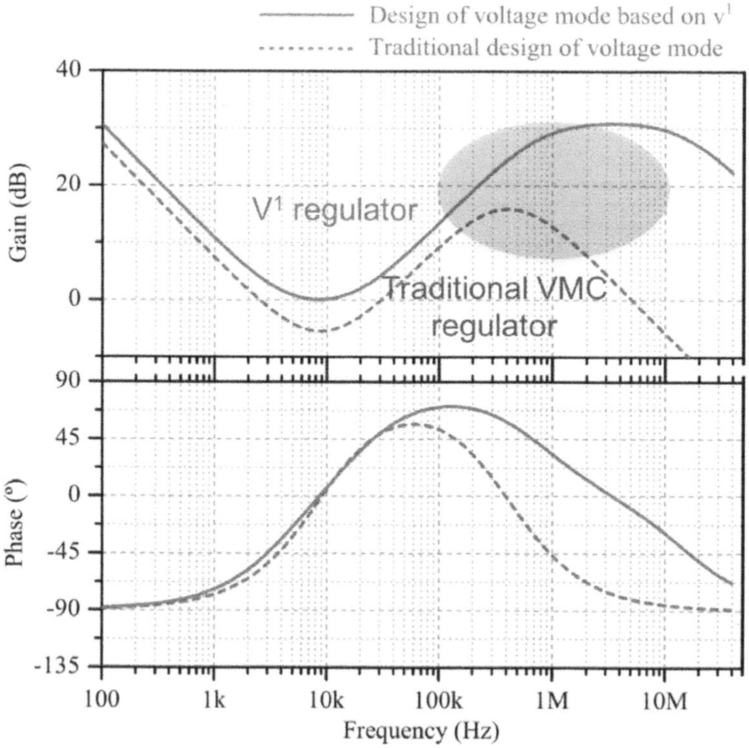

Figure 1.50. *V^1 concept regulator versus traditional VMC regulator design*

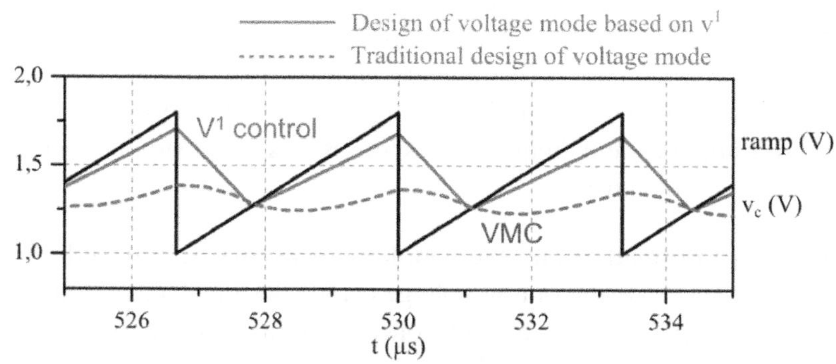

Figure 1.51. *V^1 concept: ripple information in the loop. For the color version of this figure, see www.iste.co.uk/allard/systems.zip*

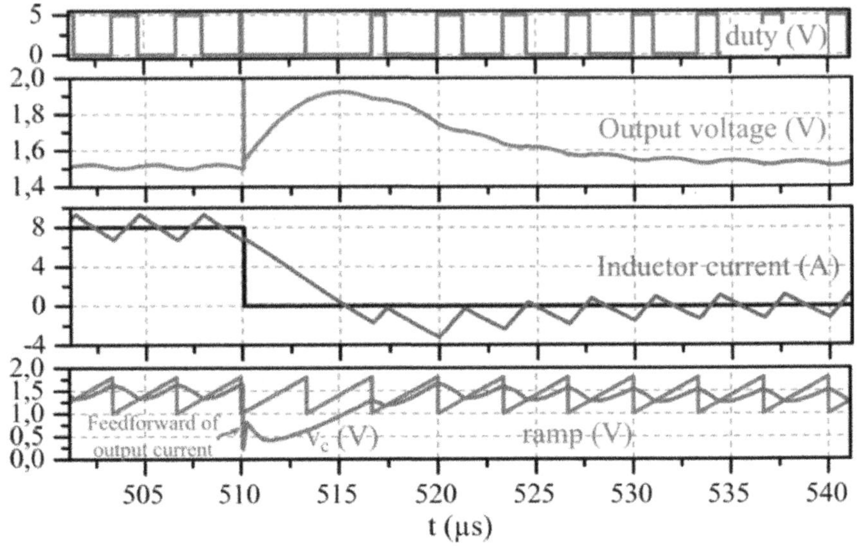

Figure 1.52. *V^1 concept: V^2I_C control implemented using a VMC with a type III regulator. Response under a load step. For the color version of this figure, see www.iste.co.uk/allard/systems.zip*

Therefore, a more accurate analysis is needed to account for tolerances of the parameters and mismatches of the current sensor. In [COR 15c], a methodology that takes into account all the above is proposed to evaluate the stability of converters based on discrete modeling and Floquet theory. Using these tools, an optimization algorithm that can design a very fast control can be developed, while assuring robustness under the whole operation region and accounting for tolerances of parameters and sensing networks [COR 14b]. Additionally, a commercial simulation program that can work with switched-mode power supplies and frequency responses is a good supplementary tool to design the V^2I_C control.

Given this information and methodology, low-cost, robust and very fast controls that only sense the output voltage can be designed and manufactured. As explained in [COR 15d], the V^1 concept can also be applied to V^2 and V^2I_C; V^2 can be designed as VMC, and V^2I_L can be designed as a current-mode control (Figure 1.1).

1.4.7. *Overview of the synchronization of asynchronous modulations*

1.4.7.1. *Hard synchronization paradigm*

The converter works in hard synchronization (HS) mode when each cycle is started by a reference clock. This synchronization paradigm does not allow asynchronous transient response but provides a better control of the generated EMI.

The first solution for phase synchronization in sliding-mode control is to use a RS or D-flop synchronization scheme to force each cycle as proposed in [RED 09]. The reference clock starts the cycle by setting the flop in a positive state and the pulse is generated until the sliding function is reached. When the comparator input reaches zero, the flop changes its state until the start of the next clock change. This operation mode is called the "fixed-frequency sliding-mode control" in [NI 09b] and "constant-frequency peak (or valley) voltage ripple regulator" in [RED 09]. Nevertheless, this synchronization scheme converts the sliding-mode controller into a current-mode-like control, with the drawback that a significant slope compensation or a complex system to switch between the peak and valley modes is required depending on the conversion ratio as proposed in [NI 09b]. Furthermore, the intrinsic sliding-mode asynchronous transient response is transformed into a current-mode synchronous transient response.

A hard synchronized converter is mandatory when the output noise and EMI have to be precisely controlled. When dealing with digital core power supply, things are very different. The output voltage has to be accurate and quickly modulated, but a particular spectrum of the supply voltage is not required.

In [RED 09], the sliding function is also modulated by a clock-synchronized ramp at the targeted switching frequency. The higher the sliding function modulation, the lower the sensitvity of the converter. If the sliding function is highly modulated, the transient has to have a high impact on the voltage error to produce a modulation variation; otherwise, the duty cycle will not be significantly changed and the perturbation will not be quickly rejected.

1.4.7.2. *Soft synchronization: compensated frequency control*

A constant on-time or a constant off-time (COT) configuration controls the converter in order to keep the advantage of asynchronous response. When the converter hits the sliding surface, the switch that is connected to the battery voltage is activated for a defined amount of time (on-time of the switch connected to the battery). The switch that is connected to the ground is also activated for a defined amount of time (off-time), respectively. The switching frequency is kept quasi-constant by adjusting the on or off-time using feedforward operation at both the input voltage and output voltage and eventually load current as in [HUA 11]. A feedforward action on the potentially controllable delays, T_{don} and T_{doff}, or hysteresis cycle, V_{hyst}, based on the measured voltage, can compensate the switching frequency variation since the input voltage and the output voltage are the most prominent external terms in the switching frequency expression:

$$f_s = \left[\frac{V_{in} V_{hyst}}{(V_{in} - V_{oout}) V_{out}} R_f C_f + \frac{V_{in}}{V_{in} - V_{out}} T_{don} + \frac{V_{in}}{V_{out}} T_{doff} \right]^{-1} \quad [1.35]$$

AOT control is well suited when the converter operates as a stand-alone solution where clock synchronization is not required, thanks to a dedicated battery decoupling network. The switching frequency is kept quasi-constant in a controlled band and preserved from any unwanted drop in efficiency. The on-time control degrades the falling transient performances, but this can be improved by the use of a ramp pulse modulation (RPM) strategy as proposed in [RED 09]. Last but not least, the sliding-mode-based asynchronous response is preserved.

The drawback that makes a feedforward synchronization-based strategy unsuitable for integration in a complex environment is the lack of phase synchronization. Instantaneous or average switching phase synchronization of the converters combined in a large power management integrated circuit is mandatory to reduce the ripple current in the battery and avoid possible interactions between the different DC/DC converters that share the same power line.

1.4.7.3. *Soft synchronization: regulated frequency control*

A more precise frequency control of the switching frequency of a converter can be achieved by the regulation of the free-running switching frequency.

Frequency locked loop (FLL) synchronization schemes for a buck converter are proposed in [FEN 09, HUE 09b] and a principle schematic is shown in Figure 1.53. The natural switching frequency (f_{out}) of the converter is sensed and converted into a voltage by a frequency-to-voltage converter. The reference clock (f_{in}) is converted into a voltage, which is compared with the observed switching frequency. The resulting error is then amplified and filtered by a compensation function prior to being used as a frequency-control input of the free-running DC/DC converter. As a result, the whole DC/DC converter is used as a VCO in the frequency loop.

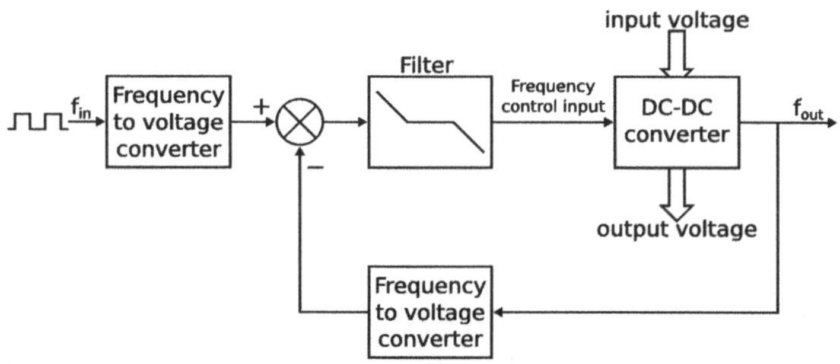

Figure 1.53. *Frequency locked-loop synchronization*

The frequency-to-voltage converters have to be properly matched to avoid unwanted offset in the frequency loop that generates a switching frequency error. The high DC gain provided by an arbitrary compensation function that stabilizes the frequency loop helps achieve no voltage error and therefore no frequency error. However, FLL synchronization still does not provide the phase synchronization effect. The converter behaves as an AOT control with a more accurate switching frequency, thanks to the feedback regulation.

The phase locked loop (PLL) synchronization operation principle is shown in Figure 1.54. The DC/DC converter operates as an oscillator in the frequency loop. The output phase is compared with the reference clock phase by a phase frequency detector (PFD). The PFD output is filtered and used to control the oscillator frequency. In [LU 14], the PLL synchronization is used in a multioutput converter.

A negligible phase error indicates a negligible frequency error between the desired switching frequency, f_{ref}, and the effective switching frequency, f_{sw}. Obviously, no phase error indicates that the switching cycles start synchronously in the steady-state operation. This ensures a good overall phase control while transient operation can start asynchronously.

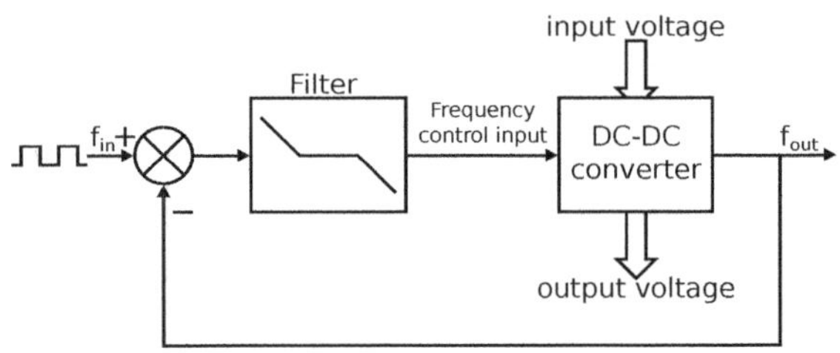

Figure 1.54. *Phase locked-loop synchronization*

1.4.8. *PFM - pulse skipping: burst modes*

The inductor current of any of the three converters can become momentarily negative at low load current. This significantly degrades the efficiency, as the RMS currents are high while the average currents are small and switching losses remain almost constant despite the low output power. Therefore, for low load, current-mode converters use some kind of discontinuous conduction mode in order to improve the low load efficiency of the converter.

These discontinuous conduction mode (DCM) modulation principles introduce a high impedance state where both switch and rectifier are turned off. The most straightforward solution is to drive the rectifier as an ideal diode so that the inductor current is always positive. Hence, a conduction cycle becomes an on-time, followed by an off-time and then by a high impedance state until the start of the next conduction cycle. This modulation scheme keeps the switching frequency of a PWM-controlled converter constant. However, switching losses do not scale down with a reduced load current and the efficiency rapidly falls. Furthermore, this scheme requires a fast zero-current-crossing detection that is difficult to design for high-frequency, multi-MHz converters.

In pulse frequency modulation, an on-time and off-time are generated followed by a high impedance state. The controller generates a conduction sequence (on- and off-time) when the output voltage hits the reference voltage. By appropriately setting the conduction times, the converter turns into high impedance mode when the inductor current is zero. The output voltage is maintained by the output capacitor and slowly discharges until the start of the next cycle. The PFM modulates the frequency of the current pluses in order to control the output voltage, and therefore the switching frequency is not controlled. However this does not significantly affect the efficiency as switching losses and conduction losses are proportional to the pulse density. Last but not the least, the bias current of the converter is the only current drawn from the battery during the high impedance state and can significantly affect the efficiency of the converter. For instance, 10 uA drawn from a 3.8 V battery represents 3.8% of a 1 mW load, such as a small digital circuit in retention mode.

In pulse-skipping mode, the converter skips one clock cycle in high impedance in order to avoid reversing current to the battery. This scheme keeps the effective switching frequency of the converter at an integer multiple of the clock frequency.

Burst mode operation increases the low load efficiency by delivering a burst of a few conduction cycles when the output voltage falls below the reference threshold. This operation must take place between the pulse-skipping mode and the PFM mode, in which the frequency of the burst is modulated but the length of the cycles is controlled.

1.5. Minimum voltage deviation controller

This section describes a hardware-efficient mixed-signal controller integrated circuit (IC) for DC/DC converter that recovers from load transients with practically minimum possible output voltage deviation. In the steady state, the IC behaves as a conventional voltage mode pulse-width modulation controller. During load changes, it enters the transient suppression mode that uses a simple algorithm, requiring no knowledge of the converter parameters and almost no processing power, to seamlessly recover back to the steady state without exposing components to a high current stress. To compensate the effects of converter losses and system delays on the controller operation, the IC also incorporates the duty ratio correction logic and dual extreme point-based detection. The simplicity of the control method enables the development and utilization of an application-specific asynchronous track-and-hold analogue-to-digital converter (ADC), with 10 times smaller silicon area and power consumption compared with state-of-the-art flash ADC. The entire IC is implemented in a CMOS 0.18μm process on a 0.26 mm^2 silicon area, which is comparable to the state-of-the-art analogue solutions. The IC enables about three times smaller output voltage deviation, when used in typical PoL 12V-to-1.8V 60W/120W applications, compared with PID compensators having a 1/10th of the switching frequency bandwidth.

1.5.1. *Introduction*

High-frequency low-power switch-mode power supplies (SMPS), used in consumer electronics, portable applications and computers, are required to meet stringent voltage regulation requirements [INT 08, PAN 01, SAG 04] using a cost-effective, small-volume implementation. The regulation is usually achieved by an application-specific on-chip integrated controller (controller IC) occupying a small silicon area [ON 11, INT 15, MAX 14]. The controller is often required to effectively minimize voltage deviations caused by load transients and, in that way, minimize requirements for the bulky output filter capacitor [ERI 01].

Conventional controller ICs [ON 11, INT 15, MAX 14] most frequently use constant-frequency voltage or current-programmed mode control [LIN 08], [TEX 07], where a linear PID compensator provides accurate output voltage regulation and system robustness over a wide range of operating conditions. To obtain a fast response and thus small output voltage deviations, a high bandwidth control loop is usually designed. However, since the validity of the averaged converter models used in compensator designs is constrained to frequencies that are significantly lower than the switching frequency [ERI 01], the bandwidth of the feedback loop is quite limited. Hysteretic-control-based analogue IC solutions, which belong to the ripple-based class of controllers [RED 09], have a very simple structure and, generally, provide a faster transient response. However, those systems usually operate at a variable switching frequency, which is not desirable in the targeted noise-sensitive applications. Also, in these systems, as well as in other ripple-based solutions, great care is required to mitigate stability problems [RED 09], such as jitter, high noise sensitivity and fast-scale instability. Constant-frequency ripple-based systems [RED 09], [MIC 08] eliminate the variable frequency and stability problems. The improvements are usually achieved by the introduction of an additional feedback loop that often slows down the transient response. Also, during transients, in hysteretic systems, the switching components and the inductor are often exposed to currents that are significantly larger than nominal [LEU 05]. As a result, the power stage is usually overdesigned. While the ripple-based solutions provide a very fast transient response, they are still not able to recover the output voltage with the minimum possible deviation. The recently developed digital controller ICs for high-frequency DC/DC converters [COR 09b], [COS 08] can further improve transient response through advanced control laws. Among the most notable examples are time-optimal [COS 08, GUA 06, GUA 07, MEY 12, ZHA 08, YOU 08, COR 09a, JIA 10, EFF 08, COR 10] and digital hysteretic [COR 09b, LEU 05, SU 08, HUE 11, COR 11, COR 08, CHE 10] controllers. For a given power stage, these solutions achieve almost the fastest possible recovery time and thus the smallest deviation. These advantages are most often demonstrated with large-scale prototypes consisting of discrete components and general-purpose digital logic. However, on-chip implementation of these controllers has not been widely adopted. This is mainly due to the challenges related to the complexity of the hardware needed for their implementation. The controllers require fairly demanding computational algorithms, to calculate the optimum switching sequence, and use complex analogue-to-digital converters (ADC) consuming relatively large power and silicon area. Furthermore, their operation is significantly affected by the imperfections of the converter circuit, i.e. converter losses and parasitic resistances, as well as by the system delays. All these characteristics make the implementation of the on-chip minimum-deviation controllers challenging and overly expensive for the targeted cost-sensitive applications. As demonstrated in [COR 09b, COS 08], an application-specific ADC for digital hysteretic and time-optimal controller occupies a larger silicon area than an entire conventional controller IC [CHE 08, TRE 11, LIO 08, CHA 10, AHN 11, WU 10].

The main goal of this section is to describe a simple minimum-deviation mixed-signal controller IC for DC/DC converters that does not suffer from the previously mentioned drawbacks and can be implemented with simple hardware. Similar to solutions presented in [GUA 06, GUA 07, MEY 12, ZHA 08], [GUA 06, GUA 07, MEY 12, ZHA 08], [COR 09a], [COR 10], the controller IC shown in Figure 1.55 has two distinctive modes of operation.

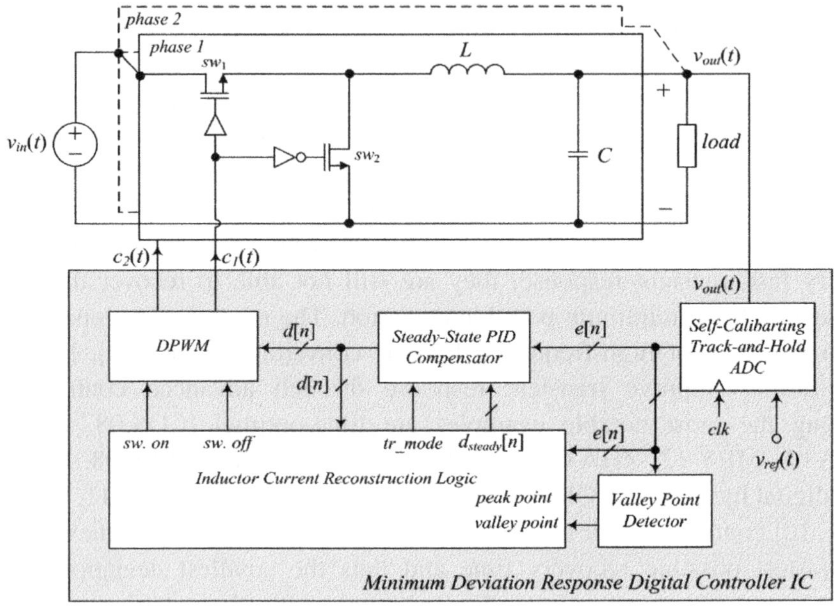

Figure 1.55. *Minimum deviation controller IC regulating operation of a two-phase buck converter*

In the steady state, the controller IC operates as a conventional constant-frequency pulse-width modulated voltage-mode controller. It uses a PID compensator to provide robust operation over a wide range of operating conditions. During transients, the controller provides recovery from load step changes with practically minimum possible output deviation by charging/discharging the inductor current to the new steady-state current value using a single on-off control action.

In this case, an accurate reconstruction of the entire current waveform, i.e. both the DC and ripple components, is very important. This is because, in modern converters, the ripple component can be as large as 40% of the maximum load current value [NXP 07a, FAI 07] and thus any ripple mismatch can have the same effect as a large load transient. The operation of the controller during a sudden light-to-heavy load change can be explained by the waveforms shown in Figure 1.56. As soon as the output voltage drop exceeds the threshold value, i.e. the detection of a transient, the controller turns on the converter main switch (SW1, Figure 1.55) causing the ramp up of the inductor current. This process continues until the detection of the output voltage valley point, by the track-and-hold ADC [RAD 13] shown in

Figure 1.55. At this point, the initial on-time of the transistor control signal, c(t), is extended by the time interval t_{on} in [1.36], followed by a turn-off period in [1.37], where D is the steady-state duty ratio value and T_{sw} is the switching period:

$$t_{on} = \frac{D \cdot T_{sw}}{2} \quad [1.36]$$

$$t_{off} = (1 - D) \cdot T_{sw} \quad [1.37]$$

As shown in Figure 1.56, this switching sequence results in a current waveform that is practically identical to that of the new steady state. As soon as this current reconstruction period is completed, the PID compensator is reactivated to perform charge recovery, bringing the output voltage to the reference value. It should be noted that the new operating conditions are well suited for the PID compensators, as the converter operates close to the steady state, where the small-signal assumption [ERI 01] is valid. As a result, smooth mode transition can be achieved despite system delays, power converter parasitics and non-idealities [RAD 13]. In addition, the simplicity of the minimum-deviation concept lends itself well to utilization in multi-phase systems [RAD13], [STR 14].

1.5.2. *Integrated circuit implementation and experimental results*

A single/dual-phase minimum-deviation digital controller IC [RAD 13], designed and fabricated in a 0.18μm CMOS process, is shown in Figure 1.57. The complete digital logic takes about 4500 gates, including 500 digital blocks used for testing and debugging, and is fully implemented through the Verilog hardware description language (HDL) and automated design tools. The main characteristics of the IC are listed in Table 1.1.

The closed-loop verification of the minimum-deviation controller IC was performed with a state-of-the-art industrial two-phase, 12 V to 1.8 V, 500 kHz, buck converter module, providing 35 A per phase [NXP 07a, NXP 07b]. The ESR of the 400 μF output capacitor is just 0.5 mΩ and each phase uses a 0.47 μH inductor. The experimental waveforms shown in Figure 1.58 indicate the transient response results of the conventional PID compensator, designed to have the bandwidth of a 1/10th of the switching frequency, during a 0A-to-30A light-to-heavy transient. In Figure 1.59, the suppression logic is active. In this case, the duty ratio correction logic is not used.

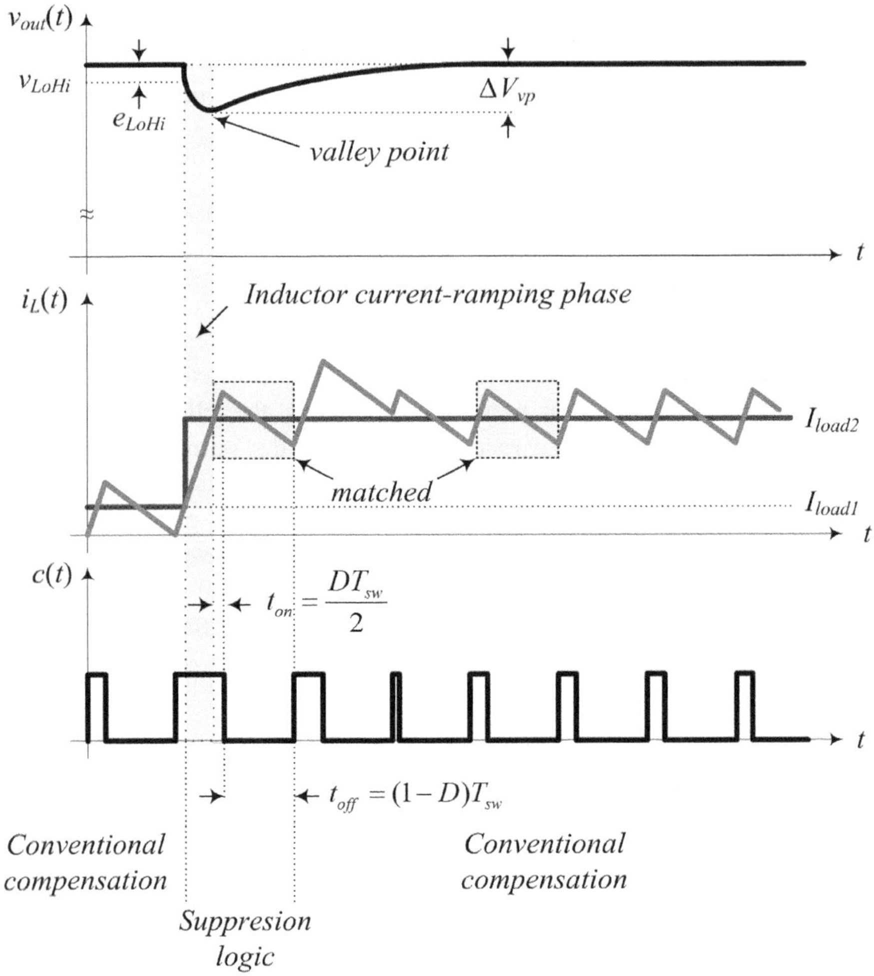

Figure 1.56. *Operation of the large–small signal digital controller during the light-to-heavy load transient with a single-phase buck. For the color version of this figure, see www.iste.co.uk/allard/systems.zip*

The results indicate that the activation of the suppression logic results in about three times smaller deviation, allowing for a proportional reduction of the output capacitor. Similar improvements can be achieved for both conventional multiphase interleaved buck topologies [RAD 13] and hybrid inductive-capacitive SMPS [NXP 07b, RAD 14].

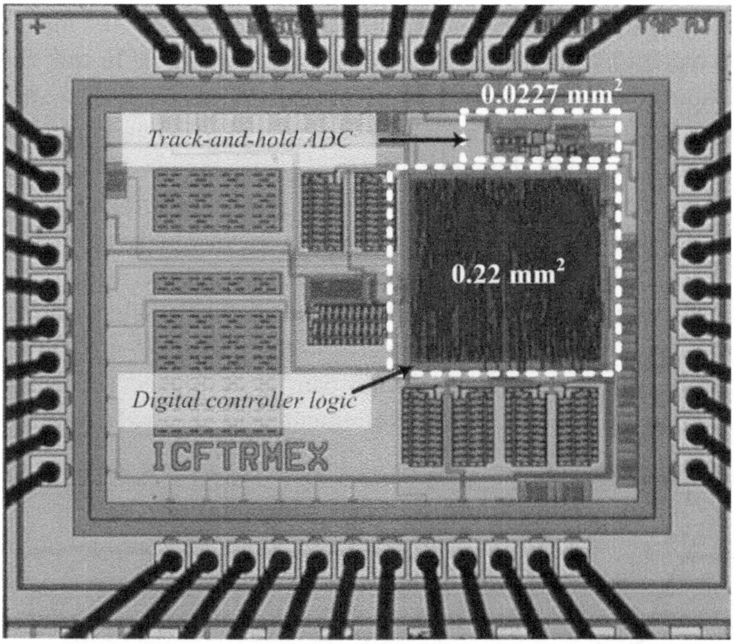

Figure 1.57. *Photograph of the minimum-deviation digital controller IC. For the color version of this figure, see www.iste.co.uk/allard/systems.zip*

Technology	TSMC 0.18μm CMOS
Supply voltage (digital/analogue)	1.8 V / 3.3 V
DPWM resolution	13 bits
DPWM nominal frequency	500 kHz
ADC area	0.02 mm^2
ADC current consumption	0.24 mA
ADC quantization voltage	4 mV
Digital logic area	0.22 mm^2
Digital logic current consumption	0.2 mA
Total area	0.24 mm^2

Table 1.1. *Minimum-deviation digital controller IC summary*

1.6. CAD tools for PwrSoC design and optimization

The design of integrated power converters on-chip can be formulated as an optimization problem that satisfies some electrical and thermal requirements subject to geometrical constraints with given technologies (semiconductors, magnetic materials, integrated capacitors and packaging). In this chapter, as an application example, we will focus on the optimization of a single-stage buck converter that will be optimized in terms of efficiency at typical load, satisfying the static and dynamic electrical requirements with constraints in the total area occupied by the passives (integrated inductor and capacitors). Finally, the optimization can be seen as a process in which many designs that comply with all the requirements are carried out, and the results are ordered according to some performance metrics that allow the selection of the best design according to the selected metrics. For example, one performance metric can be the cost, and the design can be optimized according to that metric satisfying all the electrical requirements and physical constraints.

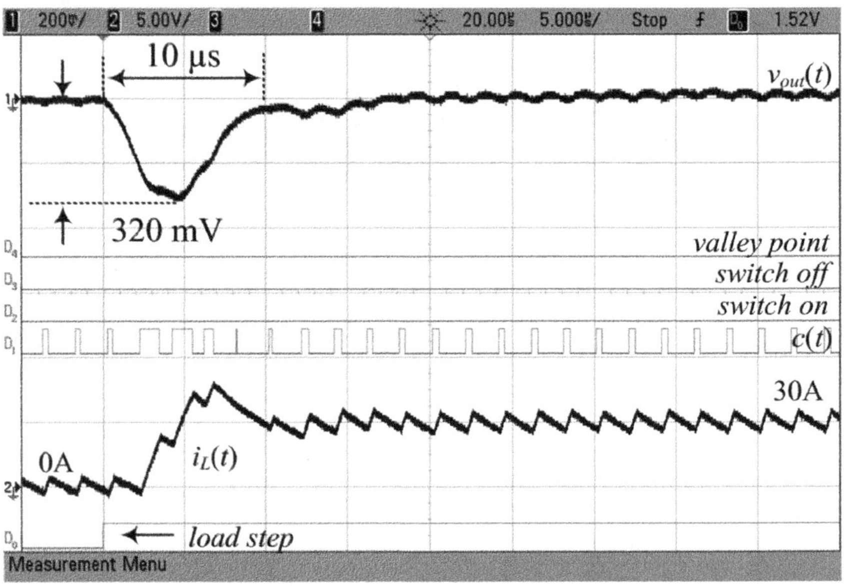

Figure 1.58. *Transient response waveforms for a 0 A-to-30 A load step (wide-bandwidth conventional compensator): Ch1, output converter voltage (200 mV/div); Ch2, actual inductor current iL(t) 30 A/div; D_0, load transient triggering signal; D_1, pulse-width modulated control signal; D_2, main switch control signal created by suppression logic; D_3, synchronous rectifier control signal created by the suppression logic; D_4, peak/valley point detection; time scale is 5 µs/div*

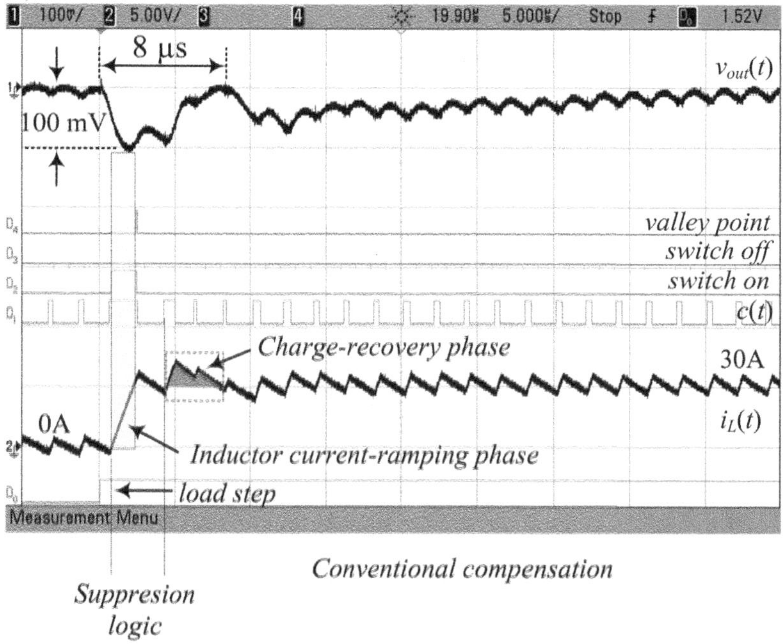

Figure 1.59. *Transient response waveforms for a 0 A-to-30 A load step (minimum-deviation controller IC): Ch1, output voltage (200mV/div); Ch2, inductor current iL(t) 30A/div; D_0, load transient triggering signal; D_1, pulse-width modulated control signal; D_2, main switch control signal created by suppression logic; D_3, synchronous rectifier control signal created by the suppression logic; D_4, peak/valley point detection; time scale is 5 μs/div*

1.6.1. *Overview of the CAD requirements*

The key element in the optimization is the design process. The design flow is shown in Figure 1.60. From the specifications, at the top of the process to the final design at the bottom, three layers can be identified.

Design options – These are the building blocks of the integrated power converter:

– technology for the semiconductor devices, the integrated inductors and the integrated capacitors. The designer can select from among different technologies for all these components that will affect the final result;

– topology. Different topologies can be used to process the energy (single-phase buck, multiphase buck, coupled inductors, switched capacitors, hybrid, etc.). The selection of the topology is coupled with the available technology and the control;

– control. As has been shown in this chapter, the control has a strong influence on the size of the LC filter; as a result, its selection and optimization will affect the final results;

– modes of operation. To improve the efficiency at light load, different strategies to modulate the converter can be applied. The trade-off between the higher complexity and the efficiency benefit can be assessed based on the optimization results.

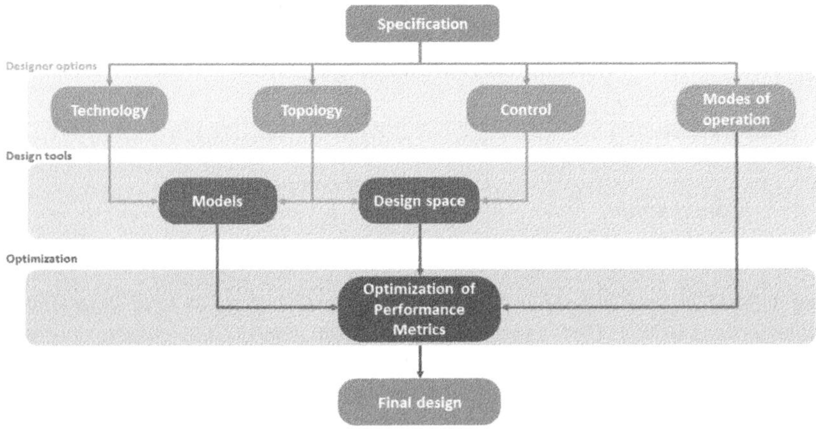

Figure 1.60. *Design layers for the optimization of integrated converters on-chip. For the color version of this figure, see www.iste.co.uk/allard/systems.zip*

Design tools – The design of the converter to meet the specifications with the designer options of the previous level has to be done based on:

– models. Different types of models (losses, static, dynamic) will be needed in the design process depending on the type of metric to be evaluated. The models will help to quantify losses, to evaluate the static behavior of the converter (waveforms in the steady state) and to design the controller to meet, if possible, the dynamic requirements of the converter;

– design space. Based on the design options selected at the design option level, and based on the specifications to be met by the converter, a design space will be identified for each switching frequency. The design space is represented by the values of the output inductor and the output capacitor, for a given switching frequency, topology and type of control, which will provide a valid solution (meeting all the requirements).

Optimization – Based on the above definition of the models and the design space, the optimization algorithm will search for the optimum solution according to some performance metrics. The optimization can be either single-objective optimization, such as optimizing the efficiency at a given operating point (meeting all the requirements), or multiobjective, such as optimizing a weighted function of several performance metrics. For this purpose, different optimization algorithms can be used. Nevertheless, even though the optimization algorithm has a strong effect on the speed of the search and to avoid local minima, the biggest effort has to be placed on the models used for the design.

1.6.2. *Loss models for integrated inductors and semiconductors*

One of the key design metrics is the losses. To carry out the design, it is necessary to have accurate loss models for the semiconductors, integrated magnetics and capacitors, as a function of the operating conditions. The models can shift from simplified analytical models that allow a first-order approximation and to gain physical insight, to highly complex 3D models based on finite element analysis calculations for the losses on the magnetic components. Although the complex models are more accurate, and allow taking into account high-order effects, the computation time required restricts its use on the optimization algorithms. They can be used, at the end of the optimization process, to validate the end results.

1.6.2.1. *Analytical models for magnetic components*

Magnetic components have to be designed in an ad hoc manner during the optimization process. As long as the value of the inductance, the switching frequency or the available area for the inductor change during the optimization, a new inductor needs to be designed for these operating conditions. The models for the inductors will help optimize their design and quantify their influence in terms of losses in the overall design. Since the optimized design of the inductor is within the power converter optimization loop, analytical models that quantify the losses will be preferable to speed up the process.

To obtain an analytical model of the inductor, 1D approximations can be used [DOW 66]. In this case, the analytical model of a basic inductor, validated by 2D FEA simulations, can be implemented based on the equations presented in [AND 13a]. The losses of the inductor are divided into DC conduction losses, $PL_{(Cu\_DC)}$, AC conduction losses, $PL_{(Cu\_AC)}$, core hysteresis losses, $PL_{(Fe\_Hyst)}$, and core eddy current losses, $PL_{(Fe\_Eddy)}$. The inductor conduction losses are estimated using the inductor DC resistance, R_{DC}, and AC resistance factor for k^{th} switching frequency harmonic, F_k; therefore, the conduction losses are calculated by using [1.38] and [1.39], where $I_{(L\_DC)}$ is the DC inductor current and I_{LK} is the amplitude of the inductor current k^{th} switching frequency harmonic:

$$P_{L\_Cu\_DC} = R_{DC} \cdot I_{L\_DC}^2 \qquad [1.38]$$

$$P_{L\_Cu\_AC} = R_{DC} \cdot \sum_{k=1}^{k_{max}} F_k \cdot \frac{I_{LK}^2}{2} \qquad [1.39]$$

Core losses are calculated by adding hysteresis and eddy current losses. Hysteresis losses, created due to the hysteretic dependence of the flux density (B) versus magnetic field (H), are calculated using [1.40], where K_H and b are material-dependent parameters, V_C is the core volume, f_{SW} is the switching frequency and ΔB_{PP} is the peak-to-peak flux density, calculated using [1.41]:

$$P_{L\_Fe\_Hyst} = K_H \cdot f_{sw} \cdot (\frac{\Delta B_{PP}}{2})^b \cdot V_C \qquad [1.40]$$

$$\Delta B_{PP} = B_{DC} \cdot \frac{\Delta IL_{PP}}{IL_{DC}} \qquad [1.41]$$

The eddy current losses of the magnetic core are calculated by using proximity loss estimation. It is assumed that the eddy currents are generated due to the proximity effect of the generated magnetic field in the core. For detailed information about the calculation, the reader can refer to [SVI 14] and [AND 13a].

1.6.2.2. *Semiconductor loss models based on combined simulation and analytical models*

To calculate the losses in the semiconductors, three typical approaches can be adopted:

– Analytical models. Analytical models are based on simplified analytical equations that describe the behavior of the devices and allow the calculation of the losses by solving a simplified equivalent circuit. One example is the calculation of MOSFET switching losses provided in this chapter. Other approaches that take into account the parasitic inductances can be found in [REN 06]. The main advantage of this approach is that it fits very well into the optimization process and its main drawback is the accuracy;

– simulation models. The use of accurate SPICE models provided by the IC manufacturer allows the use of detailed simulations to calculate losses in the semiconductor devices. It requires running a detailed simulation whenever a design parameter is changed, with the corresponding increment of computation time;

– combined simulation and analytical models. The idea is to combine both approaches, i.e. the accuracy of the SPICE simulations with the speed of the analytical models. In this approach, the device losses are computed by running an extensive set of simulations once, and the results are stored in multidimensional matrices than can be accessed very quickly during the optimization process [AND 13a].

Due to its usefulness, this approach will be explained in more detail in this chapter. The process will be illustrated using a buck converter as an example, as shown in Figure 1.61, where the high-side and low-side switches are PMOS and NMOS transistors. The process is divided into two steps:

1) characterization of the devices by means of SPICE simulations;

2) calculation of losses under any operating condition of the converter based on analytical equations.

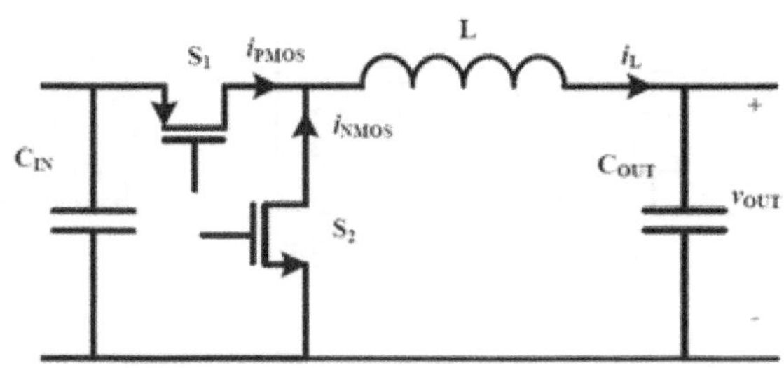

Figure 1.61. *Schematic of a buck converter*

1.6.2.2.1. Evaluation of energy losses

A discrete number of accurate calculations can be performed using SPICE simulations. From these calculations, interpolated piece-wise linear functions are constructed, allowing the estimation of the losses for all the points that belong to the input domain of the model. The computations are performed by sweeping the independent variables of the measured function (w and I and v_{GS}, if needed).

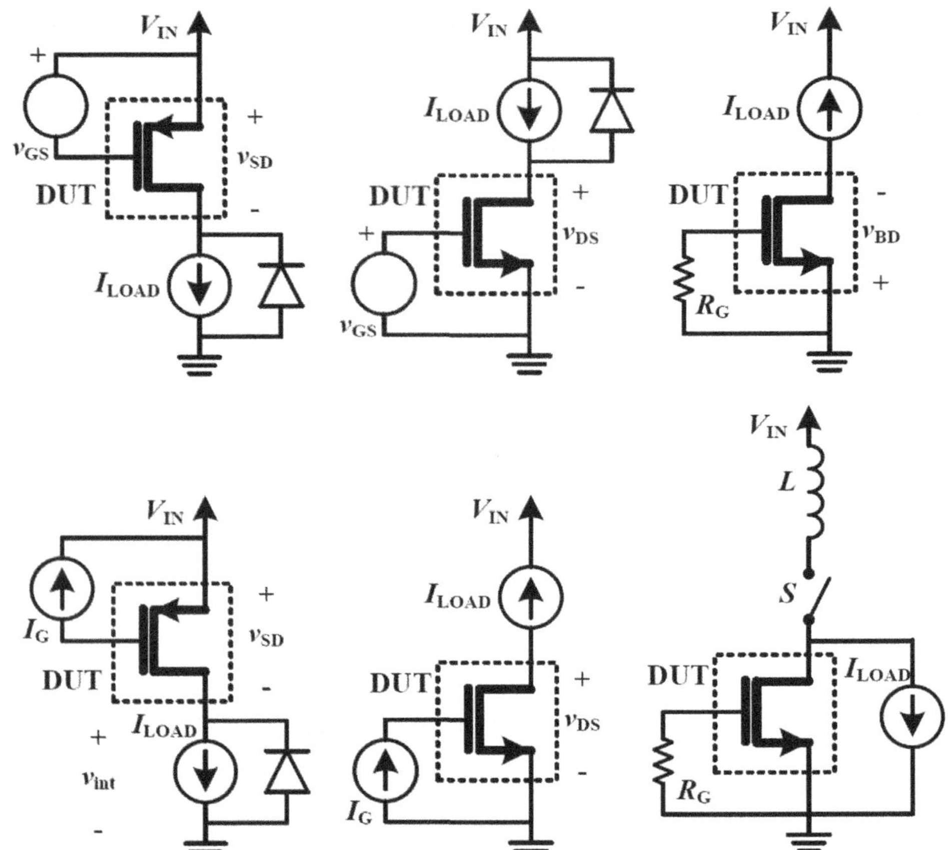

Figure 1.62. *Circuits for characterizing MOSFETs: a) PMOS on resistance, b) NMOS on resistance, c) NMOS body-diode voltage drop, d) PMOS turn-on and turn-off energies and driving gate charge, e) NMOS driving gate charge and f) NMOS body-diode reverse recovery energy loss*

To calculate the ON resistance of the PMOS, the circuit shown in Figure 1.62(a) is used varying the MOSFET width w_P, the load current I_{LOAD} and the gate-to-source voltage v_{GS}. Finally, by measuring the source-to-drain voltage drop v_{SD}, the on-resistance is calculated and the dependence $R_{PMOS}(w_P,v_{GS},I_{LOAD})$ is obtained. Similarly, the circuit shown in Figure 1.62(b) is used to obtain the on-resistance of the NMOS, obtaining $R_{NMOS}(w_P,v_{GS},I_{LOAD})$.

The circuit shown in Figure 1.62(c) is used to measure the NMOS body-diode voltage drop sweeping the MOSFET width w_N and the load current I_{LOAD}, while the gate is connected to the source via a small resistance R_G, which emulates the driver output resistance. Finally, the dependence $v_{(D_MOS)}(W_N,I_{LOAD})$ is obtained.

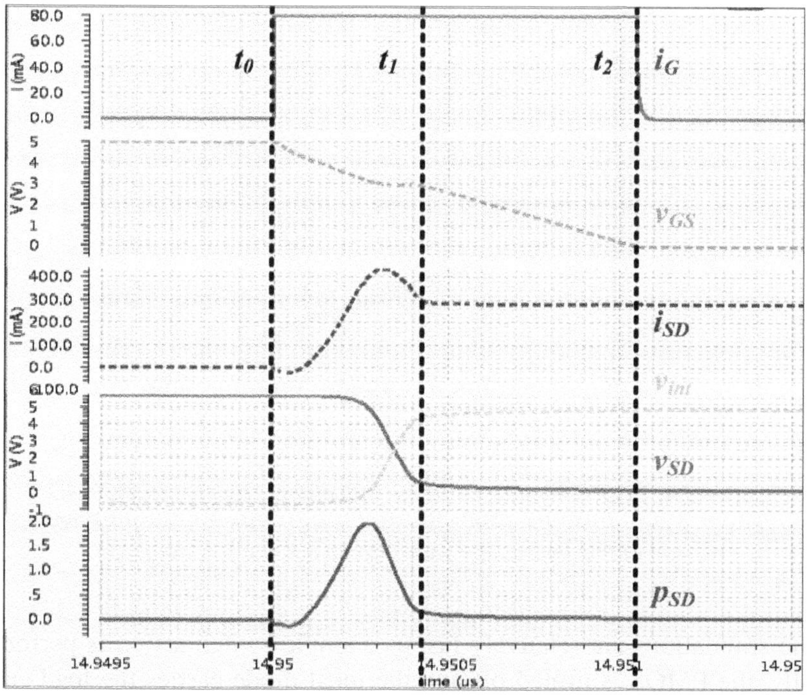

Figure 1.63. *SPICE simulation – PMOS turn-on energy loss and gate-charge calculation: gate current i_G (blue, solid), PMOS gate-to-source voltage v_{GS} (green, dotted), PMOS source-to-drain current i_{SD} (blue, dotted), intermediate voltage v_{int} (green, dotted), PMOS source-to-drain voltage v_{SD} (purple, solid) and PMOS source-to-drain power loss P_{SD} (light green, solid). For the color version of this figure, see www.iste.co.uk/allard/systems.zip*

The dynamic characteristics of the switches are obtained using circuits shown in Figure 1.62(d-f), and using the simulation results shown in Figure 1.63. To perform the SPICE calculations, the driver is modeled as the current source injecting a constant current during the transient. The circuit shown in Figure 1.62(d) is used to calculate both turn-on and turn-off energy loss and gate charge. The turn-on energy loss calculation and gate-charge calculation are shown in Figure 1.64 for a load current of 280 mA, a PMOS width of 8 mm and an input voltage of 5V.

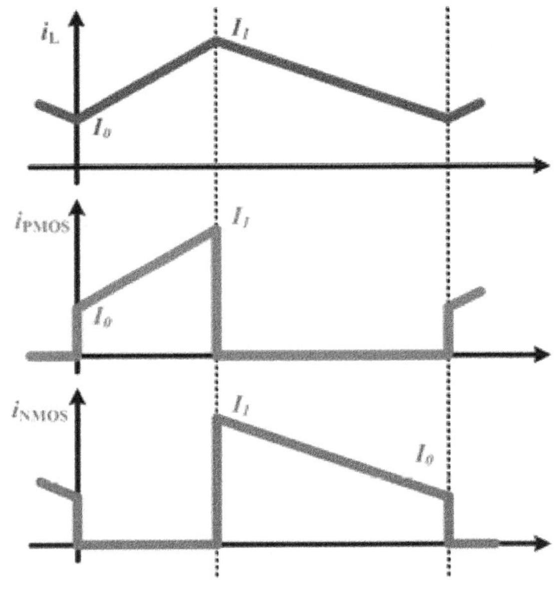

Figure 1.64. *Buck converter currents: inductor current i_L (blue), PMOS current i_{PMOS} (red) and NMOS current i_{NMOS} (green). For the color version of this figure, see www.iste.co.uk/allard/systems.zip*

The simulations are relatively fast as a single switch action is performed. Initially, the PMOS is turned-off and the ideal diode carries the load current while the PMOS blocks the input voltage. At the time instant t_0, the driver current saturates to a maximum value I_{DR} (80 mA for the present case) and the gate-to-source voltage v_{GS} starts to decrease. When the PMOS current i_{SD} reaches the load current, the source-to-drain voltage v_{SD} starts to decrease up to the voltage drop defined with the product of on-resistance and the load current at moment t_1. On the contrary, the gate-to-source voltage continues to decrease up to the targeted value v_{GSP}.

The product of the PMOS current i_{SD} and the source-to-drain voltage v_{SD} provides the instantaneous power loss p_{SD}. Integrating the power loss from t_0 to t_1, turn-on energy loss, $E_{(PMOS\_turn\_on)}(w_P,I_0)$, is calculated for a given PMOS width and instantaneous current. Similarly, integrating the gate current i_G from 0 to t_2, the gate-charge $Q_{PMOS}(v_{GSP},w_P,I_0)$ dependence is obtained for the same PMOS width and the load current I_{LOAD} and the driving voltage V_{GSP}.

Similarly, using the same circuit, the turn-off energy characteristic $E_{(PMOS\_turn\_off)}(w_P,I_1)$ can be calculated. Furthermore, using the circuit shown in Figure 1.62e, the NMOS gate-charge dependence is obtained. Initially, the NMOS is turned off and the body diode carries the entire load current. Again, the gate current is saturated at its maximal value I_G at moment t_0 and the gate-to-source voltage starts to increase. Because the NMOS is turned-on with small drain-to-source voltage (determined by the body-diode voltage drop), it can be assumed that the switching is nearly lossless (ZVS is achieved). When the gate-to-source voltage reaches the targeted value V_{GSN} at t_2, the transient is finished and the gate charge is calculated by integrating the gate current from t_0 to t_2.

Finally, the gate-charge dependence $Q_{NMOS}(v_{GSP},w_P,I_1)$ is obtained. The circuit shown in Figure 1.62f is used to calculate reverse-recovery energy loss of the NMOS body diode. Again, the NMOS is turned-off since the gate is connected to the source via a small resistance R_G. Initially, all the current is carried by the body diode. At time t_0, the ideal switch is turned on, imposing a positive voltage at the drain and thus turning off the body diode. The inductance L is used to limit the slew rate of the body-diode current, which should be selected based on the slew rate of the PMOS source-to-drain current from simulations performed with the circuit shown in Figure 1.62d. When the inductor current reaches the load current at t_1, the body diode enters in reverse recovery mode, creating energy loss. The transient is finished at t_2 when the diode current returns to zero. Integrating the product of the body-diode current and the voltage from t_0 to t_2, the reverse-recovery energy dependence $E_{(N\_rev\_rec)}(w_P,I_0)$ is obtained.

After the construction of the piece-wise linear functions, it is possible to apply analytical equations to calculate both conduction and switching losses. Conduction losses can be calculated by using the analytical equation given in Table 1.2, were $R_{PMOS}(W_P,V_{GSP},I_L)$ are the on-resistance of the devices as a function of the device width, the gate-source voltage and the current through the device obtained by means of simulations.

Type of losses	Time instant	Equations
P_{PMOS_cond}	$0 - T_{SW}$	$R_{PMOS}(W_P,V_{GSP},I_L)\cdot I^2_{PMOSRMS}$
$P_{PMOS\_turn\_on}$	$0 - T_{SW}$	$E_{PMOS\_turn\_on}(w_P, I_0)\cdot f_{SW}$
$P_{PMOS\_turn\_off}$	$D\cdot T_{SW}$	$E_{PMOS\_turn\_off}(w_P, I_1)\cdot f_{SW}$
P_{PMOS_gate}	$0 - T_{SW}$	$Q_{PMOS}(v_{GSP} w_P, I_0)\cdot v_{GSP}\cdot f_{SW}$
P_{NMOS_cond}	$0 - T_{SW}$	$R_{NMOS}(W_N,V_{GSN},I_L)\cdot I^2_{NMOSRMS}$
P_{NMOS_gate}	$D\cdot T_{SW}$	$Q_{NMOS}(v_{GSN} w_N, I_1)\cdot v_{GSN}\cdot f_{SW}$
$P_{NMOS\_rev\_rec}$	$0 - T_{SW}$	$E_{N\_rev\_rec}(w_N, I_0)\cdot f_{SW}$
P_{Ndiode_N2P}	$0 - T_{SW}$	$I_0\cdot v_{D\_NMOS}(w_N, I_0)\cdot f_{SW}\cdot t_{dead\_N2P}$
P_{Ndiode_P2N}	$D\cdot T_{SW}$	$I_1\cdot v_{D\_NMOS}(w_N, I_1)\cdot f_{SW}\cdot t_{dead\_P2N}$

Table 1.2. *Calculation of semiconductor losses*

For the calculation of switching losses, the following considerations are important. As the PMOS switch operates with hard switching, the switching losses can be divided into driving, turn-on and turn-off losses. Using the steady-state waveforms shown in Figure 1.64, the turn-on energy loss can be found to occur at the beginning of the cycle, which depends on the instantaneous inductor current I_0, the width of the PMOS, W_P, and the input voltage. As stated previously, since the input voltage is fixed, it can be omitted from the model, thus degrading the dependence to only two variables (W_P and I_0).

Similarly, the turn-off energy loss, which occurs at the end of the on-time, also depends on the instantaneous inductor current I_1 and the width of the PMOS, W_P, assuming that the input voltage has been incorporated into the model. The piece-wise linear function obtained by means of simulations for the turn-off energy is shown in Figure 1.65. The final component of PMOS switching losses is the driving loss, which occurs at the beginning of the cycle. The driving energy loss can be estimated as the product of the charge needed to turn on the device and the driving voltage. Again, the charge required depends on the width W_P, the instantaneous current I_0 and the driving voltage V_{GSP}.

Driving losses, occurring at time instant D·T_{SW}, can be calculated as a product of the gate charge and the driving voltage v_{GSN}. The charge is dependent on the width, w_N, the instantaneous current I_1 and the driving voltage v_{GSN}. The energy losses of the body-diode conduction occur at the beginning of the cycle as well at the time instant D·T_{SW}, which can be calculated as a product of the voltage drop of the diode, the instantaneous currents (I_0 and I_1) and the duration of the dead times ($t_{(dead\_N2P)}$ and $t_{(dead\_P2N)}$).

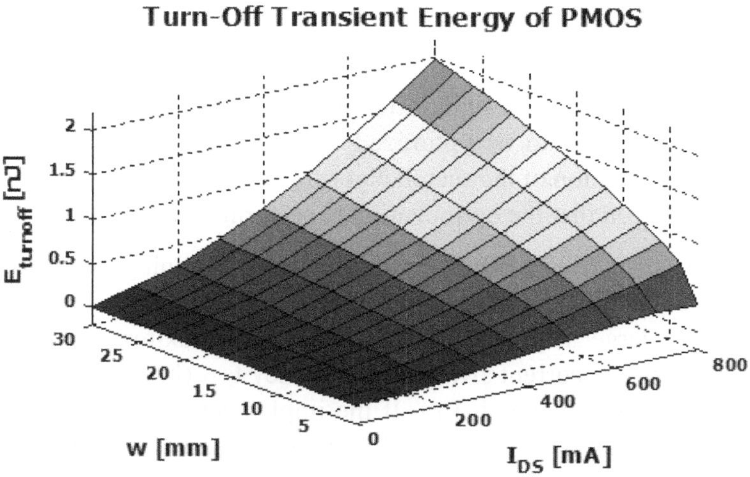

Figure 1.65. *Turn-off transient energy of the PMOS as a function of width and drain current. For the color version of this figure, see www.iste.co.uk/allard/systems.zip*

On the contrary, as the low-side NMOS operates with ZVS, the switching losses can be divided into the driving loss, the body-diode conduction losses and the reverse recovery of the body diode. Similarly, for the PMOS, the driving energy loss, occurring at time instant D·T_{SW}, can be calculated as a product of the charge and the driving voltage v_{GSN}. The charge is dependent on the width W_N, the instantaneous current I_1 and the driving voltage v_{GSN}. The voltage drop is proportional to the corresponding currents and the width of NMOS. Characterizing all the functions of the model, the total losses can be calculated using equations presented in Table 1.65.

1.6.3. *Optimization algorithms*

The optimization algorithm searches for a set of design variables (switching frequency f_{SW}, inductor area A_L, inductance L, output and input capacitance C_{OUT} and C_{IN} and widths of MOSFETs w_P and w_N) to obtain maximal efficiency at a nominal operating point defined by the typical output current I_{TYP}, the output voltage V_{OUT} and the input voltage V_{IN}. During the optimization process, the static and dynamic behavior of the system is simulated to verify that both static and dynamic constraints are satisfied under the worst-case steady-state operation and under the load steps and input voltage step transients.

To search the optimum design, different algorithms can be applied. In this case, where the metric to be optimized is the efficiency at typical load, a possible implementation of the search algorithm is shown in Figure 1.66, from which it can be seen that the optimization is performed in two steps:

1) Coarse exhaustive search of the design space obtaining the intermediate optimal point (OPT_{INT}). During this approach, the design space is searched with relatively large step in order to avoid local maxima. The algorithm starts by defining the current switching frequency $f_{SW}(k_F)$, the inductor area $A_L(k_{AL})$ and the inductance $L(k_L)$. If the implementation of the inductor is possible, the optimal design is passed downwards and the algorithm selects the current output capacitor $C_{OUT}(k_C)$ and indirectly the input capacitor $C_{IN}(k_C)$, which is defined by difference between the total implementable capacitance C_{Total} and the current output capacitance $C_{OUT}(k_C)$. The total implementable capacitance C_{Total} is defined by the capacitance density and an available capacitor area, obtained as the difference between the total area of the device and the current inductor area $A_L(k_{AL})$. After obtaining all the electrical parameters, $L(k_L)$, $C_{OUT}(k_C)$, $C_{IN}(k_C)$ and $f_{SW}(k_F)$, the converter is simulated both in the steady state and under transients. If the performance obtained satisfies the constraints of the system, the algorithm selects current NMOS and PMOS widths, $w_N(i_{(W\_N)})$ and $w_P(i_{(W\_P)})$, and estimates the efficiency at a typical operating point. After calculating all the possible combinations, the intermediate optimal design OPT_{INT} is obtained.

2) Fine gradient search from that point to the optimal design (OPT). For each variable x, the efficiency is estimated in points x-Δx, x, x+Δx. The initial step, Δx, is equal to half of the step in the exhaustive search. If the efficiency is increased for one of the new designs in any direction, the new design is taken as the optimal and the process is repeated. If there is no increase in the efficiency, all the steps are reduced by half and the process is repeated.

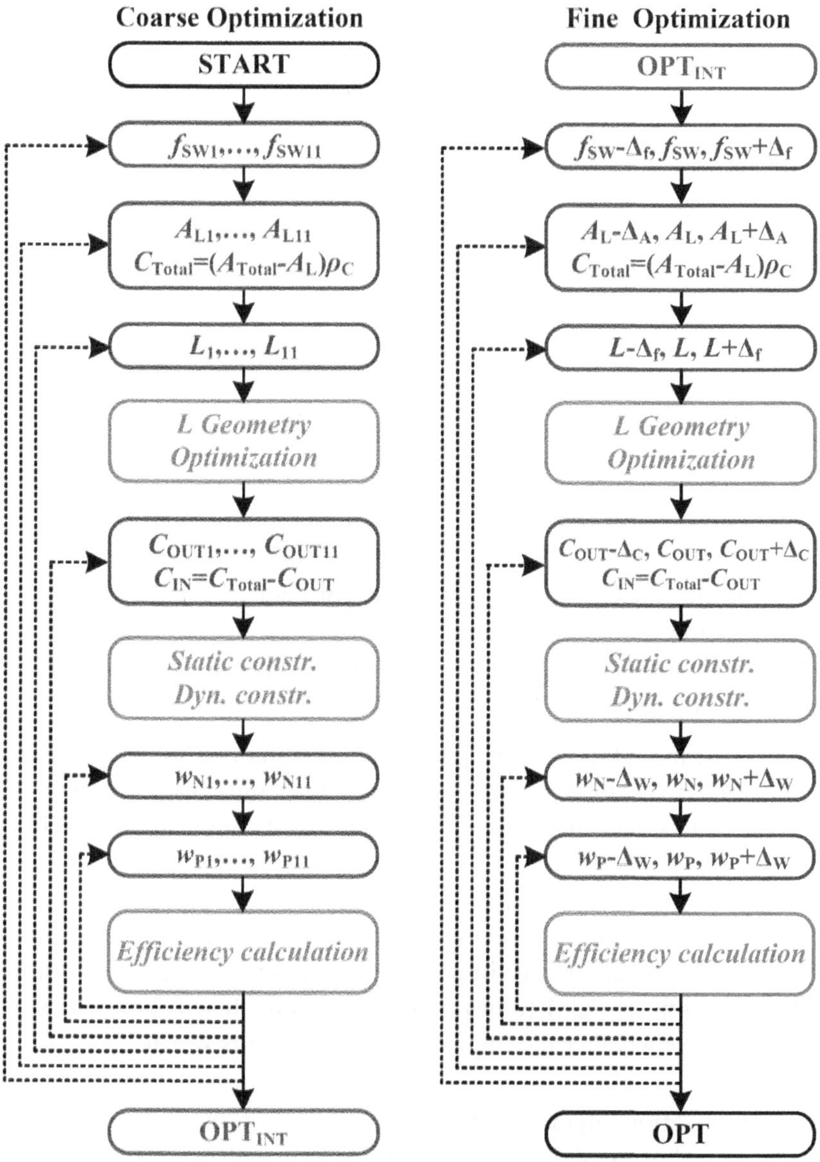

Figure 1.66. *Optimization algorithm flowchart. For the color version of this figure, see www.iste.co.uk/allard/systems.zip*

1.6.4. *Outcome of the optimization (topology, area, loss, f_{sw}, detailed design)*

As a result of the optimization process, a complete design of the power stage is obtained with detailed information about the individual optimization of all the components.

– semiconductors: optimum size of the high-side and low-side MOSFETs and their driving gate-to-source voltage;

– inductor: optimum area dedicated to the inductor, geometrical parameters (core length, width and thickness), number of turns, trace width and height. Eventually, if different core materials can be used for integration, the optimization algorithm will provide the most suitable;

– integrated capacitors: optimum area dedicated to input and output capacitors and the optimum combination of cells (with different capacitance per unit area vs. ESR per capacitance) for a given specification;

– control: the optimization algorithm will provide a design of the controller that satisfies the dynamic requirements under parameter variations of the components. The controller type (voltage mode, current mode, ripple-based) can be either an input to the optimization algorithm or selected by it;

– light load modulation: the optimization results can show the efficiency as a function of the load for different light load modulation strategies and the designer can select the most suitable one for its applications, taking into account the trade-off between the benefit (loss reduction) and the complexity of the implementation. Figure 1.67 shows the efficiency versus load results for a buck converter (5V to 1.2V) optimized for a typical load of 270 mA. It shows how burst mode operation (red) improves the efficiency at light load compared with the operation in CCM (blue) or DCM (green). It will be the designer's decision to assess whether the improvement obtained is worth.

The optimization algorithm has to provide not only the final design, optimum according to the selected performance metrics, but also how the design variables affect the results. For example, Figure 1.68 shows the breakdown of losses for an optimized design of a buck converter with an output voltage of 1.2V and an output current of 270 mA: semiconductor losses in red, inductor losses in blue, capacitor losses in black and total losses in blue. According to the metric of efficiency, or lower losses, the best design is achieved with a switching frequency of 12 MHz. Nevertheless, the difference in losses of all designs with switching frequency between 8 MHz and 20 MHz is less than 4 mW, i.e. the efficiency varies only by 1% in all this range.

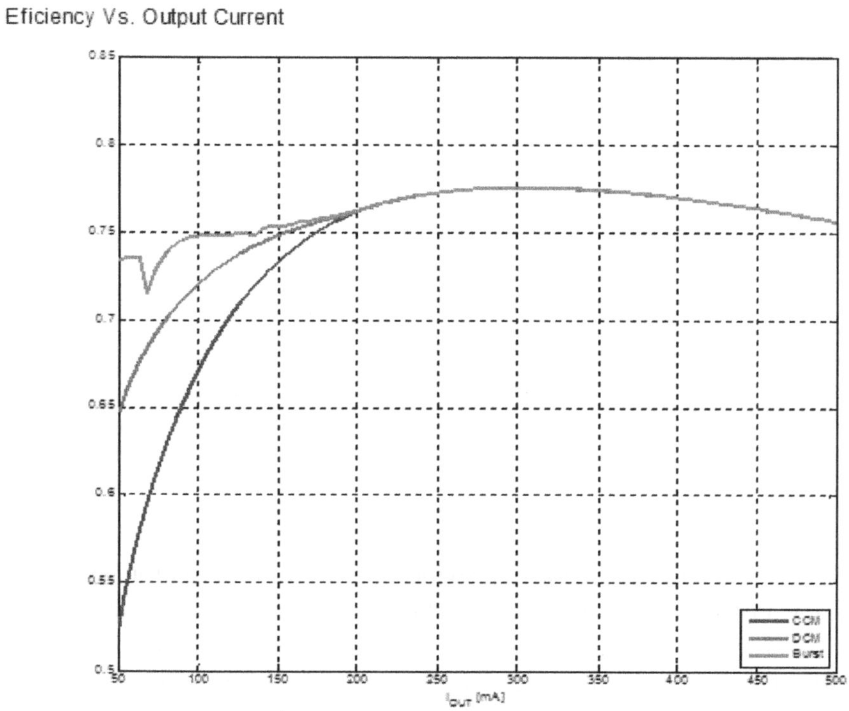

Figure 1.67. *Efficiency versus output current for different light load modulation strategies: CCM (blue), DCM (green), BURST mode (red). For the color version of this figure, see www.iste.co.uk/allard/systems.zip*

Based on this useful information, the designer can choose to operate at 8 MHz due to the lower requirements in the power management IC, knowing that the penalty on efficiency will be only 1%.

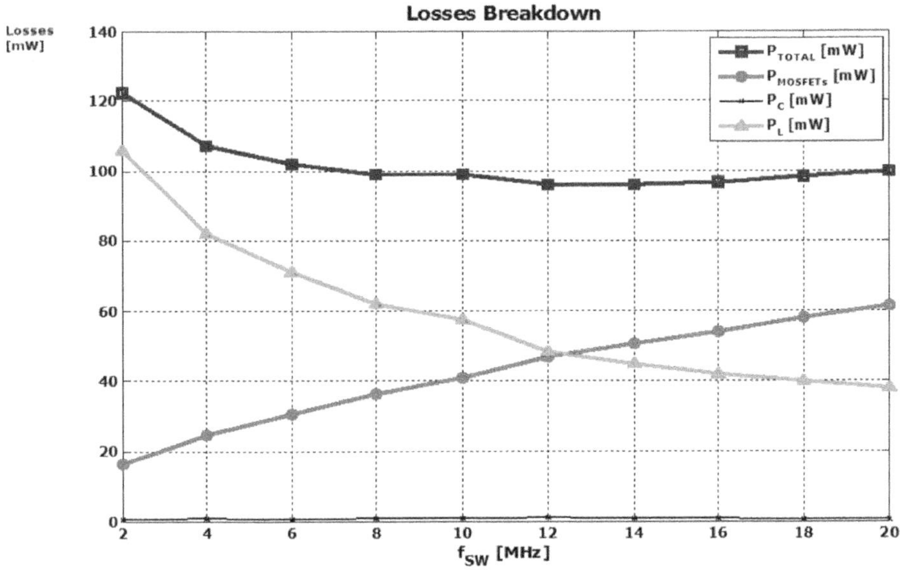

Figure 1.68. *Optimized design loss breakdown as a function of the switching frequency: total losses (blue), MOSFET losses (red), capacitor losses (black) and inductor losses (green). For the color version of this figure, see www.iste.co.uk/allard/systems.zip*

There are other plots that also provide very useful information, such as the breakdown of losses for individual technologies, shown in Figure 1.69, as a function of the switching frequency and the breakdown of losses for all the components at a given frequency, shown in Figure 1.70. In the first case, as expected, it is shown that the higher the switching frequency, the higher the losses in the MOSFETs. It can be noted that at a low frequency (2 MHz), the difference between the total losses in the PMOS or NMOS and the switching losses is about 3mW; however, at a high frequency (20 MHz), this difference is about 10 mW, i.e. conduction losses increase with frequency. This is because all of them are optimized designs at each switching frequency and at low frequency where switching losses are lower, the size of the MOSFETs can be made large to reduce conduction losses, but as the switching frequency is increased and their influence becomes more important, the optimum size of the MOSFETs is reduced to balance conduction and switching losses.

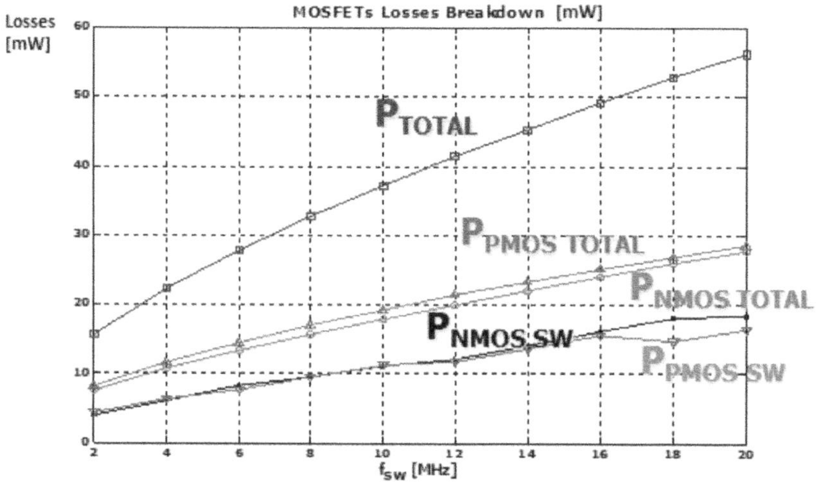

Figure 1.69. *MOSFET loss breakdown: total semiconductor losses (blue), PMOS total losses (green), NMOS total losses (red), PMOS switching losses (black) and NMOS switching losses (pink). For the color version of this figure, see www.iste.co.uk/allard/systems.zip*

The plot in Figure 1.70 shows the detailed loss breakdown of a given design at a given switching frequency, providing information about individual MOSFET losses divided into conduction, turn-on and turn-off, and driving losses and inductor losses broken down into copper losses (DC and AC) and core losses, and input and output capacitor losses. In this case, inductor losses represent close to 50% of the total losses and the other 50% is divided equally between the high-side and low-side MOSFET.

1.6.5. *Impact of technology*

One of the outcomes of the optimization algorithm is the possibility to basically evaluate the impact of technology improvements on the overall system level metrics (efficiency, area). It is possible to evaluate different scenarios to analyze where should the main efforts be placed to get the maximum improvement. For example, it is possible to evaluate the effect of having ideal semiconductor devices (with no losses), keeping the technology of the magnetics and the integrated capacitors, or the case when the inductors are ideal (the inductor occupies no space and has no losses) so the limits will be given by the Si devices.

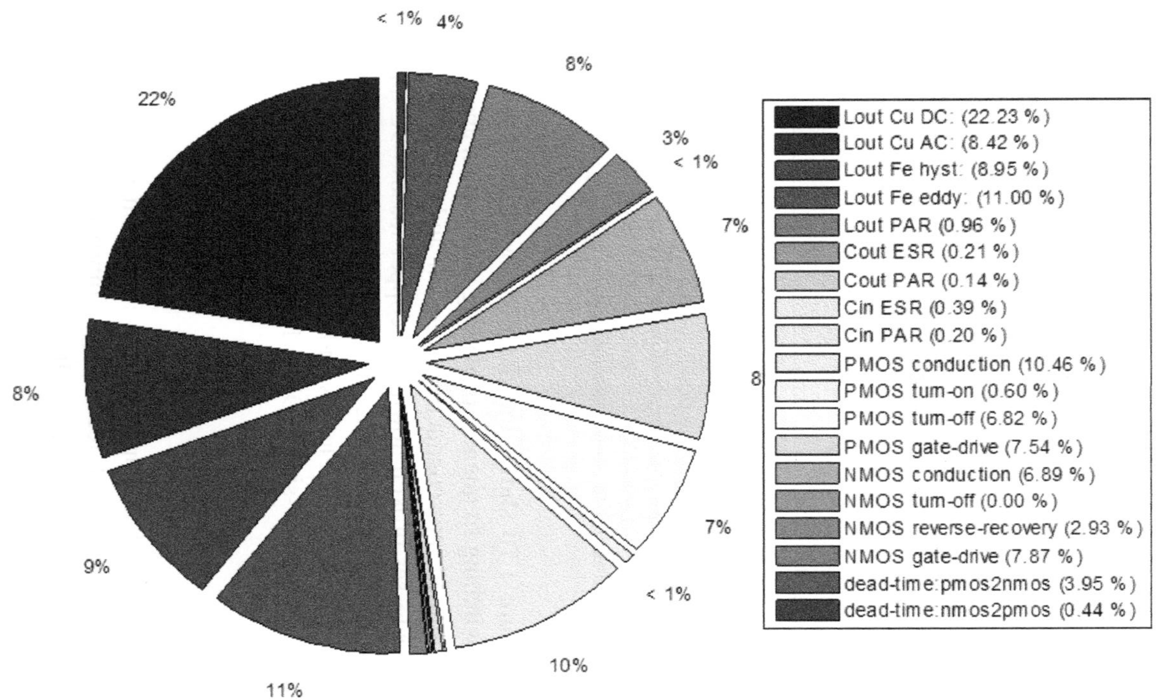

Figure 1.70. *Losses breakdown for the optimized buck converter at a typical load (270 mA) and a switching frequency of 12 MHz. Efficiency (77.5%). For the color version of this figure, see www.iste.co.uk/allard/systems.zip*

It is also possible to analyze the impact on current designs if the technology of the power devices is improved by a factor of two (for a given semiconductor width, the conduction losses and the switching losses are divided by a factor of $\sqrt{2}$). It is interesting to analyze the case when ideal integrated magnetic inductors (without losses and occupying no space) are available but the rest of the components, Si devices and capacitors keep their current limitations. The impact on the efficiency is shown in Figure 1.71, where the efficiency of the converter is plotted against the switching frequency. In this plot, for each frequency, an optimized design is obtained (a single design is not evaluated as a function of the frequency).

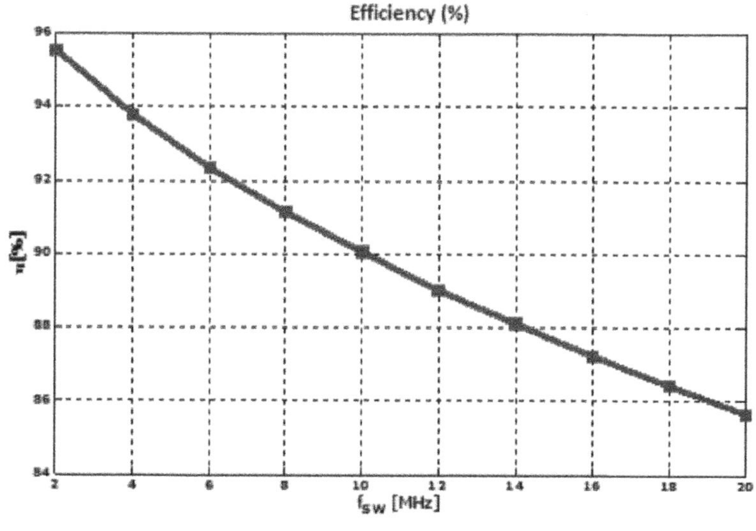

Figure 1.71. *Efficiency as a function of the switching frequency assuming ideal inductors (designs optimized at each switching frequency)*

It can be seen how the higher the frequency, the lower the efficiency as switching losses in the semiconductors increase proportionally with frequency. The result is that the better the technology of the magnetic components, the lower the switching frequency, which benefits the switching devices.

Another interesting analysis is to observe what will happen if silicon technology is improved by a factor of two, meaning that a device with the same width will have $\sqrt{2}$ lower conduction losses and $\sqrt{2}$ lower switching losses under the same operating conditions. The results are shown in Figure 1.72, where the efficiency of the optimized designs, as a function of the switching frequency, for both Si technologies is shown. Two facts can be noted: first, the differences are lower at low frequency since the switching losses have a lower impact; second, although the improvement on the technology is significant (two times), the impact on the efficiency at 12 MHz, for example, is barely 2%.

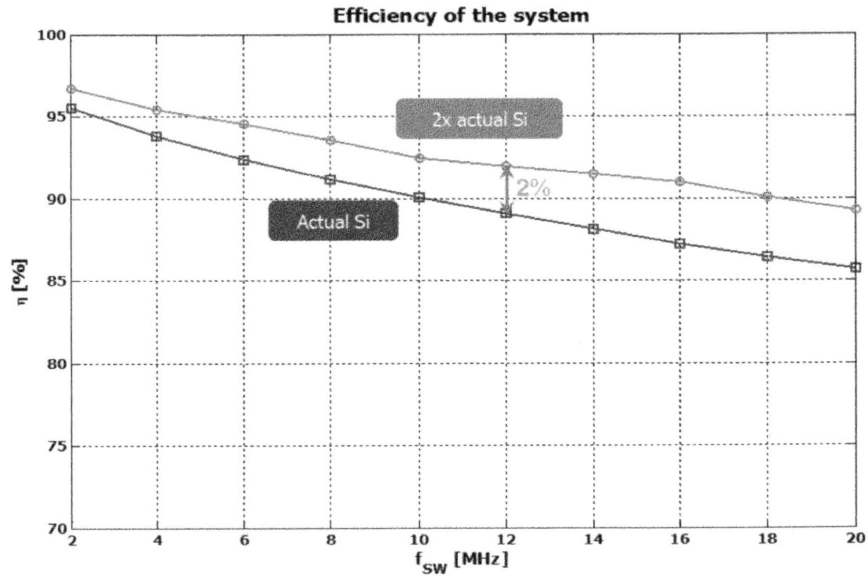

Figure 1.72. *Impact of improving Si technology by a factor of two on the efficiency: actual Si (blue), improved x2 Si (Red). For the color version of this figure, see www.iste.co.uk/allard/systems.zip*

Another interesting analysis is the impact of the integrated magnetic component technology with current Si devices shown in Figure 1.73. In the order of 10 MHz to 12 MHz, the current technology penalizes 5% of the efficiency of the converter with respect to an ideal integrated magnetic component. In this case, the comparison is not done with the same switches driving an ideal inductor and a real inductor; for each switching frequency, an optimized design with an ideal inductor and current inductor technology is carried out, so the semiconductor widths will not be the same.

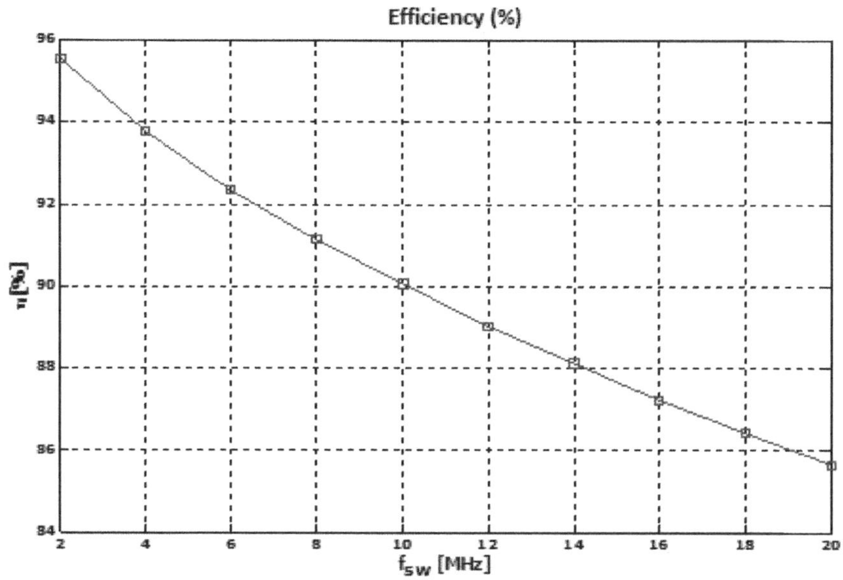

Figure 1.73. *Impact of inductor technology on efficiency: ideal inductor (blue), actual inductor technology (red)*

1.7. Conclusion

The previous section introduced the necessity for a high-level, global approach of design to decide for a trade-off between efficiency, quality of output DC energy and transient performances of a given converter architecture. Given a specification, multiple architectures may be considered and should be compared with major figures. The penalty in number of passive components and silicon area is a cursor the designer has to put somewhere in the design process but being aware of the cost on the trade-off quality.

Static metrics of the converter performances are generally put in the front at an early stage in the design process. This chapter attempted to emphasize the extreme importance of the control strategy. When more integration means higher switching frequency, it should not be forgotten that transient performances should improve as well.

This chapter has not discussed analogue versus digital implementation of controllers. Examples of both kinds are given in the following chapters.

Probably ripple-based control provides a new territory to analogue control as improved performances can be achieved using circuit complexity similar to that of the existing standard control approach. The last chapter of this book intends to show that the converter architecture complexity does not allow a choice and that digital control is the best candidate.

2

Magnetic Components for Increased Power Density

The ongoing drive towards the miniaturization of electronic products with ever increasing functionality and performance, from portable electronics to high-end computing, is placing significant challenges on power management platforms to deliver high-efficiency power at an increasing level of granularity.

In mobile phones, the use of point-of-load (POL), DC/DC, switched-mode power supplies (SMPS) as a high-efficiency replacement for low-cost linear regulators has resulted in the increase of battery lifetime. Furthermore, the increase in functionality and in the number of peripherals in portable electronics is driving the need for the proliferation of different power rails to drive various functional blocks such as RF power amplifiers, the baseband system-on-chip (SoC) and screens, as well as the increasing number of sensors and actuators such as cameras, lights and motion detection. In the case of complex SoC platforms, such as embedded micro-controllers and multi-core processors, there is an increasing need for multiple digital power rails to enable energy-efficient dynamic voltage and frequency scaling, multi-analogue rails to minimize noise as well as separate power rails for I/O drivers and RF functions. All of these functions require separate power converters that are closer to the load in order to minimize efficiency losses and reduce the effects of interconnect parasitics.

Chapter written by Santosh KULKARNI and Cian O'MATHUNA.

The above challenges in power management, together with the concerns regarding the miniaturization and reliability of power supplies, are increasingly being addressed by power semiconductor companies through their ability to deliver advanced processing and functional integration. This proliferation of functionally integrated hardware solutions is considered as an inflection point in the power supply industry that is seeing a dramatic move away from traditional power supply manufacturing (with a focus on the assembly of power supply modules or bricks from discrete components) to an increasing emphasis on power supply products derived from semiconductor and microelectronic platforms and technologies.

The POL SMPS typically consist of active silicon devices (controller, metal-oxide-semiconductor field-effect transistors (MOSFETs), drivers) and several passive components (R, L, C) to provide complete voltage regulation, as shown in Figure 2.1. When compared with linear regulators, the SMPS converter is more complex, harder to implement and more expensive as a solution. A major challenge for the further miniaturization of DC/DC converters and their integration with the load is the inability of the integrated passive components (in particular magnetics) to achieve similar electrical performance, at a comparable footprint, to discrete passive components used in today's converters operating at frequencies of 0.5 to 8 MHz.

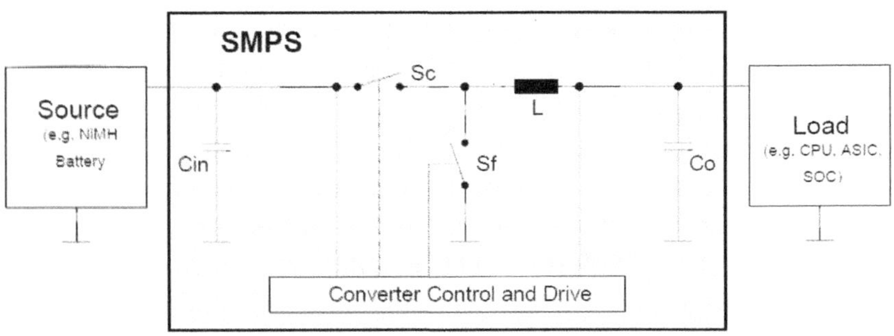

Figure 2.1. *Magnetic devices in a standard SMPS*

For a typical POL buck converter, the required inductance value and capacitance are inversely proportional to the operating frequency. The required inductance and capacitance are also affected by the load current specifications. The plot shown in Figure 2.2 gives a good illustration of how the required inductance and capacitance values change with frequency and load current. Increasing the switching frequencies in the range of 10 to 100 MHz provides the potential for the reduction of passive component values to the point where, with the right technology, their size becomes compatible with the dimensions of power management integrated circuit (IC). The ultimate target is to develop new miniaturized product formats that can be referred to as power supply-in-package (PwrSiP) and power supply-on-chip (PwrSoC).

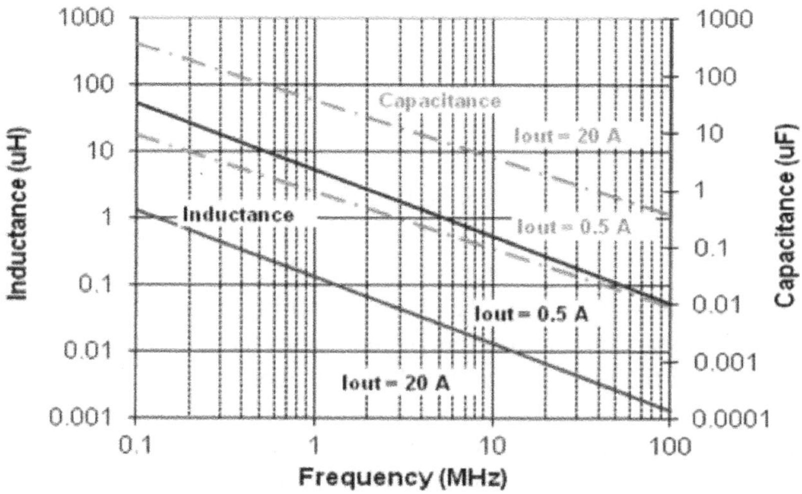

Figure 2.2. *Abacus of required inductance and capacitance values in SMPS applications. For the color version of this figure, see www.iste.co.uk/allard/systems.zip*

Qiang Li *et al.* [LI 09] from CPES recently presented a preliminary roadmap for high-frequency integrated DC/DC converters. This paper reviewed the state-of-the-art in device (switch) technologies, magnetic materials and integration techniques. Charles Sullivan from Dartmouth also presented a review of challenges of integrating magnetics for on-chip power [SUL 09]. Don Gardner *et al.* from Intel reported on on-chip inductor structures with magnetic films [GAR 09]. Other recent initiatives by PSMA, Arpa-E, the Power Electronics Ireland Group (PEIG) and Tyndall, as well as recent industry market studies, have emphasized the increasing interest in the development of the PwrSiP and PwrSoC space [REP 08, INT 08, DAR 01]

This chapter reviews the state-of-the-art in integrated magnetic devices required to realize a PwrSiP and PwrSoC. It then further analyzes and compares the ongoing research trends towards development of integrated magnetic passive components that can be packaged or monolithically fabricated as part of a switched-mode, DC/DC power management platform.

This chapter is organized as follows. Section 2.1 reviews the commercial POL SMPS products with passive components included in the package. Section 2.2 reviews the reports in the literature on demonstrated integrated passive technologies, primarily magnetic, targeted at realizing PwrSiP or PwrSoC. Key data are used, from both selected commercial products and academic research demonstrators, to illustrate the current state of the technology in terms of operating frequency, power density and actual passive integration. The review of integrated magnetic devices includes different micro-magnetic structures and materials. This work also reviews the current state-of-the-art in micromagnetic technologies using non-micro-electromechanical systems (MEMS)-based thin film processing technologies. A detailed comparison of different technologies is presented in order to highlight each of their advantages and disadvantages.

Section 2.3 assesses the performance of all the reported micro-inductors using two figures of merit (FOM). The DC performance of the inductor is evaluated in section 2.3.1 based on the plotting of L/Rdc against energy density. The AC performance is assessed in section 2.5 based on using the conventional quality factor of the inductor. An outline technology roadmap for integrating passive technology with power management ICs, along with key challenges and commercial opportunities, is discussed in section 2.4 before ending with conclusions in section 2.5.

2.1. Commercial and research trends towards PwrSiP and PwrSoC

To put this study in context, it is appropriate to refer to Figure 2.3, which shows what can be described qualitatively as the evolution from discrete power supply on PCB solutions to the ultimate monolithic PwrSoC platform, in which all the passive components are monolithically integrated on the power management IC. The following qualitative definitions are also relevant to the discussion.

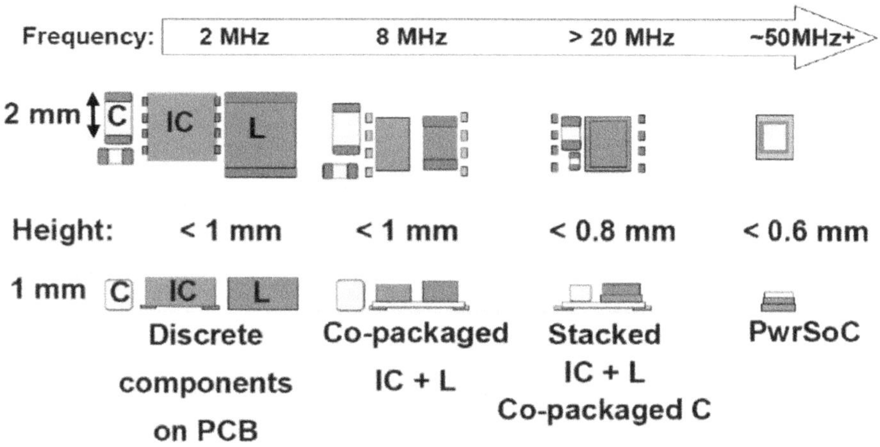

Figure 2.3. *Roadmap of the morphology of SMPS. For the color version of this figure, see www.iste.co.uk/allard/systems.zip*

Power supply module: Figure 2.4 shows a typical power module introduced by Vishay with a set of (> 10) standard discrete components mounted side by side on an FR4 substrate, and enclosed in an inexpensive plastic lid. The largest component in the converter is the inductor.

Figure 2.4. *Power module by Vishay*

Power supply-in-package: PwrSiP products typically co-package the magnetic components with the rest of the components in the converter. Most of the current PwrSiP products on the market only co-package either proprietary inductors or discrete ferrite inductors with the power management

IC to produce a single-chip solution, as shown in Figure 2.5. The passive components and the IC can be either assembled side by side or stacked on each other.

Figure 2.5. *Examples of the co-packaging of magnetic devices. For the color version of this figure, see www.iste.co.uk/allard/systems.zip*

Power supply-on-chip: To realize the ultimate PwrSoC solution, passive components will be integrated onto the power management IC, i.e. monolithic integration.

Linear regulator: Voltage regulator based on an active device generally provides a constant voltage by varying its effective resistance.

Switched-capacitor DC/DC converters: Switched-capacitor (SC) DC/DC power converters are a subset of SMPS, based on a network of switches and capacitors. They convert one voltage to another by moving the charge into and out of capacitors.

Increasingly, many companies, including Enpirion, Fuji, Micrel, National Semiconductor and TI, have reported products using a PwrSiP platform, either with one or more passive components integrated into the same package as the power management IC, in a planar or stacked form factor. Enpirion [HUE 09b] and Micrel [MIC 15] have assembled, or co-packaged, the output inductor in a plastic-encapsulated package on the same lead frame as the adjacent power management IC. This presents the benefit of removing the "pain point" for OEM companies, the inductor, which can be problematic from both a design and quality perspective. Both Fuji [FUJ 06] and National Semiconductor [NAT 09] have reported the use of a customized ceramic ferrite inductor that acts as a chip-scale substrate on which the power management IC is mounted. This dramatically reduces the DC/DC converter footprint while using the customized version of conventional magnetic component technology. More recently, TI [TEX 16] have gone a step further with this stacked chip-scale package form factor with commercial ceramic passive components, inductor and capacitors, assembled on a chip-scale PCB on which the power management IC is embedded. In all the above cases, companies have developed products with enhanced integration using creative engineering solutions to deliver the requirements of a reduced footprint and reduced component count while maintaining the overall height profile in the range of 1 mm.

In an attempt to understand the trends in this rapidly evolving area, this paper presents an overview of the progress of the performance of the typical DC/DC converter platforms outlined above. The accompanying figures provide the plots of data for a range of DC/DC converter platforms ranging from commercial power modules [VIS 05, BEL 16, DEL 16, LIN 07, MUR 16, INT 14, ROH 16, LIN 16] and PwrSiP products [ENP 16a, MIC 15, FUJ 06, NAT 09, TEX 16] to PwrSiP and PwrSoC research demonstrators, which have been reported in the recent literature [SAT 04, HAZ 05, VIL 08, SCH 07, HAY 03, KIM 02, MIN 96a, BER 09, ONI 06, WAN 10b, WIB 08, WEN 11, DIB 10, NAK 00, ALI 09, NI 09a, MUS 05]

Figure 2.6 confirms that PwrSiP products have a higher power density than conventional power modules. This is mainly due to the higher operating frequency of PwrSiP products that allows the use of smaller-value inductors, hence their smaller size and lower profile. It is worth noting that all the PwrSiP products have a profile of less than 1.2 mm, while in stark contrast, all power modules have a profile greater than 2.8 mm.

Figure 2.6 also includes key DC/DC converter research demonstrators from both industry and academia, all using various inductor technologies. While some of these demonstrators have used thin film magnetic core inductors [SAT 04, KIM 02, MIN 96a, WAN 10b, DIB 10, NAK 00] or ferrite-based inductors [HAY 03], most products have used an air-core inductor on silicon [VIL 08, BER 09, ONI 06, WIB 08, DIB 10, ALI 09, NI 09a, MUS 05] or discrete air-core inductors [VIL 08, HAY 03]. These demonstrators can be randomly divided into two groups: PwrSiP and PwrSoC. The majority strictly fall into the PwrSiP category [SAT 04, HAZ 05, VIL 08, SCH 07, HAY 03, KIM 02, MIN 96a, BER 09,

ONI 06, WAN 10b] while, currently, very few of them can be considered as truly PwrSoC demonstrators [WIB 08, WEN 11, DIB 10, NAK 00, ALI 09, NI 09a, MUS 05] in which the inductors are monolithically integrated with the power management IC.

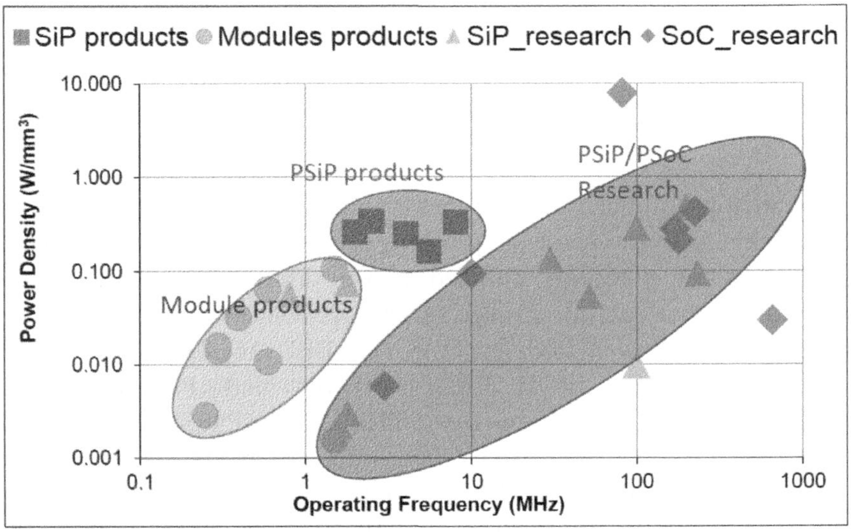

Figure 2.6. *Landscape of various products. For the color version of this figure, see www.iste.co.uk/allard/systems.zip*

The converters reported in [MIN 96a, BER 09] show relatively lower power densities compared with the commercial PwrSiP products in the same range of operating frequency. This is mainly because the commercial products use commercial ferrite chip inductors. Significantly thicker magnetic material is generally applied in commercial ferrite chip inductors than in thin film inductors, which results in much higher inductance densities for discrete ferrite inductors. For a relatively high inductance value, e.g. $1\mu H$, the size of a discrete ferrite inductor can be significantly smaller than that of a thin film inductor. However, as the switching frequency increases, e.g. above 50 MHz, when the required inductance becomes significantly low, the research demonstrators have started to show power density comparable to commercial PwrSiP platforms. To date, all the inductors applied in these research demonstrators operating at a frequency above 50 MHz are air-core inductors, either discrete-chip, air-core inductors in [HAZ 05, SCH 07] or on-chip, integrated, air-core inductors in [VIL 08, BER 09, ONI 06, WIB 08, WEN 11, ALI 09, NI 09a, MUS 05]. In [BER 09, ONI 06, WIB 08, WEN 11, ALI 09, NI 09a, MUS 05], an extra step was taken by integrating the capacitor onto the same silicon as the on-chip, air-core inductor.

Due to the use of air-core inductors in these demonstrators, the majority of the converter footprint area is occupied by the inductors. As an example, in [DIB 10], the whole converter measures 1.6 mm × 2.35 mm, with the total footprint of the four on-chip, integrated, air-core inductors being more than 2 mm^2. In their recent work, Meere *et al.*, [MEE 11] from Tyndall compared the performance of an air-core inductor with that of a magnetic-core inductor up to a frequency of 100 MHz. The authors demonstrated that the magnetic-core inductor has an advantage in terms of inductance density and efficiency at frequencies below 100 MHz, beyond which the air-core inductor becomes competitive. A detailed analysis of air-core and magnetic-core inductors can also be found in [SUL 09], which suggested that both air-core and magnetic-core inductors show promise for high frequencies, but there are more opportunities to improve magnetic-core inductors. Air-core inductors are prone to EMI issues that may ultimately justify the need for magnetic-core to actually contain magnetic fields.

In the work carried out by a group of Intel researchers [NAK 00], the size of inductors was drastically reduced by applying thin film magnetic material, i.e. permalloy, in the inductor structure. The VRM had 20 power cells with a current rating of 25 A per cell and a footprint area of 2.8 mm^2 per cell. The frequency of operation for the cells varied from 30 to 140 MHz with 16 phases. In this work, an array of thin film coupled inductors was integrated onto the same substrate as the converter IC. The coupled inductor configuration provides an advantage of DC flux cancellation so that much higher current handling capability can be achieved. A laminated, sputtered, permalloy thin film was used as the inductor core to further increase the power density of the inductors. Consequently, the resulting power density of the converter was significantly higher than the aforementioned PwrSoC demonstrators. A similar concept was demonstrated by researchers from Dartmouth and Tyndall in [PRA 04], where on-chip, integrated, coupled inductors, using a NiFe alloy core, were demonstrated for the first time, in a four-phase 5 MHz buck converter. In this case, the devices were relatively large due to the relatively low switching frequency.

To complete the competitive landscape, commercial linear regulators and silicon-integrated, switched-capacitor research demonstrators of DC/DC converters are shown in Figure 2.7. Linear regulators are widely used as power converters due to their simplicity, low noise, good regulation and low cost [IRF 16, STM 13, NAT 05, TIP 16, DIC 76]. Switched-capacitor DC/DC converters have been gaining considerable interest in recent years due to their higher efficiency compared with linear regulators, and also because they do not need an output inductor, and therefore lend themselves to full integration onto silicon using conventional technology, presenting a very attractive solution for PwrSoC [NG 09, BRE 09, MAK 99, MEY 11, LE 10, LEE 06, RIC 04, PEL 03, MEN 05, RAM 10a, SU 08b, PAT 04, ENP 16b]. Figure 2.7 compares the efficiencies of linear regulators, switched capacitors, power modules, PwrSiP and PwrSoC products, and demonstrators.

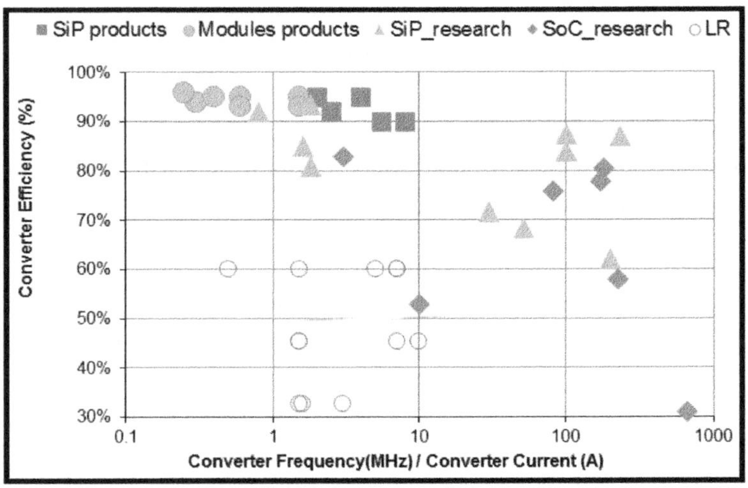

Figure 2.7. *Landscape of indicative SMPS ranked by efficiency value. For the color version of this figure, see www.iste.co.uk/allard/systems.zip*

As widely recognized, the efficiency of switched-mode DC/DC products is significantly higher than that of linear regulator products and switched-capacitor (SC) converters. Although there are no commercial SMPS products currently operating above 10 MHz, the research demonstrators have shown promising performance at very high operating frequencies (100 MHz or even higher). Converter efficiencies of up to 87% have been reported at operating frequencies ranging between 100 and 233 MHz [VIL 08, HAY 03, ONI 06] using advanced CMOS processing technology. However, achieving a higher efficiency, similar to that of the majority of commercial products operating at a multi-MHz range (i.e. $> 93\%$) remains a significant challenge.

Further DC/DC converters using Switch Cap topologies have also been gaining increasing attention. Most of these SC converters operate below 0.1 A. The reported efficiency values range between those of linear regulators and those of inductor-based converters. While the linear regulator can provide a high integration level and fast load regulation, it is not an attractive solution as far as efficiency is concerned. The switched-capacitor regulators can give a moderate efficiency ranging between 60% and 80% at a high integration level; however, it has been reported that the solution still requires a large silicon area even for 100 mA (i.e. 1 mm^2) [HAY 03]. To date, the demonstrated current handling for SC has been about 0.7 A/mm^2 [LE 10].

Currently, the SMPS is a more versatile solution in most applications for better efficiency, but with bulky passive components, namely inductors and decoupling capacitors. The benefits of moving from modules to PwrSiP platforms have already been discussed for products by Enpirion [SAN 07], Fuji [RFP 16] and National Semiconductors [BRA 03]. The current PwrSiP platforms are based on low-temperature cofired ceramic (LTCC), ferrite inductor cores, with the power management IC assembled within the package or on top of the inductor. The advantages of this technique include ease of fabrication, reduced profile (inductor), suitability for embedding in the substrate and the advantage of thick film technology for high power conversion. However, the LTCC cores require relatively high temperature processing, which is not compatible with standard CMOS processing technologies; hence, they are unsuited for realizing a PwrSoC solution.

PwrSoC is currently in the research phase and is yet to demonstrate the efficiency and footprint targets required for commercial success. To realize the PwrSoC vision, the key challenges to be addressed are increased converter operating frequency, miniaturization and system integration of passive components, while maintaining efficiency. In the next section, we review the progress made to date in addressing a key roadblock to further evolution towards a monolithic PwrSoC platform; the integration of magnetic components.

2.2. Review of magnetics

The evolution of micro-inductors and the development of magnetic thin films can be traced back to the early work by Olivei in the late 1960s. However, significant developments in the area did not occur until the early 1990s. Different research teams have been working on the development of integrated power converters for the last 20 years, with particular emphasis on integrated planar magnetic components. They use different materials, processing methods and structures to fabricate the micro-magnetic devices. The reported inductor technologies can be classified into two different categories, using standard, clean-room, low-temperature, CMOS-compatible processing techniques or using different package/assembly technologies without clean-room requirements. The inductor technology subsequently determines which technology platform is applicable for the power converter, PwrSiP or PwrSoC.

2.2.1. *Micro-inductor structures*

The main question in the design of a micro-inductor device is the best arrangements of winding and core, taking into account the inductance, efficiency, size/footprint and fabrication process. Power inductors typically use a magnetic core to enhance the inductance. Depending on the arrangement of conductors with respect to the magnetic core, the construction of a micro-inductor can be categorized into two different approaches. In the first approach, the planar coils are enclosed with a magnetic material. In the second approach, the conductor is wrapped around a planar magnetic core using a multi-layer metal scheme. The main structures using the first approach are spiral inductors (including racetrack inductors) and stripline inductors. Typical examples of structures using the second approach are solenoid and toroidal inductors. Key parameters of interest include high inductance value per unit area, low DC resistance for high current and high Q-factor for high efficiency.

2.2.1.1. *Spiral inductor*

The spiral-type structure (Figure 2.8) is one of the most appropriate structures to form a planar inductor because of its high inductance value per unit area, low DC resistance and high Q-factor [KAT 00, AHN 98, WAN 05]. The copper windings are completely surrounded in the magnetic material to achieve the required inductance and to minimize electromagnetic interference. The typical shape is either square or circular. However, the main drawback of the square or circular spiral structure is that it is difficult to induce anisotropy in the magnetic core during core deposition. This may limit its application frequency (<10 MHz) as it compromises the ability to minimize core loss at high frequencies. The integrated spiral inductors for power supplies are typically limited to switching frequencies of <10 MHz with a current of less than 1A.

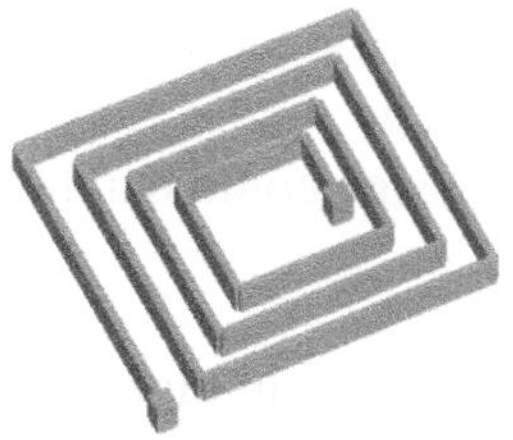

Figure 2.8. *Geometry of a spiral inductor*

Figure 2.9. *Example of a racetrack inductor*

A racetrack-shaped spiral inductor, as shown in Figure 2.9, can be formed by stretching a circular spiral. The magnetic core will typically be used to wrap the straight part of coils, hence taking advantage of the uniaxial anisotropy of the magnetic material [MEE 09b, WAN 07b, WAN 08, SUL 93]. The core material for the device can be deposited in the presence of a magnetic field to induce a uniaxial anisotropy along the longitudinal direction, so as to generate a simple magnetic axis in the direction of current flow in the copper winding. Racetrack micro-inductors have been demonstrated in various buck converters at high frequencies up to 100 MHz [WIB 08, MEE 09a, SUL 96]. Compared with the square- or circular-shaped spiral inductors, racetrack micro-inductors are more suited for power supplies with switching frequencies in the range of 10 to 100 MHz where inductances of approximately 10 to 200nH are required.

Sullivan and co-workers compared two micro-inductor structures, solenoid and racetrack, in terms of the distribution of external magnetic fields [SUL 93, KIM 02]. The E-core shape used in the racetrack-shaped magnetic field was shown to be the better design because the shape enclosure shields the windings, resulting in low external magnetic field. Using an analytical study, they also demonstrated that the E-core design could handle more power in the same area than the toroidal structure.

An alternate approach to developing high power spiral inductors, with low profile and increased packaging densities, is to embed the inductors in the substrate, i.e. PCB or silicon. Zhang and Sanders at the University of California, Berkeley [WU 13, WAN 07a] reported a design and prototyping process for integrating magnetic components within a multilayer printed circuit board. Bak from Wroclaw University [MIN 96b] also reported electrical properties of surface and embedded planar PCB inductors with square, spiral and meander structures. Ludwig *et al.* from Tyndall demonstrated inductors using plated NiFe alloys on PCB, which also incorporated NiFe through holes or vias to close the magnetic core circuit [PAR 98a].

Others have investigated inductors embedded onto silicon substrates. Pan *et al.* from the University of Minnesota [PAR 03a] reported the fabrication of silicon-embedded, high-Q, integrated inductors. They reported a Q-factor of over 60 for a 2 μH air-core inductor at a frequency of 30–40 MHz. Wang from the University of Florida [LOT 98] reported the characterization of an integrated power inductor in a silicon substrate. The embedded inductor had a footprint of 3×3 mm^2 and a profile of 0.6 mm with measured inductance of 390 nH and DC resistance of 117 mΩ at 6 MHz for a current of 500 mA. More recently, R. Wu from Hong Kong University [MEY 10] reported a coreless, power inductor embedded onto CMOS silicon, with a footprint of 0.8 mm^2, a DC resistance of 87 mΩ and an inductance of 13.1 nH for a current of 0.6 A. Bae from Samsung [YAM 93] investigated Ni-Zn ferrite inductor embedded onto polymer substrates. He reported an inductance of 1.6 μH and DC resistance of 1.25 Ω for a current of 900 mA. This inductor had a profile of 0.21 mm.

2.2.1.2. *Stripline inductor*

The stripline micro-inductors have a magnetic material wrapped around a single turn winding, as shown in Figure 2.10. Sullivan and his team at Dartmouth College reported a V-groove, micro-inductor design [PEN 13, XIN 13] for use in a high current converter. The V-shaped conductor was deposited into a wet-etched groove in the silicon substrate, which was surrounded above and below by a $Co - ZrO_2$ thin magnetic film. The large cross-section area of the V-shaped conductor led to ultra-low DC resistance, which was one or two orders of magnitude lower than other stripline inductors realized using conventional metal schemes [NAK 00, WAN 15]. However, this large cross-section resulted in a long magnetic path length, hence having less inductance enhancement. The V-groove inductor was also significantly longer than other conventional stripline inductors for a similar inductance. The stripline inductors are generally applied for high frequency and high current where inductances of the order of 10 to 50 nH are required.

Figure 2.10. *Geometry of a stripline inductor. For the color version of this figure, see www.iste.co.uk/allard/systems.zip*

One drawback of the stripline inductor is its limited inductance. However, researchers have focused on improving the inductance density by increasing the overall thickness of the magnetic core by the deposition of laminations [WAN 15, ZHA 99, ZHA 04]. One advantage of the stripline inductor is its relatively simple structure, which allows the realization of a multi-array, coupled inductor. Although the saturation current of each individual conventional stripline inductor is low [ZHA 04], it has been shown that to cancel the DC flux, a coupled inductor configuration can be used to significantly boost the current handling capability from 150 mA to higher than 1 A [NAK 00].

2.2.1.3. *Toroidal inductor*

The toroidal structure, shown in Figure 2.11, is another widely used structure to fabricate micro-inductors, using multi-level metal schemes to wrap metal windings or conductors around a magnetic core. The toroidal structures are more suited for power supplies with switching frequencies of < 10 MHz. In 1994, Ahn and Allen from the Georgia Institute of Technology [BAK 08] implemented a toroidal micro-inductor structure with magnetic material for use in a DC/DC converter. The design of this inductor consisted of interconnected metal coils wrapped around a ferromagnetic bar-type core. Here, the electroplated metal interconnects and vias were utilized to minimize heat and DC resistance loss. The inductor achieved an inductance of 400 nH at 10 kHz, with a DC resistance of 300 mΩ. The main feature of this design was the use of thick copper windings to carry currents of up to 400 mA. The main drawback of the design was the via contacts used, which may introduce a relatively high contact resistance [LUD 03] and considerably increase the complexity of the process. In addition, an induced uniaxial anisotropy field is difficult to achieve for toroidal cores [PAN 05].

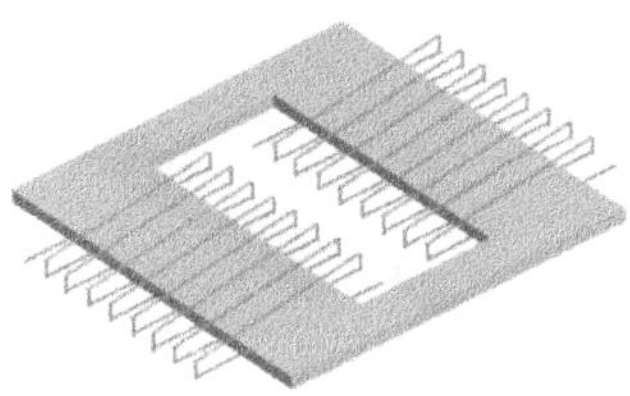

Figure 2.11. *Geometry of a toroidal inductor*

2.2.1.4. *Solenoid inductor*

The solenoid structure is similar to the toroidal structure, except that it uses a bar-shaped magnetic core as opposed to a toroidal core as shown in Figure 2.12. In 2008, Lee from Stanford University published his work on small-area (i.e. 1 mm^2) solenoidal inductors for high-frequency operations up to 100 MHz [WAN 10a]. A bar of magnetic material was placed in the center of the device, and the windings were placed around this. A fabricated 70 nH device had a measured DC resistance of 670 mΩ. The solenoid structure also takes advantage of the anisotropic magnetic core that allows this structure to be applied at higher frequencies than that applied for toroidal micro-inductors.

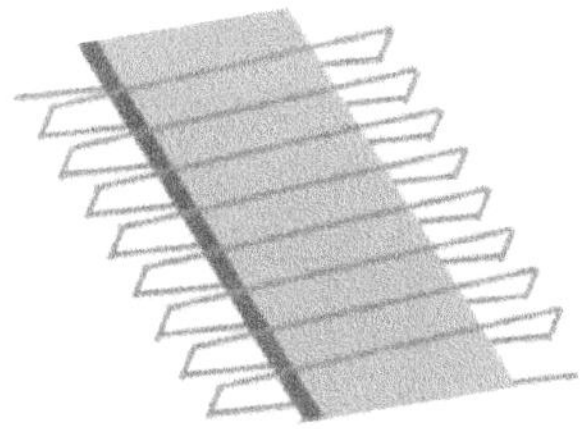

Figure 2.12. *Geometry of a solenoid inductor*

Figure 2.13. *Examples of wire-bond inductor with or without a magnetic material*

John Shen from the University of Central Florida [WU 11, BAE 05, PRA 03] proposed a simple but elegant approach to form solenoidal, on PCB, in-package or on-chip, magnetic structures utilizing existing bond wires encapsulated in a ferrite-filled epoxy core. Fabricated bond-wire inductors fabricated onto PCB are shown in Figure 2.13. The advantage of this approach is that wire bonding is readily available in all power management IC packages, which can be easily adapted for realizing on-chip or in-package inductors using aluminum or gold bond wire of diameter 1–15 mils (25–280 μm). This technique also lends itself to use with magnetic core materials that are deposited using sputtering or electroplating. This technology has been demonstrated in a 5 MHz buck converter [PRA 05]. The reported measured efficiency is about 52% and the low efficiency is attributed to the power MOSFET losses.

The choice of the structure is critical in the design of a power micro-inductor in order to achieve a high inductance enhancement, minimize resistive losses at high frequency and carry the largest current possible without saturating the magnetic core. Table 2.1 summarizes the pros and cons of each inductor construction.

Conventional micro-inductors	Advantages	Disadvantages
Solenoid and toroidal	– High saturation currents – Ease of magnetic core deposition and lamination – Uniaxial anisotropy applicable in solenoid cores	– Complex fabrication process for windings – Uniaxial anisotropy for closed-core structures difficult to realize for toroidal cores
Planar and spiral	– Ease of winding deposition – High inductance densities – Suitable for low-medium current applications	– Higher DC resistance – Low saturation current – Needs a two-layer magnetic core for higher inductance – Uniaxial anisotropy for closed-core structures difficult to realize
Racetrack	– Ease of winding deposition – High inductance densities – Suitable for low-medium current applications – Uniaxial anisotropy applicable in racetrack cores – Higher operational frequency	– Higher DC resistance – Low saturation current ($< 1A$) – Needs a two-layer magnetic core – Core lamination is more difficult to realize
Stripline	– High inductance densities – Low DC resistance – High current handling – Easy of winding deposition – High operational frequency	– Needs a two-layer magnetic core – Core lamination is more difficult to realize – Difficult to achieve high inductance within a reasonable aspect ratio structure

Table 2.1. *Comparison of different magnetic inductor structures*

2.2.2. *Magnetic materials and processing for thin film integrated micro-magnetic devices*

Soft magnetic material is typically used as a magnetic core in power inductors and transformers in order to achieve a reasonable enhancement of the air-core inductance. This provides the desired inductance for inductors, and provides a high magnetizing inductance and a good coupling factor between the primary and secondary windings for transformers.

2.2.2.1. *Key metrics defining the performance of soft magnetic core materials in inductor cores*

The growing necessity for smaller and more efficient micro-inductors has created the need for magnetic materials that exhibit most, if not all, of the following characteristics at frequencies in the range of 10 to 100 MHz and higher: lower core loss, higher saturation flux density, higher anisotropy field, lower magnetostriction and higher permeability. They are described as follows:

1) core losses are mainly composed of two different contributions, eddy current losses and hysteresis losses. Low core losses require soft magnetic materials with high resistivity and low coercivity field;

2) resistivity (ρ): the magnetic core material must have a high resistivity, which will allow for the control of eddy current effects in the core. Eddy currents represent a dissipative loss of energy within the core and define the maximum frequency of operation for the inductor. Additionally, a high resistivity material will have an increased skin depth to ensure that the magnetic field intensity does not vary with the thickness of the magnetic film;

3) coercivity (H_c): the material must have a low coercivity, which will minimize the hysteresis loss at high frequencies;

4) saturation magnetization (Bsat): the material should have high saturation magnetization in order to enhance the current handling capability of the inductor;

5) relative permeability (μ_r): the material must have relatively high permeability along with high anisotropy field, and should remain constant at high frequencies for increased inductance enhancement and stable high frequency performance of the micro-inductor;

6) anisotropy field (H_k): the material must have a high anisotropy field. This will increase the operational frequency and current handling ability of the inductor.

Among the above-listed properties, resistivity and saturation magnetization are intrinsic to the material. While permeability, anisotropy field and coercivity can be varied, they will also depend on the magnetic material deposition process and the shape of the magnetic core, as reported previously by authors from Tyndall [GAR 08].

In addition to an appropriate magnetic material, it is also important to develop processing techniques that are cost-effective and compatible with MEMS and high volume fabrication for the incorporation of magnetic materials in micro-magnetic devices. Various approaches to deposit the magnetic core layer have been investigated and demonstrated over the last 20 years. The advantages and disadvantages of applying these magnetic materials and their implementation using different deposition methods will be discussed in the following section. The electrical performance of micro-fabricated magnetic devices will also be discussed.

2.2.2.2. *Deposition techniques for magnetic core materials in micro-inductors*

The most commonly used techniques for the deposition of thin film magnetic cores are the following: screen printing, sputtering and electrodeposition, each of which have their own advantages and disadvantages. Screen printing is ideally suited for the deposition of non-metallic thin films, as the core materials are typically suspended in a polymer matrix for deposition. NiZn and MnZn are the most popular soft ferrites, which were deposited as magnetic cores for micro-fabricated inductors using screen printing as early as 1998 [YAO 11]. Because the technique presents a good compromise between the deposition of the magnetic material with high resistivity (i.e. $> 1\ \Omega m$) and the simplicity of processing the polymer, it has been adopted by other researchers in their work in realizing micro-fabricated inductors [MOR 11, AHN 94b, FLY 06]. While screen printing provides a comparatively simple process, along with deposition of low-loss cores, the requirement for a high temperature annealing for realizing the required magnetic phase makes this process incompatible with standard silicon-based MEMS fabrication technology.

The second approach for the deposition of thin film magnetic cores is sputter deposition. This technique has the advantage of depositing a wide range of magnetic materials, including alloys and oxides with resistivities in the range of 100–1000 Ωcm, which greatly assists in limiting eddy current losses. Sputtering has been investigated since the early 1990s for different types of magnetic materials for micro-inductors, including CoZrNb [ORL 06], FeCoBC [LEE 08, SHE 07, JIA 08], FeHfO and CoFeHfO [LU 10, JAM 10], FeZrO [PAR 98b] and Fe59Co20B14N7 [KOW 05]. Since 2000, sputtering has been widely used for depositing magnetic alloys with high resistivity, including CoHfTaPd thin film [AHN 98, FUK 03], and multilayer CoZrTa/SiO2 [WAN 10a]. The sputtered magnetic alloys have the advantage of higher saturation flux density and permeability, and their deposition is compatible with low-temperature CMOS processing technology.

Sullivan and his group from Dartmouth College focused on developing granular nanocrystalline films as core materials for micro-inductors. Sputtering is the usual method used to deposit these films. They investigated and reported the use of a variety of different core materials, including CoMgF2 [MAN 05] and CoZrO [XIN 13, TOM 96]. More recently, in 2011, they reported the use of sputtering to deposit laminated CoZrO thin films with ZrO2 insulation in order to reduce eddy current losses within the core [INO 98]. In their recent work, Dibene *et al.* [NAK 00] and Pat Morrow *et al.* [SAT 96] from Intel used sputtering to deposit laminated cores of permalloy to demonstrate a fully integrated DC/DC converter with the micro-inductor fabricated onto the top of the power management IC.

Sputtering is a line-of-sight technique providing a uniform and controlled deposition process, along with the option of depositing laminated thin films as the core material. The technique is ideal for depositing thin films up to a thickness of a few hundred nanometers. However, for thicker films, sputtering is an expensive and slow process.

Electroplated magnetic films are a good option for achieving the desired thick core layers for micro-inductors. The process is relatively inexpensive and compatible with standard IC fabrication, while lending itself to the processing of three-dimensional (3D) device structures. The most commonly used electrodeposited magnetic material is permalloy (typically 81% Fe and 19% Ni). The combination of relatively high permeability, low hysteresis losses and near-zero magnetostriction (i.e. stresses in the devices will not have an impact on magnetic performance) has driven its use in macroscale and microscale sensors, actuators and power converters. Perhaps the most significant reason for its common use in magnetic MEMS is that this material is used in magnetic recording heads, and the technologies necessary for depositing and micromachining have been developed by the well-established recording-head industry.

In 1994, Ahn and co-authors reported a fully integrated, solenoidal micro-inductor realized onto a silicon wafer using a multi-level metallization technique [SAT 97]. NiFe permalloy was electroplated as the core material. Permalloy has also been used in realizing different micro-inductor structures including spiral [KAT 00, SAS 97, SAS 00], toroidal [LUD 03, SAT 00] and solenoid [NAK 00].

Electrodeposition has also been reported in the deposition of magnetic materials such as supermalloy (NiFeMo) [COO 00] and orthonol (Ni50Fe50) [LU 07] to fabricate micro-magnetic devices. Additionally, electroplating techniques for laminating the thin film core have also been investigated [MOR 11]. Since 2000, researchers at the Tyndall National Institute have reported the use of electroplating for the deposition of both permalloy and Ni45Fe55 thin film cores in their micro-inductor structures [MEE 09b, WAN 07b, AHN 94a]. C thin films have the advantage of higher saturation flux density, high anisotropy field, relatively low coercive field and higher resistivity when compared with Ni80Fe20 permalloy thin films. There are other interesting electroplated materials including electrodeposited nanocyrstalline alloy, Co65Ni12Fe23 [BRA 00, WES 00, GAO 05], and electroplated amorphous alloy, CoFeP [PAR 04], both showing relatively high saturation flux density (>1.8 T).

The following section compares the properties of different magnetic materials when used as micro-inductor cores.

2.2.2.3. *Comparison of different magnetic materials*

Table 2.1 summarizes the properties of different magnetic materials used as micro-inductor core materials in the literature. From the table, it is clear that there is no single material that has the required properties of high resistivity, high flux density, high permeability and low coercivity, and can be used as the core material in micro-inductor structures for all DC/DC converter applications. At low frequencies (1 MHz), ferrites are most commonly used as core materials for power applications. Although these materials have low saturation flux density and permeability, films in the millimeter thickness range can be used, due to their significantly higher resistivity, which keeps the eddy current losses to a minimum at increased thicknesses. These materials are typically deposited using techniques such as screen printing, due to the requirement of greater thicknesses. As shown in the table, these materials have higher coercivity, leading to significant hysteresis losses at higher frequencies.

At higher frequencies ($> 5MHz$), magnetic thin films with higher flux density and lower coercivity are better suited (all materials are listed in Table 2.1). These materials are also compatible with low-temperature silicon integration technologies, such as sputtering and electroplating. In the early years of thin film inductor research, permalloy ($Ni_{80}Fe_{20}$) was the magnetic alloy most widely used as the core material. Permalloy has significantly higher flux density (1 T) than ferrites, along with ultra-low coercivity (hard axis). However, permalloy as a metallic alloy has low resistivity, limiting the allowed core thickness to achieve low loss at high switching frequencies. In the quest for a higher resistivity and higher flux density core material, researchers have investigated various different combinations of magnetic alloys (as listed in Table 2.2). The key trade-off in the material properties is between saturation flux density and resistivity. A higher composition of magnetic elements such as Fe, Ni and Co is required for higher permeability; however, higher metal composition results in lower resistivity and thus the composition of metallic elements has to be diluted with non-magnetic elements, such as oxygen or phosphorous, for improved resistivity. This, in turn, lowers the saturation flux density. Hence, there has to be a compromise, based on the application requirements, between the composition and properties of the metallic thin film.

As mentioned previously, for switching frequencies up to 100 MHz, inductors using magnetic thin films as core materials provide an efficient solution as passive technology. However, eddy currents within the magnetic thin film structure become a significant concern as the switching frequencies of DC/DC converters increase above 100 MHz. Eddy currents within the magnetic thin film structure become a significant concern. The key material property at these elevated frequencies is resistivity. Currently, the issue of eddy current losses in thin films at high frequencies is addressed through laminations.

	Thin film material	Deposition technique	Structure	B_s (T)	H_c[1] (Oe)	μ_r	ρ ($\mu\Omega.cm$)	Thickness (μm)
#1	$Ni_{80}Fe_{20}$ [FLY 06]	Electroplated	Solenoid	0.8	3	2000	20	10
#2	CoTaZr [LEE 08]	Sputtered	Solenoid	1.44	0.4	500	100	2.2
#3	NiFeMo [PAR 99]	Electroplated	Solenoid	0.8	0.715	2500		5
#4	$Ni_{80}Fe_{20}$ [PAR 99]	Electroplated	Toroidal	1	0.402	1300	20	12
#5	$Ni_{80}Fe_{20}$ [ORL 06]	PVD	Toroidal	1		900	20	16 [2]
#6	CoZrO [LU 07]	Sputtered	Toroidal	3	80	2000	6	
#7	CoHfTaPd [KAT 00]	Sputtered	Spiral	1	0.5		170	9
#8	Ni-Zn [KOW 05]	Screen printed	Spiral	0.2		6	108	300
#9	NiZnFeO [FUK 03]	Screen printed	Spiral	0.23	17.5	25	108	100
#10	NiZnFeO [MAN 05]	Screen printed	Spiral	0.65	70	60	108	55
#11	FeBN [KIM 02]	Sputtered	Spiral	1.95	30	4600	100	3
#12	CoZrO [PRA 05]	Sputtered	Stripline	1.3	1	80	600	10
#13	CoZrTa [GAR 08]	Sputtered	Stripline	1.52	0.015	300	99	2 [3]
#14	CoZrO [YAO 11]	Sputtered	Stripline	1	1	100	300	4 [4]
#15	$Ni_{80}Fe_{20}$ [MOR 11]	Sputtered	Stripline	1	1	1000	20	3.4 [5]
#16	$Ni_{45}Fe_{55}$ [WAN 05, MEE 09b, WAN 07b]	Electroplated	Racetrack	1.5	0.5	280	45	3.5-5
#17	CoZrTaB [WU 13]	Sputtered	Racetrack	1.2	0.1	1000	115	1 [6]

Table 2.2. *Summary of magnetic materials used as core materials in inductor structures*

From Table 2.1, it can be seen that lamination allows a thicker magnetic core, while keeping the eddy currents to a minimum. These laminations are 2D with alternate layers of magnetic and dielectric thin films. While sputtering is better suited as a deposition technique for laminations compared with electrodeposition, researchers at Tyndall reported a novel technique for laminating magnetic cores using electrodeposition [PAR 03b]. Other techniques investigated for core laminations include the use of a sacrificial layer developed by J. Park *et al.* from the Georgia Institute of Technology [BRU 02, OSA 02].

Recently, researchers have focused on developing a new generation of nano-engineered magnetic materials for addressing this significant challenge. These nano-engineered materials include nano-granular [OSA 00, OSA 98, BRU 04] and core-shell nano-composite [PAR 02, PAR 03a, DHA 04, OHN 95] magnetic materials. The nano-granular materials consist of randomly deposited nano-grains of magnetic metal/alloys in a ceramic matrix, commonly prepared by sputtering the transition elements/alloys in the presence of gases, such as O2 and N2, while the core-shell nano-composite materials are typically prepared either with nano-sized magnetic particles deposited onto an insulating spherical core (micrometer-sized) of SiO2, Al2O3, etc. or by oxidizing nano-magnetic particles to create an insulating shell around them. These two types of materials provide a significant improvement in resistivity, as they are more engineered for an optimized structure having necessary magnetic interactions. This approach allows for what can be described as 3D laminations where individual nano-particles are laminated within the thin films. Additionally, these magnetic particles being smaller than their single domain size, based on the random anisotropy model proposed by Herzer [HER 90], should have very low coercivity that varies as D6 (where D denotes the grain size).

Nano-granular ferromagnetic thin films were reported by Sullivan from Dartmouth as a potential core material for high-frequency micro-inductor structures [SUL 09]. These nano-granular materials consist of nano-particles of metallic magnetic material separated by dielectric thin films. The material has the advantage of increased resistivity due to dielectric thin films, while the metal grains are nano-scale (typically 2–10 nm), minimizing eddy currents losses at high frequencies. In his recent work, Sullivan showed that this material can be potentially used up to a frequency of 1 GHz [DAS 02].

However, one issue to be resolved with this material is the small processing window for depositing high performance material in a repeated manner [SUL 09].

2.3. Figures of merit for performance of integrated magnetics

Much of the work reported in the literature has focused on the comparison of micro-inductor performance based on inductance density and inductance enhancement. A detailed comparison of on-chip micro-inductors with magnetic materials was presented and examined in [GAR 09] based on inductor quality factor versus inductance density. However, to truly evaluate a power micro-inductor for PwrSiP or PwrSoC applications, the range of parameters that needs to be considered includes inductance, DC resistance, Q-factor, saturation current, operating frequency, component footprint and height profile. Since it is impossible to arrive at a single FOM, this paper assesses the performance of all the reported micro-inductors using two FOM. The DC performance of the inductor is evaluated in section 2.3.1 using a plot of inductance per unit DC resistance against the energy density. This section also assesses the AC performance of the inductor based on the conventional quality factor of the inductor.

2.3.1. *Figure of merit – DC performance*

The key DC parameters for an inductor structure are inductance (L), DC resistance (Rdc), saturation current (Idc) and footprint area. For a given footprint area of an inductor, inductance can generally be increased by increasing the number of winding turns (N), using a magnetic material with a much higher relative permeability or using a thicker magnetic core. However, the increase in the number of turns will subsequently result in a higher winding resistance.

The inductance per unit DC resistance, L/Rdc, has been widely accepted as a key parameter for evaluating the performance of an inductor. For micro-inductors, with a constrained footprint and limited conductor aspect ratio, increasing the number of windings or turns for an inductor may restrict the conductor thickness, which subsequently undermines L/Rdc. The parameter L/Rdc can be improved by increasing the magnetic core thickness (if all the other dimensions are kept the same). However, if the magnetic material is conductive, the thickness of the core will be limited by the material skin depth. Lamination techniques can be applied to deposit a thicker magnetic core, but with increased process complexity and cost. The L/Rdc can also be improved by depositing thicker conductors, which can again be constrained by the fabrication process. It can also be increased by adopting magnetic materials with higher relative permeability. However, this approach has its own disadvantage, namely lower current handling capability, as the saturation current of the micro-inductor is inversely proportional to the effective permeability of the magnetic material.

As stated earlier, the saturation current and the footprint area are two important parameters for evaluating the performance of power micro-inductors. Generally, the comparative DC performance of power micro-inductors can be made by plotting L/Rdc against current density. However, this is not applicable for stripline inductors unless the inductance value is fixed. For the stripline micro-inductor, the L/Rdc will remain the same if only the length of the stripline is shortened. This is because the saturation current of a shortened stripline inductor does not change while the current density is increased. A reasonable L/Rdc with a very high current density can be created by simply shortening the stripline inductor. To avoid this, the comparison with energy density is introduced. The energy density of a power inductor can be expressed as [2.1], where E is the energy stored per unit area in the core of an inductor, L is the inductance, i is the current and A_r is the footprint area.

$$E = \frac{Li^2}{2A_r} \qquad [2.1]$$

From basic magnetics theory, inductance is proportional to the relative permeability of the magnetic core, μ_r, while saturation current is inversely proportional to μ_r^2. Reducing μ_r will generally allow a higher saturation current, and hence a higher energy density, while the inductance becomes smaller. Alternatively, increasing μ_r can be used to achieve a higher inductance, but with a lower energy density. It is also worth noting that the most effective way to increase the energy density is to increase the thickness of the magnetic core.

Using L/Rdc versus energy density, Figure 2.14 compares the performance of different thin film micro-inductor structures with non-thin film inductors and commercial products. The inductors using ferrite or ferrite composite, such as commercial inductors, integrated ferrite inductors and wire-bond inductors show very high energy density with a relatively high L/Rdc. This is due to the large volume of the magnetic material generally used in these inductors, while the conductor windings can be constructed using thicker copper with fewer process constraints. However, these inductors are only suited for PwrSiP applications and are unlikely to be compatible with silicon integration for PwrSoC applications. Some toroidal inductors show relatively high L/Rdc in the plot mainly due to the use of higher relative permeability materials, which, in turn, results in relatively lower energy density. The most effective approaches to improving L/Rdc while maintaining high energy density include using a laminated core and applying thicker conductor, where possible.

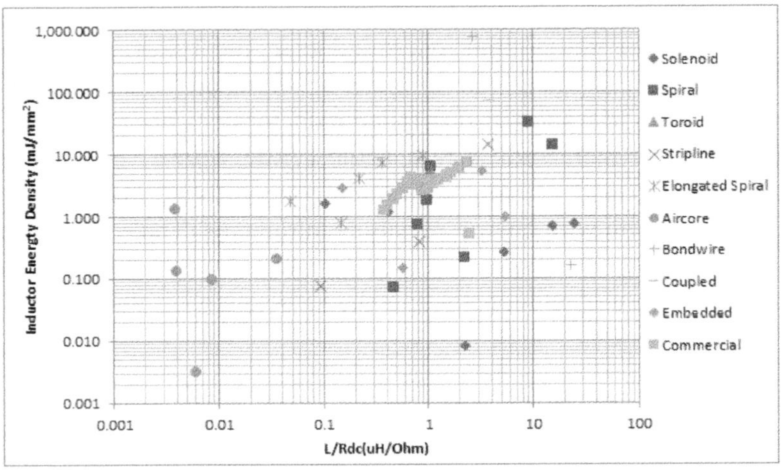

Figure 2.14. *Typical abacus of energy density versus L/Rdc factor. For the color version of this figure, see www.iste.co.uk/allard/systems.zip*

In terms of the performance of a DC/DC converter, the L/Rdc factor also indicates the electrical performance of the output inductor. For a buck converter, the required inductance value is dependent on the switching frequency (f), output voltage (Vo), duty cycle (D), output current (I_0), and current ripple ratio (I_r), as expressed in [2.2]:

$$L = \frac{v_0(1-D)}{f.I_0.I_r} \qquad [2.2]$$

By taking into account the DC resistance, [2.2] can be expressed as [2.3], which can be further expressed, using the output power of the converter (P_0) and inductor's DC conduction loss (powerloss$_{dC}$), as in [2.4]:

$$\frac{L}{R_{dc}} = \frac{v_0(1-D)}{f.I_0.I_r.R_{dc}} \qquad [2.3]$$

$$\frac{L}{R_{dc}} = \frac{P_0(1-D)}{f.I_r.Powerloss_{dc}} \qquad [2.4]$$

For a converter with 5 V input, 1 V output and 0.3 ripple current ratio, there will be a certain required L/Rdc to ensure that the efficiency of the micro-inductor is greater than 95% (the typical efficiency of commercial chip inductors) before taking other losses into consideration. For example, for 10, 20, 40 and 80 MHz switching frequencies with the above given input voltage, output voltage and ripple current ratio, the minimum required values of L/Rdc to achieve an inductor efficiency higher than 95% are 5.07, 2.53, 1.27 and 0.63 $\mu H/\Omega$, respectively. For any converter specifications, estimation of the achievable L/Rdc will give a good indication of how efficient the inductor can be.

Referring again to Figure 2.13, micro-inductors using a magnetic core have much higher inductance per unit area than air-core inductors. Among the magnetic-core micro-inductors, solenoid inductors exhibit relatively high L/Rdc with low energy density. Compared with spiral inductors, solenoid inductors have the advantage of increasing the number of winding turns without increasing the winding mean length turn, which results in higher L/Rdc. With constrained saturation flux for the magnetic material, a large number of winding turns leads to lower saturation current and hence lower energy density. The spiral structure is generally considered as a structure that can yield high inductance density and high energy density. Therefore, many of the air-core inductors demonstrated use this structure. Due to the complexity of inducing anisotropy in the magnetic core, as mentioned previously, spiral inductors are generally aimed at low operating frequencies (<10 MHz), for which a high inductance value is required by the converter. The highest demonstrated frequency of square or circular spiral inductors using a thin film magnetic core is 8 MHz [MIN 96a].

For toroidal- and racetrack-shaped micro-inductors, the trade-off between L/Rdc and inductor energy density still exists. The toroidal- and racetrack-shaped inductors can achieve similar performance to that of solenoid and spiral inductors, i.e. higher L/Rdc results in lower energy density. However, racetrack and solenoid inductors are more capable of operating at high frequencies than toroidal and spiral inductors due to the ability to induce anisotropy in the core. Solenoid and racetrack-shaped inductors have been demonstrated in a frequency range from a few MHz to tens of MHz, where high or moderate inductance values are required by the converters.

The stripline micro-inductor is generally used for achieving low inductance values because it is a single turn device, resulting in a low L/Rdc [WAN 15]. However, the reported stripline micro-inductors showed random distribution of FOM values because of the significant difference between the processing technologies used. The L/Rdc of a stripline micro-inductor can be significantly increased by either increasing the conductor cross-sectional area [PEN 13], laminating the magnetic core [NAK 00] or using both [ZHA 99]. Due to the nature of the stripline structure, most stripline micro-inductors have very low inductance, which implies that they need to operate at high frequencies, typically higher than 60 MHz [NAK 00]. The air-core micro-inductors show the lowest L/Rdc with various energy densities. As the operating frequency increases, the difference in L/Rdc between thin film magnetic core inductors and air-core inductors becomes smaller unless advanced magnetic materials or core lamination techniques are used [BAN 00].

At switching frequencies of several hundreds of MHz, the inductance values required are of the order of nH or tens of nH. It might be considered that with such small inductance values, magnetic material is not required to achieve the inductance. However, the use of a suitable magnetic material always increases the inductance achievable per unit area. The inductance enhancement can be directly translated to a saving in silicon area as well as lower DC resistance. For power micro-inductors, the reduction of EMI through the shielding effect of the magnetic core may also be a very important consideration. However, the level of inductance enhancement is only possible if the resistivity of the magnetic material is sufficiently high so that relatively thick layers of the material can be used at the frequency of interest, or the magnetic core can be laminated to control the eddy current loss.

2.3.2. *Figure of merit and AC performance*

The AC performance of micro-inductor structures, which is primarily an evaluation of the magnetic core material, can be quantified by the ratio of inductive to resistive impedances and is termed the quality factor (Q) expressed as [2.5], where R_{ac} is the AC resistance of the inductor and f is the frequency:

$$Q = 2\pi f L / R_{ac} \qquad [2.5]$$

The Q-factor can be seen as a measure of the AC loss within the inductor structure at different frequencies of operation. The AC loss of the inductor using a magnetic core includes AC winding loss and AC core loss, where the AC core loss is determined by the core material properties as well as the flux density.

Figure 2.15 compares the performance of micro-inductors using different deposition techniques described in section 2.2. From the graph, the screen-printed materials show a higher Q-factor at frequencies below 10 MHz. This is due to the deposition of thicker, screen-printed core material with higher resistivity. The electroplated thin films have comparable Q-factors compared with sputtered films at frequencies of about 10 MHz. This is because at these frequencies the core thickness required can be more easily achieved by electroplating, which is well within the skin depth. However, at higher frequencies where the core thickness required is lower and core laminations can be used, sputtered thin films show a better performance compared with electroplated cores.

2.4. Technology roadmap and challenges

In section 2.1, Figure 2.3 shows a schematic roadmap for the evolution of SMPS. It shows a possible route from power modules to PwrSiP to the ultimate PwrSoC, whereby the passive components are monolithically integrated with the power management IC without increasing the footprint of the power management IC chip. As discussed in the subsequent sections of this chapter, a central requirement to achieve this level of miniaturization of the passive components is to increase the switching frequency of the DC/DC converter in the range of 10 to 100 MHz.

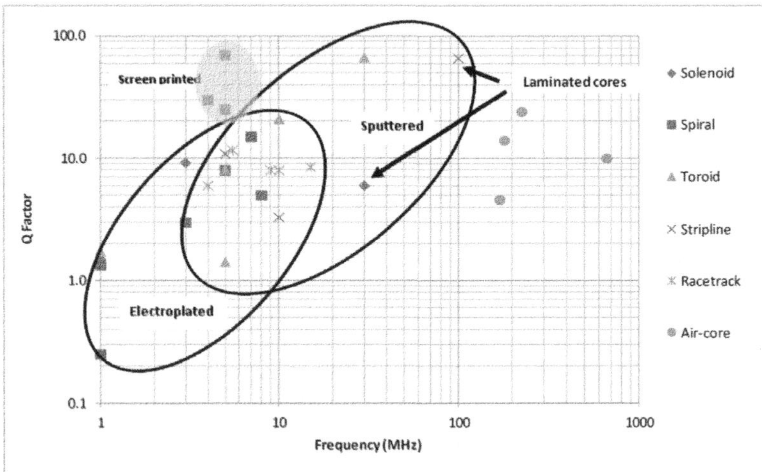

Figure 2.15. *Landscape of Q-factor versus operating frequency of the inductor. For the color version of this figure, see www.iste.co.uk/allard/systems.zip*

This section presents a perspective on some of the key issues that need to be addressed in progressing along this roadmap. First, market drivers based on potential "killer applications" are presented. Second, supply chain issues need to be addressed, which will facilitate high-volume manufacturing capability for thin film magnetic materials on silicon. Third, key challenges need to be addressed to deliver the materials and technology platforms.

2.4.1. *Market drivers*

A number of key market opportunities can be identified as driving the progression of the PwrSoC roadmap. First, the technology opens up the possibility of delivering the efficiency of a high-performance switched-mode

DC/DC converter with the same footprint and cost of a linear regulator chip. While micro-inductors on silicon have not yet achieved the efficiency performance of commercial chip inductors (i.e. 95 to 96%), efficiencies as high as 93% have been reported for current levels of 500 mA [GOD 11]. With this in mind, in replacing a linear regulator with a typically low efficiency, there is a significant scope for existing micro-inductor technologies to deliver substantial efficiency improvement. This even applies with multi-MHz switch and control circuits that are still not optimized from an efficiency perspective, compared with state-of-the-art commercial power management ICs operating in the 3 to 6 MHz space. Achieving the competitive cost targets of linear regulators will still be a major challenge.

At the other extreme from the discrete converter solution is the ultimate opportunity presented by the PwrSoC concept: delivering multi-rail arrays of on-chip power supplies to complex SoC applications, including multi-core processors. As mentioned in section 2.1, such SoC platforms currently require off-chip power management ICs to deliver multiple digital and analogue power rails as well as dedicated rails for I/O and RF functions. Bringing these functions directly into the SoC package and, ultimately, onto the SoC chip itself will provide significant performance and cost benefits. Dibene and colleagues at Intel have already demonstrated this concept with an array of coupled on-chip micro-inductors to demonstrate a granular power supply capable of delivering up to 400 A to a microprocessor. This concept is increasingly being referred to as a granular on-chip power management system architecture for SoC.

While both of the above applications present what might be considered the ultimate "holy grails" in the PwrSoC space, a more near-term commercial opportunity is needed to develop industry confidence in the market. Enpirion have announced the launch of a power converter product with a stacked micro-inductor incorporating a single layer of thin film magnetic material. While this layer can, at the absolute maximum, only double the inductance per unit area, it does allow the company to bring up a dedicated manufacturing foundry capability and facilitate the qualification of a relatively easy fabrication process.

Another more ambitious example of micro-magnetics on silicon is the air-core micro-transformer technology from companies such as Analog Devices [HER 90] and Infineon [HER 93]. This technology is silicon based and allows micro-transformers to be integrated in the same package as the active silicon, achieving a relatively small footprint. Analog Devices has reported that this technology has delivered significant business for the company in isolated power and signal transfer for instrumentation applications. In this case, the integrated micro-transformer technology presents a significant added value to the product offering, unlike the integrated micro-inductor, which is currently seen to be competing against a very low cost, small footprint, commercial chip inductor.

2.4.2. *PwrSoC supply chain challenges*

The biggest impediment to the development of a commercially viable magnetic material on silicon technology is the perception that the supply chain infrastructure required to deliver it does not exist and would be too expensive and take too long to put in place. This clearly does not stand up to scrutiny given the fact that the technology has been used in the high-volume, cost-sensitive, magnetic disk-head industry for over three decades [LI 05, MEE 11]. This then raises questions about the nature and ownership of a viable manufacturing facility: should it be captive or is a commercial foundry model feasible?

For a captive production facility, because of potential contamination issues from magnetic materials within a CMOS facility, a dedicated back-end-of-line (BEOL) facility presents an option for existing power semiconductor companies. Given that many of these companies already have copper electroplating and associated photolithography capabilities for high current power semiconductors, the addition of magnetic processing capabilities should not be onerous. In the case of a commercial foundry option, this could either be established within a MEMS foundry or be a packaging/assembly contract house. Both facilities already have considerable expertise of post-processing CMOS wafers.

While the discussion here has focused on the supply chain infrastructure for magnetics on silicon for micro-inductors and micro-transformers, it should be recognized that there is also considerable opportunity for such a production facility to target other market opportunities, such as processing of magnetic sensors on silicon for on-chip compassing and current sensing [VAL 69, ROM 70, CHI 00].

Other issues that need to be addressed in terms of the supply chain infrastructure include:

– the development of a micro-magnetic component design environment that interfaces with standard IC CAD environments, both in terms of circuit simulation and mask-making;

– a magnetic material characterization capability as a development and quality control tool;

– the implementation of a wafer-level test capability for micro-magnetic components.

2.4.3. *PwrSoC technology platform*

The third area requiring discussion relates to the technical challenges that still need to be addressed to deliver the ultimate PwrSoC technology platform. While the main focus of this discussion relates to thin film magnetic materials and associated processes, other issues that are pertinent are also discussed.

The main issue for thin-film magnetic materials in power micro-magnetics is related to the need for a low-loss material system that can deliver a high-efficiency micro-inductor component (i.e. greater than 95%), while being compatible with 3D stacking or monolithic integration within the footprint of the power management IC chip. The main requirement here is for a material with high saturation flux density, high resistivity and low coercivity. As seen in section 3, while sputtered magnetic materials can have higher resistivity than electroplated materials, the lower cost of electroplating is a major driver in material selection. This leads to the consideration of laminated cores, as was addressed in section 2.2.

Other issues that still need to be addressed relate to the electrical and thermo-mechanical interaction of the micro-magnetic component with the underlying power management circuitry, whether in a stacked or monolithically integrated format. While the magnetic core will provide shielding, EMI issues and associated containment need to be considered where stray fields from windings without a core may induce noise. From the thermo-mechanical perspective, magnetostriction in the magnetic core materials may result in stress-related changes to the micro-magnetic component performance [SHA 08].

One of the barriers in using magnetics on silicon for DC/DC converters is the ongoing need to have on-chip I/O to access both input and output filter/storage capacitors. Furthermore, despite the fact that discrete commercial ceramic chip capacitors are at least an order of magnitude cheaper than chip inductors, a number of academic and industry researchers have reported significant progress in delivering high-density, 3D, on-chip, trench capacitor technologies with capacitances per unit area in the range of 100 to 400 nF/mm^2 [TIP 05].

The availability of high density capacitors raises the possibility to consider a passive LC interposer, comprising the input and output capacitors and the output inductor for a switched-mode DC/DC converter. This passive interposer could then be stacked with a power management IC to deliver a complete PwrSiP platform. Furthermore, an LC interposer consisting of an array of multiple LC combinations could also be stacked onto a large SoC chip or multi-core processor, thereby enabling the previously mentioned concept of a high granularity power supply system architecture delivering multiple, isolated voltage rails for digital, analog, I/O and RF functions. Delivering such an integrated solution would potentially require the application of wafer level bonding and through-silicon vias.

2.4.4. *Integrated magnetic devices for PwrSoC – opportunities*

A market study by US-based Anagenesis (http://www.anagenesis-inc.com) entitled "PSiP & PwrSoC: The Ultimate Power Density Challenge, 2008–2015" predicted dramatic growth in the PwrSiP/PwrSoC market over the next 5 years. It was predicted that by 2015, the market would expect to be of the order of £1.5 bn with a compound annual growth rate estimated at more than 50%. This is validated with different products presently available in the market. Intel have demonstrated a PwrSiP solution for granular power management of their Haswell processor. They use a PCB-based air-core magnetic component and claim a resulting increase in battery lifetime by 50%. Analog Devices have developed the only PwrSoC solution currently available in the market using their stacked air-core transformer for power transfer over an isolation barrier; this product is called the "icoupler". With this product, Analog Devices' share of the digital isolation market is about 60%. The isolation market is estimated to be about $800 M, with digital isolators making up to 10% of this market. Analog Devices predict a doubling of their power-on-chip product between 2015 and 2018. Figure 2.3 (ref. 5–34) shows the advantage of power component integration with the highest power density reported for a full PwrSoC solution from Intel. They demonstrated this performance based on a granular approach to delivering power to the load (microprocessor). In this chapter, we propose to implement this granularity in nano-scale structures, further improving the power density to address different application requirements.

Based on the market trends and status of the technology, this program will develop a platform magnetic passive technology to target two different power management applications:

1) micro-processor power delivery;

2) isolated power transfer.

This particular application was first identified by Intel as having the potential for fully integrated power delivery, where the key challenge is a requirement for efficient power delivery over a wide current range of 500 mA to 100 A. Additionally, with microprocessor loads, there is a significant requirement to provide fast transient power with low output ripple. As explained earlier in this section, Intel's proposed solution is to use air-core PCB inductors in a multi-phase coupled implementation. However, the use of air-core technology limits the power density, particularly if implemented on silicon; hence Intel have demonstrated this product in a stacked PCB structure.

Due to the limitations of copper tracking on silicon, realizing this technology in a fully integrated system requires the inclusion of a magnetic core to couple the multiple phases of the power delivery system. Furthermore, Tyndall (with collaborators) have demonstrated a highest efficiency of 86% for a three- or five-phase integrated coupled topology operating at a frequency of 20 MHz. Within the PowerSwipe project, INSA Lyon, UPM, IPDiA and Tyndall demonstrated PwrSiP technology on a passive interposer with an efficiency of > 80% at 100 MHz. The on-silicon coupled inductor showed an efficiency (estimate) of 90%. This is the highest efficiency for a coupled inductor for PwrSoC reported in the literature. Table 3.1 compares the different on-silicon coupled inductors reported in the literature.

One of the key constraints with integrated magnetic inductors is the winding conduction loss. This is due to the fact that planar coils have small cross-sectional dimensions due to limitations in photoresist patterning and electrodeposition processes. This results in higher DC conduction losses in integrated inductor structures than for their discrete counterparts. This issue is not a major concern in integrated transformer applications, where only AC current is transferred between the two windings. This was confirmed in our previous publication where a stacked copper design had higher efficiency compared with a single-layer interleaved design with twice the cross-sectional area of copper. This is due to lower AC losses in the stacked copper design compared with single-layer copper windings. Table 2.4 compares the different on-silicon transformer research reported in the literature.

	Inductance (nH)	DCR (mΩ)	CL footprint (mm^2)	Frequency (MHz)	CL efficiency (%)
Dartmouth Tyndall 2004	14	40	80	7	70
Intel 2011	25	14	2.8	> 60	80
Columbia IBM/ Ferric 2013	7.4	480	10.9	100	89
Columbia IBM/Duke 2012	12.5	270	2.5	> 40	80
Tyndall 2014	45	27	3	5	85
NUIG Tyndall 2015	23	71	5	20	86
PowerSwipe 2016	47	340	1.25	100	90.7

Table 2.3. *Comparison of state-of-the-art in integrated coupled inductors*

2.5. Conclusions

As presented in section 2.1 on the survey of commercial POL, SMPS converters and research demonstrators, the technological shift has already been seen for low power devices from discrete component power modules to on-chip component-based PwrSiP platforms. Older generations of PwrSiP products have a similar footprint area to the power modules, but with a much smaller profile, which results in great power density (mW/mm^3). More recent

PwrSiP products, adopting a 3D stacked approach, are achieving higher power density. To further increase the power density, the operating frequency needs to be increased so that smaller inductors can be used. As has been reviewed in this work, there are many types of magnetic technologies which can be applied to further improve the performance of PwrSiP products. In the near future, the main challenge to achieve this is to reduce the inductor size and resistance while maintaining the current handling capability. Each reported magnetic technology provides an attractive alternative for use in future PwrSiP products, either because of its good electrical performance or due to the simplicity of the fabrication processing technology.

	Inductance (μH)	DCR (Ω)	Freq (MHz)	Gain (dB)	Coupling	Size (mm^2)	Efficiency (%)
Racetrack [WAN 07a]	0.4	0.48	1-20	-1	0.93	24	72
Solenoid [MIN 96b]	0.8	7.7	3-40	-3	0.93	25	
Solenoid [PAR 98a]	0.8	1.1	7-11	-4	0.9	4	
Square spiral [PAR 03a]	1.3	3.5	25	-5	0.85	12.9	
Square spiral [LOT 98]	1.4		5	-3		9	
Square spiral [MEY 10]	0.046	2	125	-4	0.63	2.25	78
Circular spiral [YAM 93]	0.5	20	5		0.95	7.44	67
Circular spiral [PEN 13]	0.06	66	10-100	-3	0.7	0.5	
Toroidal [XIN 13]	0.053		20-50		0.9	4.8	
Racetrack [WAN 15]	0.27	1.22	10-30	-1	0.9	3	78

Table 2.4. *Comparison of state-of-the-art in integrated transformers*

Although, as yet, there has been no PwrSoC product introduced to the market, different integrated magnetic technologies have been successfully used and demonstrated in many PwrSoC prototypes reported in the literature, achieving reasonable performance and efficiency.

Integrated magnetic devices will become a key enabler and differentiator in the development of future PwrSoC products. Other than the challenges of how to improve the performance of magnetics, there are more challenging issues to be addressed to realize PwrSoC products, including system design and technology compatibility issues. Compatibility issues to be addressed include processing, performance degradation due to component interference and parasitics, packaging/integration issues and long-term reliability. From design perspectives, the system co-design will include modeling and optimization of all critical components. A system co-design and verification methodology and a CAD tool for system simulations with high abstraction level models will also be required.

2.6. Acknowledgments

The authors wish to acknowledge the funding from the European Union, Enterprise Ireland, Science Foundation Ireland, MCCI (Microelectronics Centre of Competence Ireland) and PSMA (Power Sources Manufacturers Association), which has facilitated much of the research reported by the Tyndall National Institute in this chapter.

3

Dielectric Components for Increased Power Density

Capacitors, despite their banal appearance, are devices of great technologies. An IoT device does not offer many possibilities for selecting capacitor technologies. The frequency, linearity and reliability performances of this capacitor are closely linked to the dielectric selection and its integration process. Through a detailed description of the mechanisms of charge storage in dielectric materials, options for the implementation of an ideal capacitance are discussed. An almost completely satisfactory technology available for mass production is then detailed, along with the way to practically layout optimal capacitors.

3.1. Introduction

As a trivial definition, capacitors are devices used to locally store energy in an electronic circuit. This happens by local buildup of charges on the electrode of the capacitor, as an image of the polarization of the dielectric material that is built-in. Many different types of capacitor technologies are available in the ecosystem, mostly related to the nature of the mechanisms (thus dielectric materials) used for charge storage (Figure 3.1). This yields

Chapter written by Frédéric VOIRON.

large differences in terms of the capacitor's electrical performance including breakdown field, specific capacitance, linearity (Hz, deg. C, V), losses, reliability figures, etc. The physical link between the nature of the dielectric material and the electrical properties is discussed in the next section.

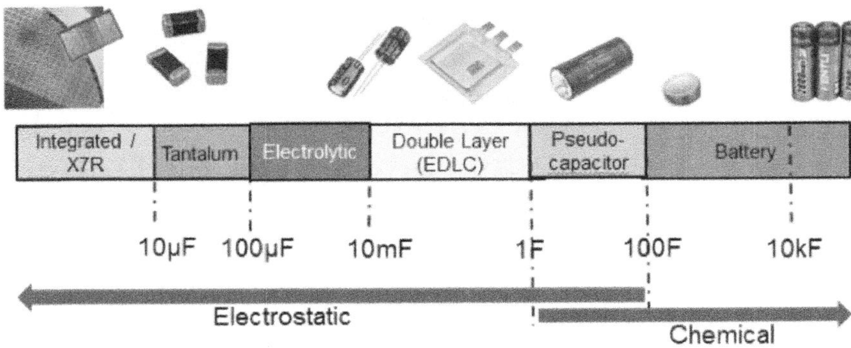

Figure 3.1. *Industrial landscape of capacitor technologies*

In section 3.3 we will focus on a technological approach that is nearly optimal for the applications requiring ultra-thin and flexible form factor, combined with high frequency switching capabilities. This technology is based on the integration of the capacitors within 3D structures, embedded onto a silicon substrate. This offers the advantage of very high specific capacitance, combined with nearly defect-free processes, bringing large advantages in terms of structure stability and reliability. Starting from the construction of a broadband electrical model, we will also explain the methodology to layout optimized capacitors with very low resistive losses and higher frequency stability. The technical discussion presented in this section will be illustrated by examples based on the PICS platform supported by IPDiA. However, it can be generalized to any other platform implementing an equivalent technological stack.

3.2. Basics of dielectric physics

3.2.1. *Forewords*

This introductory section is dedicated to a general description of the physical mechanisms taking place in a dielectric subjected to an electric field. Although this is part of the knowledge of each device engineer, it is important to recall the basic principles that are the foundation for the integrated capacitance architecture that will be presented in the next section. We will recall the relationship existing between the susceptibility of a dielectric, the polarization vector and the capacitance. The physical origin of the susceptibility is also introduced with reference to a simplified example (electric polarization in a static field) and further discussed for real dielectric materials. Finally, the notion of loss is briefly discussed at the end of the section.

3.2.2. *Polarization, dipole and capacitance*

For most of the capacitors, the insulating layer consists of a dielectric material. This material has the peculiar property of preventing free charge motion when subjected to an electric field. In contrast to conductors where charge displacement towards the outer surfaces (electrons to the anode and holes to the cathode) would exactly screen the electric field in the volume, charges partially build up on the dielectric interfaces, leaving a significant electric field across the dielectric layer (Figure 3.1). The origin of this charge buildup is related to the dielectric polarization mechanism. As a very general explanation, we consider that dielectric materials consist of a combination of charges (nucleus, electrons) that are combined as atoms and bind as molecules. When a dielectric is subjected to an electric field E, the respective centroids of the Q+ and Q- charge cloud are shifted, leading to the formation of an apparent dipolar structure. This phenomenon is known as polarization P, which is expressed as the dipolar moment per unit of dielectric volume. At any point in the dielectric, the P magnitude and orientation is related to the E field by relation [3.5], where χ is the electric susceptibility of the dielectric and ϵ_0 is the vacuum permittivity. In the case of a parallel plate capacitor, the polarization inside the material corresponds exactly to the density of charge that is built-up on the dielectric surface (Figure 3.2):

$$P = \chi \epsilon_0 E \qquad [3.1]$$

Integrating Gauss's law over the surface indicated by the dashed line in Figure 3.2 yields the following relation [3.2]:

$$V = E.d = \frac{(Q^+ - P).d}{\epsilon_0} = \frac{Q^+.d}{\epsilon_0.(1+\chi)} \qquad [3.2]$$

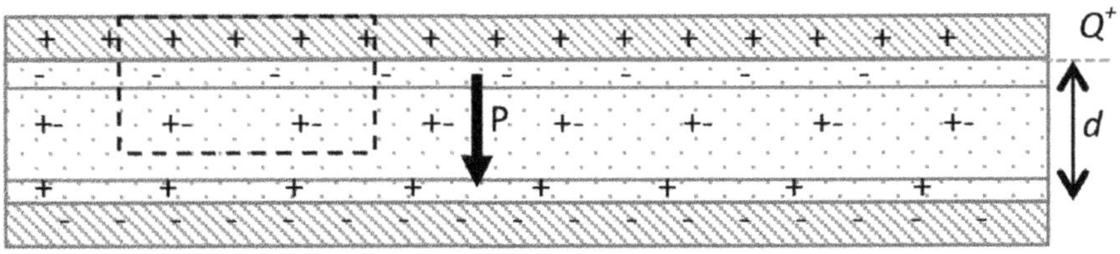

Figure 3.2. *Representation of the polarization within a dielectric buildup in a plane capacitor. For the color version of this figure, see www.iste.co.uk/allard/systems.zip*

For a capacitor of a surface S, the charge Q^+ is proportional to the potential V through the capacitance C [3.2], which yields relation [3.3] for the capacitance. Through relation [3.3], we can observe a capacitance that is obtained from a dielectric confined between conductive plates. It strongly depends on the ability of the dielectric to polarize under a given electric field, E. The fundamental mechanisms leading to the polarization are closely related to the structure of the dielectric material, and are of great influence on the performance (storage capacity, linearity) of the capacitive element. This consideration will be detailed in the following sections:

$$C = \frac{S\epsilon_0(1+\chi)}{d} \qquad [3.3]$$

3.2.3. *Polarization mechanisms in dielectrics*

Dielectrics correspond to a large set of material states; however, we will concentrate on solid-state dielectrics that are of interest for silicon integrated capacitors that will be treated in the following chapter. By definition, dielectric materials have an electronic structure preventing free conduction in their volume that allows electric field to settle. In the previous section, we have explained how capacitance relates to the polarization and ultimately to the electric susceptibility of the dielectric material. In this section, we will detail the mechanisms linking the susceptibility to the structure of the material, and discuss the impact on dielectric performance.

3.2.3.1. *Theoretical approach*

Electric polarization refers to the creation of a dipole moment by the impact of an electric field on the atom. The effect of the field E on the electron cloud results in a non-symmetrical shift in the centroid of charges, thus inducing the dipole moment. Also, the electronic binding force can be considered as "stiff", and the displacement of the electron is sufficient to create an electric polarization. The theoretical calculation of the polarization of a non-ionic molecule requires the consideration of the complete quantum-mechanical theory. However, a simplified approximation can be obtained by considering that the centroid of electric charges reaches the equilibrium, when the force resulting from the interaction of the E field on electrons is compensated by the restoring force inherent to the electronic binding in the solid state [DEX 53]. In [3.4], m is the mass of an electron, P is the polarization and ω_0 is the optical absorption pulsation of the dielectric material:

$$m\frac{d^2P}{dt^2} + m\omega_0^2 P = q^2 E(\omega) \quad [3.4]$$

The solution of [3.4] gives the expression of the polarization in [3.5], which enables us to identify the susceptibility in [3.6]. The latter approach is ultra-simplistic because common dielectrics are not single atoms or non-polar molecules but complex combinations of different atoms. As a result, they do not have a single but multiple absorption lines, reflecting their complex electronic and molecular structure. However, relation [3.5] indicates the direct relation between the susceptibility of a material and its internal structure represented by the absorption frequency spectrum:

$$P = \frac{q^2 E}{m\omega_0^2}\chi = \frac{q^2}{\epsilon_0 m\omega_0^2} \quad [3.5]$$

$$\chi = \frac{q^2}{\epsilon_0 m \omega_0^2} \qquad [3.6]$$

3.2.3.2. *Polarization in real dielectrics*

In most dielectric materials, atomic and ionic polarizations are superposed on the electronic contribution. Atomic polarization occurs in materials consisting of several non-ionic atoms. The underlying mechanism is exactly similar to the electronic contribution but takes place at the atomic binding level. Electrons involved in the binding will not be shared equally and be shifted towards the more electronegative atoms, resulting in a permanent dipole moment. If an external electric field is applied, the atoms position will shift, such as to align the dipole and the field. This results in a shift between centroids of charges inducing polarization (see [3.2] and [3.3]). Ionic polarization is similar to atomic polarization but involves the displacement of ionic molecules under the electric field. This can lead to a higher apparent value of the relative dielectric constant, above several thousands. However, the mobility of ionic specifies within solid-state dielectric is low. As a consequence, applications for such material are limited to the lower range of the frequency spectrum. This is illustrated in Figure 3.3, which sketches the frequency range of operation (related to the ω_0 range, see equation [3.5]) for different polarization mechanisms. From this observation, we can conclude that the internal structure of the dielectric has a great influence on the frequency linearity of the capacitors.

Finally, to complete the qualitative description of "real" dielectric characteristics, we can differentiate two main families of dielectrics: paraelectric and ferroelectric. Ferroelectric materials do not lose their ionic polarization when the electric field is released. Due to the additional restoring force resulting from the crystal lattice in ferroelectrics, the ions can reach configurations that do not relax back to their original state. As a result, residual polarization might be left after the field is removed, with polarity depending on the direction of the last field. It is important to note that some ferroelectric materials (i.e. BaTi03, Ba(Sr)TiO3) do exhibit very large dielectric constants in specific crystal configuration (texture, grain size, thickness) that are dependent on the processing conditions [EZH 00]. This is related to the displacement of Ti under the effect of the electric field in the crystal lattice, resulting in the amplification of the net polarization. However this amplification only manifests across a limited voltage and temperature spectrum, resulting in a highly nonlinear behavior. In contrast, paraelectric materials are quite insensitive to the processing and biasing conditions, thus yielding very stable capacitors when used as a dielectric material. This is illustrated in Figures 3.6(a) and 3.6(b) with a comparison of oxy-nitride and ferroelectric dielectric-based capacitors for linearity.

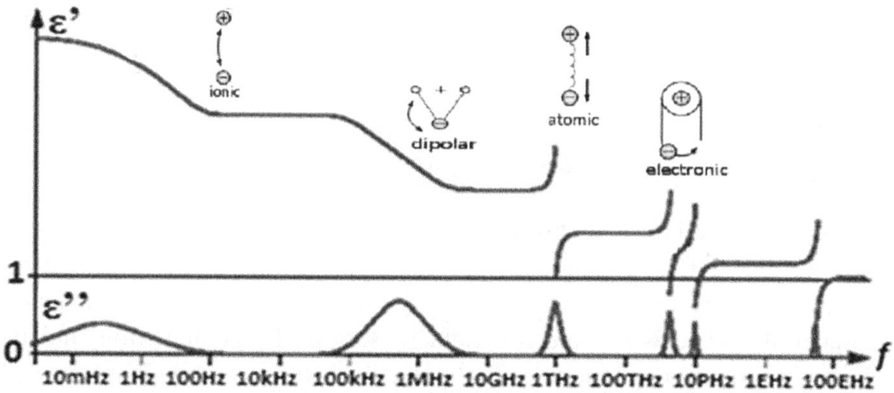

Figure 3.3. *Illustration of the frequency response of different polarization mechanisms. The vertical axis represents the magnitude of polarization for a given value of the E field. For the color version of this figure, see www.iste.co.uk/allard/systems.zip*

3.2.4. *Losses in dielectrics*

Relative permittivity can be expressed in the complex form, with an imaginary part corresponding to the losses in dielectrics. The ratio between the dielectric losses and the permittivity is sometimes quantified as the loss tangent or $tan\delta$. This is an important factor of performance as it reflects the ratio of energy that is dissipated in the capacitor. Losses in a perfect dielectric result from the inability of the polarization process to follow the pulsation of the oscillating electric field. This is a consequence of the relaxation time that corresponds to the delay needed for the dipoles to return to their original states, under the effect of restoring forces induced by the crystal lattice (see [3.4]). If the relaxation time is faster than the pulsation of the electric field, we can observe a minimum of loss. This situation corresponds to the usable frequency range for a dielectric built in a capacitor. When the pulsation of the electric field is faster than the relaxation time, the polarization cannot build up, and energy is absorbed and dissipated as heat in the material. This occurs in the microwave region (Figure 3.3) for faster polarization mechanisms. The higher region of the frequency spectrum involves the excitation of electrons. The dielectric constant shows a maximum before a symmetrical drop over a certain frequency. The frequency at which the turning point occurs is the absorption frequency, ω_0, illustrated in [3.5], and corresponds to the maximum of dielectric loss, ϵ'. The lower region of the frequency spectrum corresponds to the other types of polarization (i.e. atomic, molecular, ionic), each having their own specific relaxation time.

It must be noted that charge conduction through the oxide also contributes to losses, through a dependency on the apparent resistivity, ρ, and the complex dielectric value in [3.7], where τ is the scattering time of carriers in the material and ω is the electric field pulsation. The conduction into the dielectric is greatly dependent on the dielectric nature (i.e. bandgap), as well as on the integration scheme (mainly electrode/dielectric work function difference) and the processing conditions that might be the source of local defects. This point will be further discussed in the following section:

$$\epsilon'' = \frac{1}{\epsilon_0 \rho \omega} \cdot \frac{1}{1 + (\omega \tau)^2} \quad [3.7]$$

3.3. Silicon integrated capacitors

In this section, we will focus on the implementation of silicon integrated capacitors [MUR 06]. The advantage of this type of capacitor technology, for highly integrated applications, is related to the ultra-low thickness ($< 100\mu$m) and the flexible form factors that can be obtained. Furthermore, for certain application domains (typically for low voltage $<$ 15V), silicon integrated capacitors can outperform pre-existing technologies (see Table 3.1) in terms of specific capacitance [ULR 00], [ROB 11].

3.3.1. *Integrated capacitors for enhanced performance*

To support the technical discussion developed in this section, we will focus on the *Passive Integrated Common Substrate* technology supported by IPDiA (www.ipdia.com), which is an open, accessible and fully qualified passive platform. We will review the key technological steps to build the PICS integrated capacitor and discuss their influence on intrinsic electrical and reliability performances. Also, techniques to model and optimize the PICS switching capability will be reviewed, and a comparison in terms of switching performances (resonance frequency and resistive losses) will be presented.

Dielectric type	Dielectric constant	Thickness (μm)	Specific capacitance (nF/mm^2)	Typical size for 100nF/cap (mils)
Spin on BCB	2.7	2	1.2	3600
Spin on glass	3.7	1	3.3	2200
SiO_2	3.7	0.2	16	970
Al_2O_3 (anodic)	9	0.2	40	640
Ta_2O_5	24	0.2	110	380
TiO_2	50	0.2	220	270
Ferroelectric thick (paste)	100	1	89	420
Ferroelectric thin (sputtered/SolGel)	2000	0.2	8900	40
PICS3	6	0.002-0.5	25000	24

Table 3.1. *Comparison between specific capacitance and size for the conventional type of passive technologies and PICS 3D silicon integrated capacitor*

3.3.1.1. *An example of silicon integrated capacitors, the PICS technology*

This is an integration platform featuring all complementary passive components that are required by common analogue applications. It allows embedding multiple flavors or resistors ($13.5\Omega/\Box$ and 0TC), high-quality copper inductances, customizable diodes and high-density capacitors. The high-density capacitor [MUR 06] device combines a high aspect-ratio micro-metric 3D structure with a multi *Silicon In Silicon capacitor* (see

Figure 3.4). The passive devices can be insulated from the substrate by using a deep NWELL strategy, enabling the cohabitation of multiple voltage domains (differential voltage up to 70 V) and frequency domains (typical rejection > 50 dB in the DC-5 GHz frequency range) on a common silicon substrate.

Figure 3.4. *SEM micrograph showing the high aspect-ratio structures etched in the silicon substrate (tripods with an aspect ratio of 50:1)*

A schematic representation of the PICS3 structure, with the double SIS structure sunk into the 3D trench formed by the inter-spaces of pillars, is shown in Figure 3.5(a). The bottom capacitor electrode (represented as HL in Figure 3.4) is obtained by a high-temperature phosphorous drive-in into the space left between the tripod structures. As a result, a high doping level is obtained, contributing to the improvement of voltage linearity and the control of bottom electrode resistance. The lower electrode conjointly forms the NWell underneath the 3D structure, isolating the capacitor from the silicon substrate. The upper electrodes are the *in situ* phosphorus-doped polysilicon layer (represented as Ps/Ps2 in Figure 3.5(a)) deposited by a high-temperature CVD process. Their thicknesses are optimized for a trade-off between electrode resistance and hole filling capability.

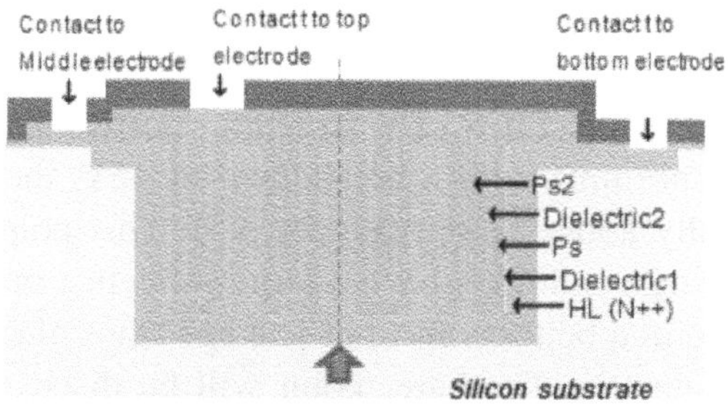

a) Schematic cross-section

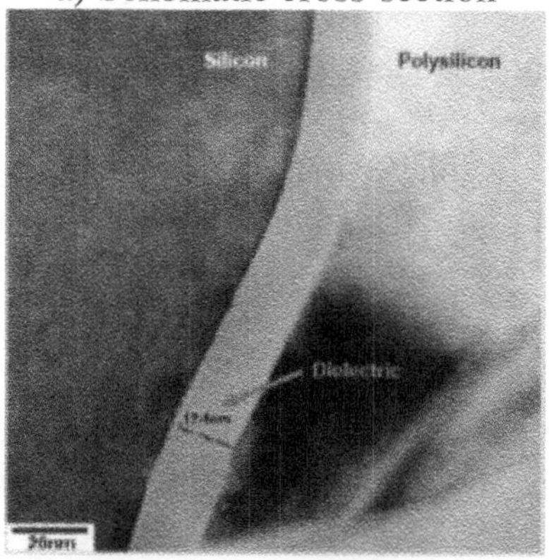

b) SEM cross-section of the mIM structure at the mesa of the trench

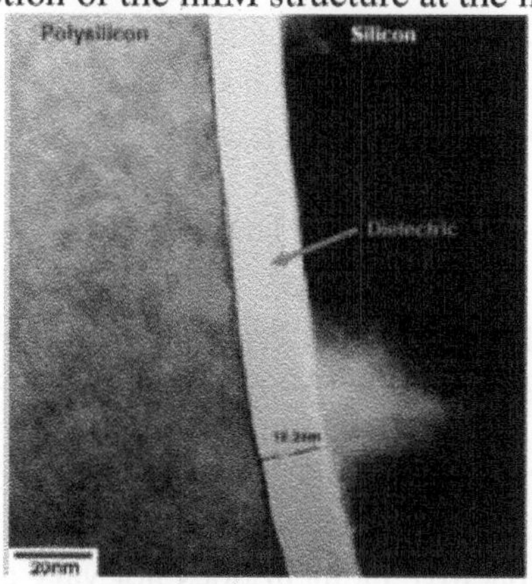

c) SEM cross-section of the mIM structure at the bottom of the trench

Figure 3.5. *Overview of the PICS3 structure. For the color version of this figure, see www.iste.co.uk/allard/systems.zip*

Both dielectric 1 and dielectric 2 are oxy-nitride paraelectric composites, formed by a combination of nitridation and TEOS oxidation high-temperature LPCVD processes. For the PICS3-LV structure, corresponding to a specific capacitance of 250 nF/mm^2 with a breakdown of 15V, the thickness of the dielectrics is typically about 17 nm (Figure 3.5). This point is important as long as the processing conditions have a large influence on the reduction of defectivity, resulting in a better control of leakage (thus electrical losses) and reliability of the dielectric. The latter point will be discussed further in the following section.

Finally, PICS elements (single caps, network or Integrated Passive Devices) are available in thicknesses up to $80\mu m$ for applications requiring ultra-low profiles. This is a key advantage as long as decoupling/filtering capacitance can be placed very next to the IOs, thus preventing parasitic impedance to the load, which is induced by board routing parasitic inductances.

3.3.1.2. *Increment hierarchy position*

As described in the previous section, the capacitor SIS structure is based on a high-temperature ($> 700°C$) defect-free process, combining a paraelectric dielectric with degenerated electrodes. This combination of material aims to deliver voltage and temperature linearity. Indeed, as discussed in Chapter 1, paraelectric materials (especially SiO_2) are mostly insensitive to voltage, temperature and frequency variation. In addition, degenerated polysilicon electrodes behave as quasi-metal and are therefore insensitive to transverse electric fields. For comparison purposes, the linearity of a 100 nF PICS3 capacitor is characterized over its usual voltage/temperature spectrum and compared with an equivalent X7R low ESL commercial MLCC capacitor. The voltage linearity is shown in Figure 3.6(a). Remarkably, the voltage de-rating over the voltage spectrum is extremely low and remains $< 0.5\%$. A similar comparison is performed with respect to temperature de-rating. Remarkably, the de-rating over the temperature spectrum remains $< 2\%$. This provides an advantage for highly integrated applications: as the capacitive element does not vary across temperature and voltage excursions, it can be implemented without margins, contributing to reduced component size and costs.

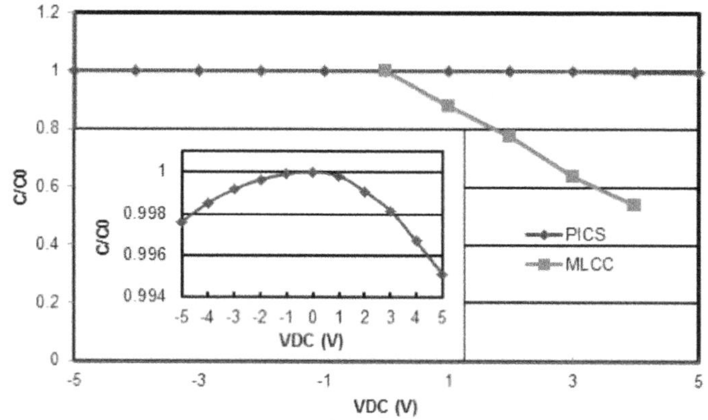

a) Voltage linearity of impedance. Measurement at 1 MHz, 25 deg C.

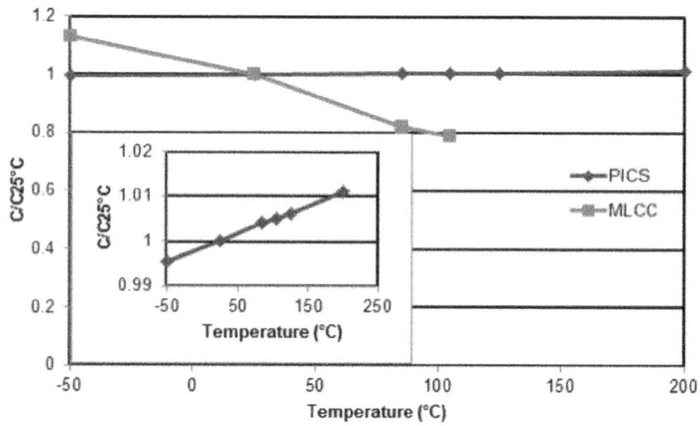

b) Temperature linearity of impedance. Measurement at 1 MHz and 3.5V DC bias.

Figure 3.6. *Linearity of a 100nF capacitor impedance. The blue line corresponds to a PICS capacitor (paraelectric) and the red line to an X7R commercial MLLC component. For the color version of this figure, see www.iste.co.uk/allard/systems.zip*

3.4. Integrated capacitors for enhanced reliability

3.4.1. *Dielectric processing*

The most critical defect mechanism for capacitors corresponds to the dielectric breakdown.

It results in a low-resistive path in between the capacitor electrodes, usually leading to an immediate and unrecoverable system failure. As for any MIS capacitor, the breakdown is activated by thermally and voltage-induced stress [MOR 91], [SIM 71]. Other failure mechanisms, such as current-related aging effects on electrodes and interconnections, can be managed at design, by adding margins to the physical layout of the capacitors (line width, redundancy, metal level selection, etc.). In most of the application cases, margins do not induce any limitation in terms of integration density or electrical performance.

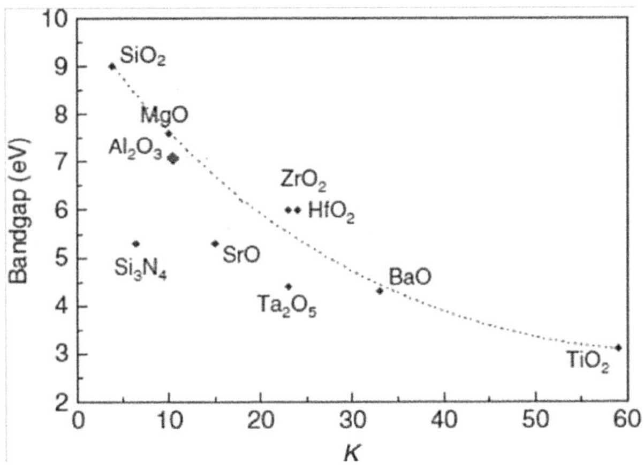

Figure 3.7. *Representation of dielectric bandgap as a function of dielectric permittivity*

The dielectric strength and leakage properties are greatly dependent on the dielectric nature. In general, dielectrics with a large bandgap are preferred for applications where reduced leakage and/or dielectric strength are required, as long as they are contributing to reduced charge injection from the electrodes, while high-K materials are preferred to increase the capacitance density. However, it can be observed that those characteristics do evolve oppositely [JED 04], as shown in Figure 3.7. To obtain an optimal trade-off between capacitance density and dielectric strength, composite dielectrics are often considered:

– implemented with a barrier material having a large bandgap to reduce charge injection and contribute to the overall dielectric strength improvement;

– combined with a high-K material to enhance the equivalent permittivity of the composite stack.

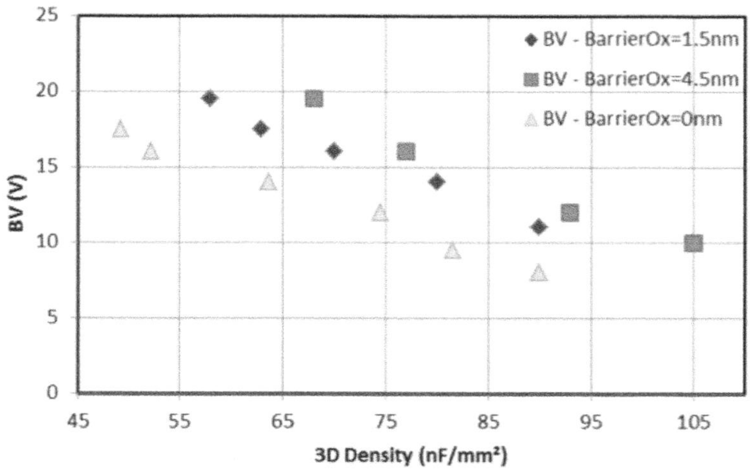

Figure 3.8. *Evolution of the dielectric strength with respect to the composite dielectric buildup*

This implementation on an oxy-nitride composite dielectric is shown in Figure 3.8, where the relative thicknesses of the barriers vary with respect to the high-K dielectric material [JAC 13]. It can be observed that for equivalent apparent permittivity, dielectric strength increases with the thickness of the barriers (represented as breakdown voltage on the vertical axis of Figure 3.8). Different variations in the relative thicknesses of the oxy-nitride composite are currently used for the PICS technologies, scaled to address a wide voltage spectrum.

In the dielectric layer, structural defects resulting from the deposition process also play an important role in leakage and dielectric reliability [MOR 91], [SIM 71]. Indeed, dielectric aging under a high electric field is assisted by charge circulation across the layer. For carriers having sufficient energy and not recombining, this would result in the creation of additional defects, ultimately driving a catastrophic failure by runaway mechanisms.

Conduction mechanisms in the dielectric have been widely studied and reported in the literature [MOR 91, SIM 71, JAC 13, DOG 05]. For relatively thick (> 15 nm) oxy-nitride dielectric composites of interest in the case of the PICS technologies, it can be observed that the leakage is driven by Poole–Frenkel conduction [JAC 13]. It can be recalled that the Poole–Frenkel effect corresponds to a trap-assisted conduction mechanism where the thermal emission of charge carriers from trap levels into the conduction or valence band of an insulator is driven by the applied electric field. The leakage magnitude relates to trap characteristics through the normalized relation [3.8], where a is a proportionality constant (related to trap density and section capture r), J is the leakage density, E is the electric field, β is a constant related to dielectric permittivity and φ is the barrier height of the traps (Figure 3.8):

$$\frac{J}{E} \propto a.e^{\frac{-q\varphi}{kT}}.e^{\frac{\beta\sqrt{E}}{kT}} \qquad [3.8]$$

Figure 3.9 shows a typical Poole–Frenkel plot for a 20 nm PICS3 oxy-nitride composite, showing an almost linear behavior in the typical voltage spectrum used (i.e. 3.5 V to 5 V) with a remarkably low intrinsic leakage level. As predicted by [3.8], the evolution of the leakage is exponential to the temperature.

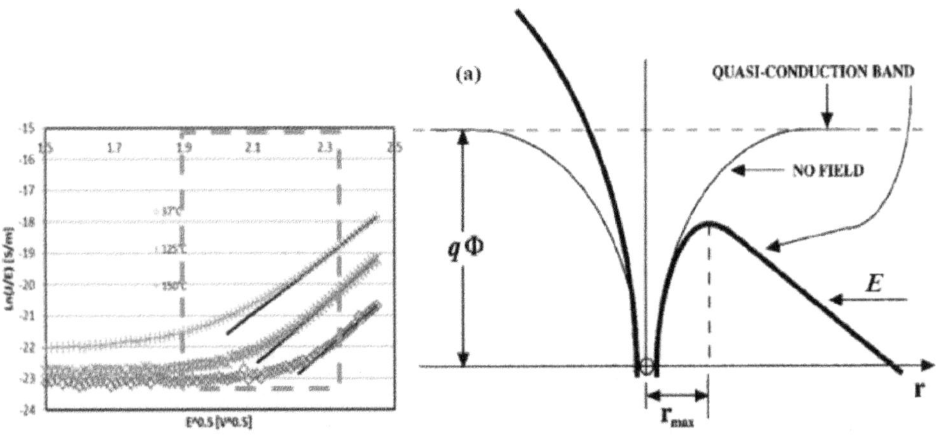

Figure 3.9. *Poole–Frenkel plot of the leakage for the PICS3 dielectric composite at different temperatures (left). The plain black lines represent the PF model according to equation [3.8] and the dashed contour represents the typical application voltage window. Representation of Poole–Frenkel conduction mechanisms through trap hopping (right). For the color version of this figure, see www.iste.co.uk/allard/systems.zip*

Several causes leading to the presence of traps in oxy-nitride dielectrics [DOG 05], [HAB 91] have been reported in the literature. It is generally observed that the Poole–Frenkel conduction level is greatly related to the dielectric stoichiometry and processing conditions such as deposition temperature, byproduct outgassing and heating or quenching ramp. For the PICS technology, optimized leakage and strength properties are obtained through the reduction of initial defects in the oxy-nitride composite. To obtain this result, a high-temperature, low-defect deposition technique is used, including a thermal processing in dry O2 for the first oxide barrier, LPCVD (low pressure chemical vapor deposition) from SiH2C12 and NH3 for the nitride and LPCVD from TEOS (tetraethyl orthosilicate) for the second oxide barrier. In [3.9], γ represents the field acceleration factor and Ea represents the thermal activation energy, with the other parameters having their usual meaning. The parameters for the voltage/temperature acceleration model are extracted from a time-dependent dielectric breakdown experiment [JED 04]:

$$TTF \propto e^{-\gamma E_{ox}}.e^{\frac{Ea}{kT}} \qquad [3.9]$$

Typical lifetime for the PICS3 technology is presented in Table 3.2. Remarkably, the TTF model remains very linear at high temperatures and is validated up to a temperature of 225°C.

Vg (V)	TTF for PIC3-LV			
	$37°C$	$100°C$	$150°C$	$225°C$
3.6	57 y	16 y	8 y	4 y
4.5	9 y	2.6 y	1.3 y	202 days
5	3.2 y	338 days	163 days	72 days

Table 3.2. *Lifetime predictions based on the TDDB E-model (PICS3) at 0.1% cumulative failures for the PICS3 technology*

3.4.2. *Lifetime considerations*

As presented in the previous sections, integrated capacitor technologies are using a high temperature process to enhance capacitor linearity and reliability. As a consequence, polysilicon electrodes are used that are, by nature, limiting the intrinsic performance in terms of access resistance to the capacitor. To overcome this issue, a layout methodology has been developed to ease predictability, and optimization of the capacitor resistive (ESR) and inductive (ESL) parasitic. This technic is leveraging the front-end patterning capabilities that are available to silicon based integration technologies.

3.5. Integrated capacitor optimization for power switching

As presented in the previous sections, integrated capacitor technologies use a high-temperature process to enhance linearity and reliability performances of the capacitor. As a result, polysilicon electrodes are used, which, by nature, limit the intrinsic performance in terms of access resistance to the capacitor. To overcome this problem, a layout methodology is developed to ease predictability and the optimization of the capacitor resistive (ESR) and inductive (ESL) parasitic. This technique leverages the front-end patterning capabilities that are available to silicon-based integration technologies.

3.5.1. *Regular layout*

Integrated capacitances are built with a massive parallelization of basic building blocks that are cascaded along a global power grid. Table 3.3 gives a detailed representation of the parallelization process from the microscopic (tripod) to the macroscopic (global capacitance) scale and the topology of the corresponding RLC model in the PICS technology.

It can be observed that the equivalent model for the interconnected cell corresponds exactly to a propagation line, which is well known by RF engineers. By this simple consideration, we can readily foresee the potential benefit to build up massive serial, parallel or serial-parallel networks. As a result, localized or distributed filtering RLC functions can be built, covering a broad range of frequencies and applications.

3.5.2. *Broad band modeling*

The RC for a tripod (Table 3.3) can be calculated using the usual sheet resistance and capacitance calculation formulas. The $R_C C_C$ scaling to the cell level corresponds to a parallelization of the RC obtained at the tripod level with addition of the resistance induced by contact and local routing. The LI calculation for the global routing corresponds to the terms of self- and mutual inductance between adjacent wires [ROS 08] and Rl to line resistance. The interconnection transfer matrix [ABCD]s is constructed by the interconnection impedance while the cell transfer matrix [ABCD]p is constructed by the cell admittance. These definitions yield the following transfer matrix [3.10] and [3.11]:

$$\begin{bmatrix} A_S \; B_S \\ C_S \; D_S \end{bmatrix} = \begin{bmatrix} 1 \; Z \\ 0 \; 1 \end{bmatrix} \qquad [3.10]$$

$$\begin{bmatrix} A_P \; B_P \\ C_P \; D_P \end{bmatrix} = \begin{bmatrix} 1 \; 0 \\ \gamma \; 1 \end{bmatrix} \qquad [3.11]$$

Tripod with double SIS	Cell	Cell with local grid	Global capacitance in a “finger” shape with n=4 cells
	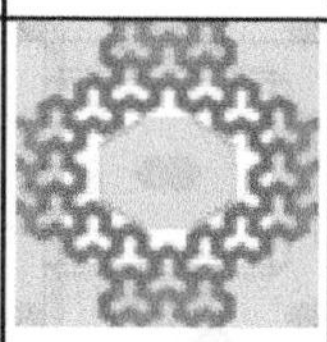	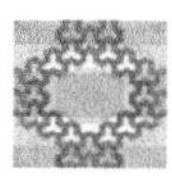	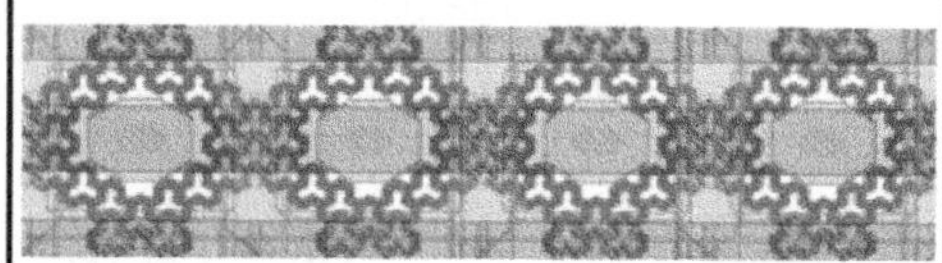
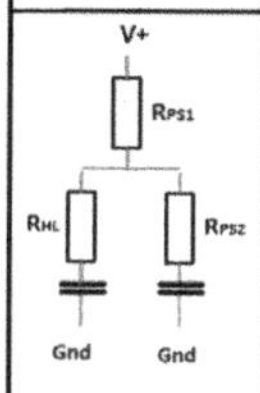	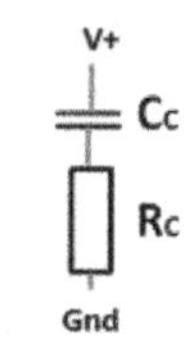	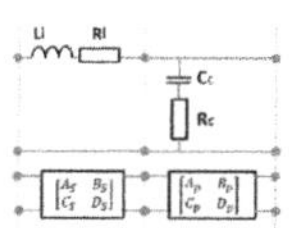	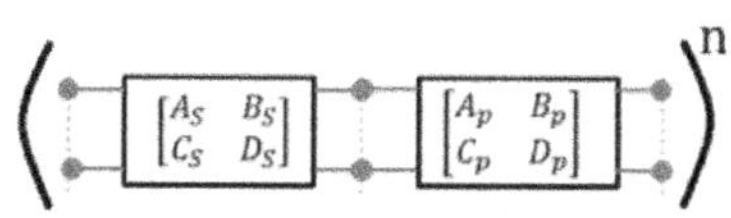

Table 3.3. *Electrical representation of cascading from the tripod to the cell level and corresponding electrical models. The green, blue, brown and red layers represent, respectively, the SIS structure sunk into the interspace between pillars, metal for local interconnect, contact to the electrode and thick metal for global interconnect. For the color version of this table, www.iste.co.uk/allard/systems.zip*

Using the matrix cascade, the interconnected cell transfer matrix can be expressed as [3.12]:

$$[ABCD]_{cell} = [ABCD]_s \bullet [ABCD]_p \qquad [3.12]$$

A cascade of n cells, yielding a finger as presented in Table 3.3, can be expressed as [3.13], where $[ABCD]_{as}$ and $[ABCD]_{ag}$ represent the terminations of the fingers, respectively. The terminal impedance Zf of the fingers can be deduced from the expression of the line impedance Zcf and attenuation γ_{cf} [3.14]:

$$[ABCD]_f = [ABCD]_{as} \bullet ([ABCD]_c)^n \bullet [ABCD]_{ag} \qquad [3.13]$$

$$\begin{cases} \gamma_{cf} = \frac{a\cosh(A_f)}{l_f} \\ Z_{cf} = \frac{B_f}{\sinh(\gamma_{cf}/l_f)} \end{cases} \cdots Z_f = Z_{cf}.\coth(\gamma_{cf}.l_f) \qquad [3.14]$$

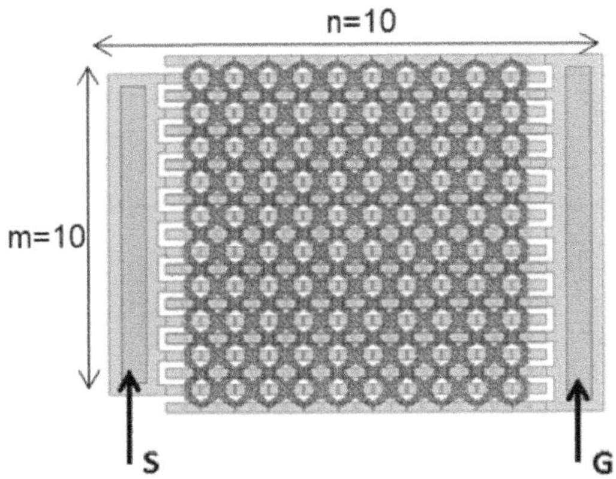

Figure 3.10. *Representation of a group of n=10 x m=10 cascaded cells in a serial-parallel network. The left termination corresponds to the signal line and the right to the GND line. For the color version of this figure, see www.iste.co.uk/allard/systems.zip*

Finally, for a parallelization of m fingers, as illustrated in Figure 3.10, we obtain the impedance in [3.15] for the global capacitive network:

$$Z = \frac{Z_f}{m} \qquad [3.15]$$

The equivalent RLC behavior of the PICS capacitor can be estimated from the Z impedance analysis.

3.5.3. *Capacitance parasitic suppression*

The modeling approach presented in the previous section gives a precise estimation of the capacitor behavior across a wide frequency spectrum. Furthermore, through relations [3.13] and [3.15], we can observe the following parasitic canceling scheme:

– the cell intrinsic Rc, and possibly Lc (not accounted in the model), are reduced by a factor m x n. As a result, the contribution of the polysilicon electrode intrinsic and contact resistance can be reduced to any level by properly selecting the capacitance of the cell and the repetition factor. The tuning of cell capacitance is obtained by addition or subtraction or tripods;

– the power grid intrinsic Rl and Ll can be adjusted by varying the m parameter. Obviously, for a given capacitance (m x n fixed), a larger m would result in the drastic reduction of the ESL and the moderate reduction of the ESR (i.e. the power grid is implemented onto a low-resistive interconnection level).

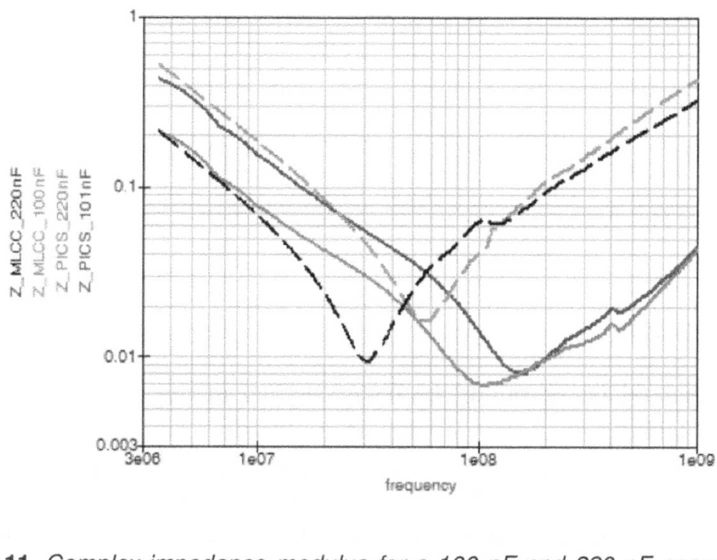

Figure 3.11. *Complex impedance modulus for a 100 nF and 220 nF capacitor. The plain line corresponds to the PICS and the dashed line corresponds to the low-ESL commercial MLCC. Impedance measurement is performed at 0 dBm, 25°C. For the color version of this figure, see www.iste.co.uk/allard/systems.zip*

As a result, for a targeted capacitance value, both the ESR and the ESL can be set to any value by tuning the individual cell and properly adjusting the form factor of the global capacitance grid (i.e. m x n). To illustrate the benefit of this approach, a comparison between a PICS optimized with the parallelization approach and an X7R commercial MLLC capacitor that is optimized for fast switching is shown in Figures 3.11 and 3.12. It can be observed that the self-resonance of the PICS component is much higher than that of the MLCC, as a benefit of the reduced ESL. Also, the ESR for the PICS is much lower at the resonance, while remaining more stable across the whole frequency range.

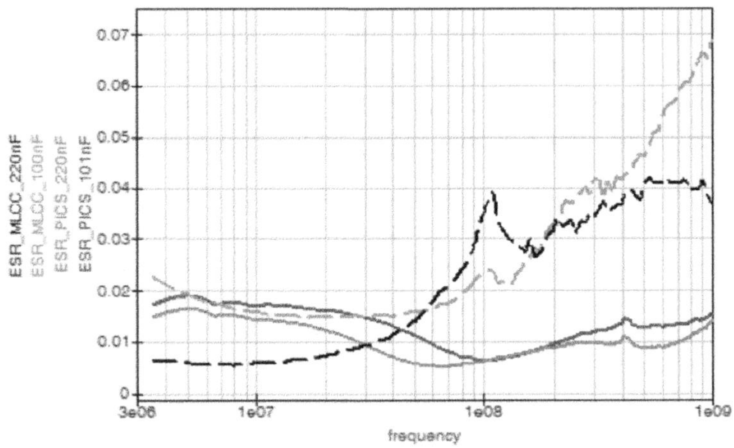

Figure 3.12. *Real part of the impedance for a 100 nF and 220 nF capacitor. The plain line corresponds to the PICS and the dashed line corresponds to the low-ESL commercial MLCC. Impedance measurement is performed at 0 dBm, 25° C. For the color version of this figure, see www.iste.co.uk/allard/systems.zip*

3.6. Conclusion

As a general background, we have recalled in this chapter, the general principles leading to charge storage in dielectric materials. The theoretical relationship between the polarization, the dielectric susceptibility and the capacitance has been recalled, for a static electrical field. The general explanations linking the dielectric susceptibility to the material absorption spectrum have been presented, and consequences on relaxation/frequency linearity have been explained. The case of real dielectrics is discussed for the different polarization mechanisms (electronic, atomic, dipolar and ionic), and several examples are given.

In the second part of this chapter, we focused on silicon integrated capacitors, which corresponds to a nearly optimal solution for the applications requiring ultra-thin and flexible form factors. Details have been given on the process used to fabricate those devices, and corresponding advantages (linearity, reliability) and intrinsic limitation (mainly serial resistance) are detailed and discussed. To overcome the intrinsic electrical limitations, a novel layout approach is presented. The benefits of this approach are discussed and compared with concurrent ceramic technologies (MLCC).

4

On-board Power Management DC/DC Inductive Converter

This chapter discusses the overall challenges for developing an intermediate on-board DC/DC converter for an IoT device. The proposed methodology follows a typical V-type development cycle. System-level specifications are discussed. The proposed list is not exhaustive and application-specific requirements may occur. Some items may not be relevant for a particular design. However, defining the most obvious requirements alleviates the development risks as quickly as possible. A clearly specified objective discards most unsuitable solutions and emphasizes the possible issues.

Again, this does not intend to be an exhaustive list of solutions but most IoT-related cases should be covered. Only fundamental equations are presented and discussed. References are given for more details while this chapter focuses on giving an overall description of the trade-offs between solutions and requirements.

A complete example is provided to illustrate the application of current-mode sliding-mode control, as introduced in section 1.4.4.2.

4.1. Specifications

Power supply design requires many compromises between several quantities such as cost, efficiency and volume, which are described in this

---

Chapter written by Benoît LABBE and Bruno ALLARD.

section. This list is unforunately not exhaustive as the converter requirements are not described in this section. The usual standard requirements are considered as detailed in the Introduction.

4.1.1. *Load-related requirements*

An extensive knowledge of load is the most obvious starting point for a power supply design. This should extend beyond the steady-state current and voltage characteristics of the load, and sometimes requires further investigation prior to any power supply design.

The DC characteristics of the load are a matter of some importance. The characteristics define the range of output voltage to be delivered by the converter. The actual output voltage of a DC-to-DC converter can be fixed by design or can be adjusted while the converter operates. The first case covers the supply of most analogue subsystems or inter-chip communication voltages, such as the *de facto* standard 1.8 V I/O voltage in a mobile platform. The second case covers the supply of high-performance digital subsystems using a dynamic voltage scaling methodology or the supply of radio-frequency power amplifiers using the average power tracking method, for instance. The output voltage is fixed by interval. The length of the interval varies but is long enough to be considered as DC. Hence, the DC/DC converter has to be designed and characterized not only for a single typical output voltage, but also for a possible wide range of output voltages.

The maximum current drawn by the load is often the main parameter that describes a converter. However, it is unlikely that the load requires a constant current. For instance, the current drawn by a digital core will vary with the activity factor of the digital glue and the actual clock frequency if a dynamic frequency scaling is used. Most modern circuits have different power states such as standby, hot standby and normal mode with different operating currents that can extend the range of possible output currents for the power converter. However, knowing the load current range is not enough to fully optimize the system's power supply. The supplied system is likely to operate some significant portion of the time in a determined region and only a small portion of the time in another region. A different use case occurs when the operating points are evenly distributed across the operation range. The ideal power supply is designed to maximize the power efficiency while meeting each system specification when the system is working in a real environment. When designing the above-mentioned first use case, the converter efficiency should be optimized for the most frequent operation region and a slight degradation for infrequent points is acceptable. The second use case requires the optimization of the efficiency flatness of the converter at the expense of lower peak efficiency. As will be discussed later, this has consequences for the control method and the power stage design.

Transient behaviors of the load have an influence on the design of the converter. This includes the load current variations as well as the rate of change of the voltage when required. Load current behaviors can be analyzed in the harmonic and time domain. The harmonic domain is used to specify the small signal output impedance of the converter. It is also well suited to check the resonant modes of the complete power delivery network (PDN). However, most modern loads present large current transients where nonlinear factors come into play. Frequent nonlinear factors are the slew rate limitation of the inductor and the duty cycle limits. Hence, the time domain is used to analyze the circuit performances, and extensive time-domain simulations or experiments are required. Highly transient loads are supplied using some safety margins. For instance, the supply voltage of a high-performance digital core must be between a low bound and a high bound at any time while keeping a lowest possible value to minimize the power consumption. Therefore, the output voltage of the converter is set to the minimum voltage that guarantees the proper operation of the digital circuit plus a sufficient margin that includes, among others, the transient response of the converter, the ripple amplitude and the DC drop voltage across the PDN.

Noise can be a problematic issue for some loads. It includes not only the peak-to-peak output voltage ripple, but also its frequency and harmonic content. As switching in a DC/DC converter is a noisy process, it must be ensured that the noise levels and bandwidths of interest are well defined at an early stage of the project.

4.1.2. *System-related requirements*

The input-side or line-side of the converter adds several elements to the requirement list. First, the line voltage can be controlled or highly variable. The former can be the case of an intermediate voltage bus and the latter can be the case of a high impedance battery. Second, variations of the input voltage must be rejected. As for an amplifier, a DC/DC converter rejects the line variations with a characteristic that can be far from ideal.

Start time and transition time between different states is also a parameter that can influence the design of the board-level converter. Output decoupling capacitors must be charged up during the start sequence of the converter. The start current can exceed the maximum load current when large output decoupling capacitors and fast start time are required. Furthermore, when the entire device is powering up, or recovering from a deep sleep mode, other subsystems are likely to draw the current from the main power source. Using a power source with some significant output impedance can create a line voltage drop and lead to awkward constraints.

Mechanical constraints also need to be considered at the specification stage. The thickness of the solution can be a challenging specification, especially for mobile or wearable devices. Cooling also depends on the mechanical constraints for the converter. Nowadays, the cooling of the converter can be less challenging than that of the load itself, such as a high-performance system-on-chip (SoC), thanks to the high power efficiencies.

Electromagnetic interferences (EMI) and electromagnetic compatibility (EMC) between the components across the board should be carefully considered. There is often no EMC standard for an on-board converter as standards apply to the overall system. However, this does not discard a proper analysis of the EMI generated by the converter for the sake of the overall system. For the sake of clarity, we assume that EMC deals with the line side of the converter while noise deals with the load side of the converter. However the two latter concepts are collectively addressed in literature as EMC issues.

Switched waveforms can cause reliability issues if the converter is not properly designed. This involves the component choices as well as the board design. Lifetime prediction and failure rate are not detailed in this chapter.

Other context-based such as cost, component sourcing and manufacturing capabilities come into play. These parameters are not addressed in this chapter.

4.1.3. *Power delivery network*

Defining the board configuration between the power source and the load is part of the specification of the system and should not be an afterthought. A typical power delivery path between the output of the converter and a SoC is made up of various parasitic inductors and capacitors. A careful board layout limits the power path impedance and resonances to low values. As several loads can be connected to the intermediate converter, the PDN may be modeled as a multidimensional network instead of a single filter [GUP 07].

A simplified DC/DC to silicon-die impedance model is shown in Figure 4.1. The low-frequency regulation, up to a few hundreds of kHz, is performed by the power supply and an array of decoupling capacitors provides high-frequency currents. Various parasitic inductances limit the current slew rate of the converter as well as the capability of the converter to deliver fast current edges to the load.

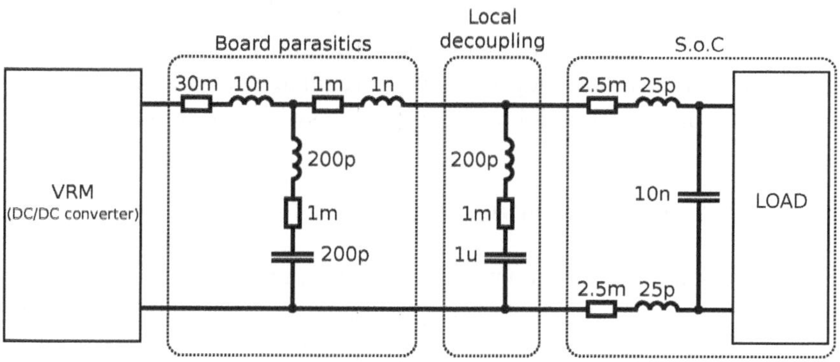

Figure 4.1. *Simplified model of a PDN*

One possible way to overcome the limitations of the power delivery path is to integrate the power supply on-chip with the load. This approach is described in this book. However, an intermediate board-level DC/DC converter is still required to maintain the on-chip input voltage at a known value.

4.2. Current-mode sliding-mode control implementation

The control of a buck converter that supplies a high-performance digital core is proposed. The load presents large current transients as the digital core operates in deep-sleep mode when computing resources are not used and when the user uses the device at full power. Furthermore, the board area must be reduced to the bare minimum; therefore, all the active parts of the circuit must be integrated on a single silicon die.

A sliding-mode approach is selected for its asynchronous transient response that helps to minimize the size of the output capacitor while keeping a good transient response. A quasi-V^2 structure is a suitable solution for high power continuous conduction mode of the converter. The converter is configured as a current-mode sliding-mode control when operating in PFM mode. This allows the power consumption of the control part to be reduced by deactivating the slow loop and reusing the main comparator at the cost of a few analogue switches.

To keep the efficiency high and predictable, an average switching frequency control by means of a phase-locked loop (PLL) is added to the sliding-mode controller. Hence, the power stage can be designed and optimized following the previously discussed recommendations. The PLL regulates the propagation delay of the sliding-mode loop in order to regulate the switching frequency. The longer the propagation delay, the lower the switching frequency.

4.2.1. *System analysis: voltage regulation loops*

The proposed quasi-V^2 converter without the synchronization loop is shown in Figure 4.2. As in the current-mode sliding-mode converter, the passive network formed by R_f and C_f combines the voltage across the inductor to produce a scaled representation of the inductor current I_L. This representation is only pertinent in small-signal analysis since the current-to-voltage transfer function is defined by [1.29].

An outer loop integrates the output error to produce a compensation voltage, V_{comp}. An operational amplifier is used in an integrator configuration, and the integral time constant has the common expression:

$$\frac{V_{comp}(s)}{V_{ref} - V_{out}(s)} = \frac{1}{R_i C_i s} = \frac{1}{\tau_i s} \qquad [4.1]$$

From a qualitative point of view, the inner loop regulates the fast transient operation. When a load positive-step transient occurs, the resulting undershoot is fed-back to the negative input of the comparator through the current emulation capacitor (path 1 in Figure 4.2). As a result, the comparator error, ϵ, increases suddenly and the power stage switches on if it is in the off-state, otherwise it remains in the on-state. Under a negative-step transient, the starting overshoot is fed-back through the capacitor, C_f. The comparator input error diminishes and the power stage ties the node V_{LX} to the ground, if required. As the fast transient regulation is mainly performed by the inner loop, the bandwidth of the outer loop (path 2 in Figure 4.2) can be relatively small compared with a voltage-mode PWM compensation function.

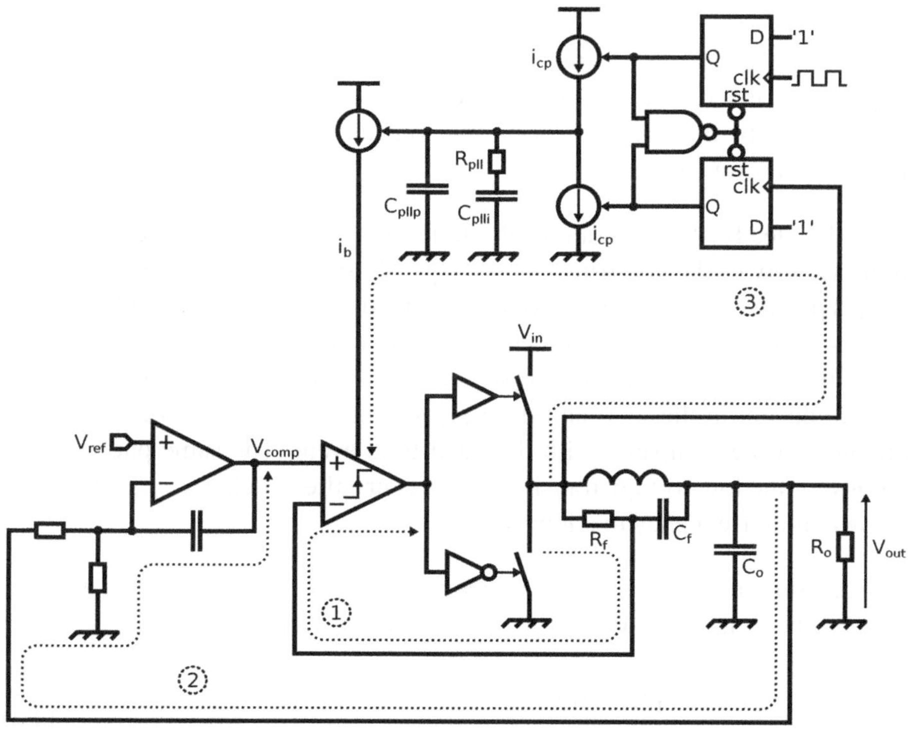

Figure 4.2. *Quasi-V^2 structure used for this example*

The comparator makes the inner-loop DC feedback voltage, $\overline{V}_{C_f} + \overline{V}_{out}$, equal to the requested DC compensation voltage, $\overline{V}_{comp}$. The inner-loop feedback voltage is the DC value of the power stage switching node, $\overline{V}_{LX}$, such that:

$$\overline{V}_{comp} = \overline{V}_{C_f} + \overline{V}_{out} = \overline{V}_{LX} = R_l\overline{I_{out}} + \overline{V}_{out} \quad [4.2]$$

Using a high-gain amplifier with a DC gain A_0, we obtain:

$$\overline{V}_{out} = \frac{A_0}{A_0 - 1}\left(\frac{R_l\overline{I}_{out}}{A_0} + \overline{V}_{ref}\right) \approx \overline{V}_{ref} \quad [4.3]$$

In contrast to the simple inner-loop converter output voltage in [1.32], the proposed V^2 converter output voltage presents a theoretically negligible DC output resistance. Furthermore, the comparator input offset is also compensated by the integral action of the outer loop. This singularly eases the design of the comparator.

In the steady-state operation, the inner-loop feedback voltage is situated in the vicinity of the compensation voltage. When the comparator input error, $\epsilon(t)$, is positive, the V_{LX} node voltage is pulled to V_{in} and the inductor current rises as well as the comparator negative input, until $\epsilon(t) = 0$ V. However, due to the intrinsic deadtime and propagation delay of the power stage, a delay, T_{doff}, appears in the loop and the V_{LX} voltage is tied to zero after a small amount of time compared with the ideal time event. The same phenomenon applies for the low-to-high transition that occurs after a delay, T_{don}. This limits the switching frequency to a low value.

According to the sliding-mode theory, the control law associated with the sliding function, $\epsilon = s\left(V_{comp}, V_{C_f}, V_{out}\right)$, is:

$$V_{LX} = \begin{cases} V_{in} & \text{if } \epsilon > 0\text{ V} \\ 0 & \text{if } \epsilon < 0\text{ V} \end{cases} \quad [4.4]$$

It can be verified whether the hitting condition is straightforward in this case: when operating in the subspace $\epsilon < 0$ V, the converter is associated with the control law $V_{LX} = 0$ V, so it asymptotically reaches $\epsilon = V_{in}$, i.e. in the subspace $\epsilon > 0$ V, because the integral action of the outer voltage loop saturates and the output voltage asymptotically reaches 0 V. When operating in the subspace $\epsilon > 0$ V, the converter is associated with the control law

$V_{LX} = V_{in}$ and asymptotically reaches $\epsilon = -V_{in}$, i.e. in the other subspace $\epsilon < 0$ V. In each case, it is obvious that $\epsilon = 0$ V is reached.

Once the converter reaches the sliding surface $\epsilon = 0$ V, it has to be ensured that the trajectory of the converter is maintained on it, i.e. whether the system complies with the existence condition defined in [TAN 05, UTK 09]:

$$\epsilon\dot{\epsilon} < 0\, V^2.s^{-1} \tag{4.5}$$

In the subspace $\epsilon > 0$ V, the existence condition implies that $\dot{\epsilon} < 0$ V.s^{-1}, thus:

$$\dot{V}_{comp}(t) - \dot{V}_{C_f}(t) - \dot{V}_{out}(t) < 0\, V.s^{-1} \tag{4.6}$$

Assuming that the triangle signal across C_f is small and globally centered on V_{out}, we obtain:

$$\dot{V}_{C_f} = \frac{V_{in} - V_{out}}{\tau_C} \tag{4.7}$$

Introducing [4.1] and [4.7] in [4.6] gives:

$$\frac{V_{ref}(t) - V_{out}(t)}{\tau_i} - \frac{V_{in} - V_{out}(t)}{\tau_C} - \frac{I_L(t) - I_{out}(t)}{C_{out}} < 0\, V.s^{-1} \tag{4.8}$$

Now let us consider the converter at the switching instant when $\epsilon > 0$ V. Then, the inductor current, I_L, verifies:

$$I_L(t) - I_{out}(t) = -\frac{V_{out}}{L}T_{don} \tag{4.9}$$

where I_{out} is the output current. This leads the existence condition to become:

$$T_{don} < \frac{C_o L}{V_{out}}\left[\frac{V_{in} - V_{out}}{\tau_C} - \frac{V_{ref} - V_{out}}{\tau_i}\right] \tag{4.10}$$

The integration time constant, τ_i, is set approximately equal to the capacitive branch time constant, τ_C, in the order of magnitude of a few switching cycles. Therefore, since the output error is three orders of magnitude smaller than the input-to-output voltage in typical conditions (a few mV compared with a few V), the existence condition can be approximated by:

$$T_{don} < \frac{LC_o}{\tau_C} \cdot \frac{V_{in} - V_{out}}{V_{out}} \quad [4.11]$$

Using the same methodology for the other sub-space $\epsilon < 0\ V$, we obtain:

$$T_{doff} < \frac{LC_o}{\tau_C} \cdot \frac{V_{out}}{V_{in} - V_{out}} \quad [4.12]$$

Depending on the conversion ratio, the most stringent conditions are defined in [4.11] or [4.12]. Nevertheless, in the standard steady-state operation, these two conditions imply that the ideal output voltage ripple is non-negligible with respect to the amplitude of the voltage across V_{C_f}. Operation outside the existence domain does not mean that the converter is strictly unstable, but it does not follow a sliding-mode operation principle and does not operate with the delay-to-frequency characteristic used thereafter. Once the converter operates in sliding mode, its average trajectory remains on the sliding surface, i.e. $s = 0$ and $\dot{s} = 0$, respectively, where s is the average value of ϵ when neglecting the chattering[1] effects. Then, averaging the sliding motion in [1.31] gives:

$$\overline{V}_{comp} = \overline{V}_{C_f} + \overline{V}_{out} \quad [4.13]$$

Higher-order development terms give the average dynamic control-to-output transfer function of the inner loop:

$$\frac{\overline{V}_{out}(s)}{\overline{V}_{comp}(s)} = \frac{1}{1 + \frac{R_L}{R_o}\frac{1+\tau_L s}{1+\tau_C s} + R_L C_o s \frac{1+\tau_L s}{1+\tau_C s}} \quad [4.14]$$

1 In [UTK 09], "*Chattering, [...], describes undesired system oscillations with finite frequency caused by system imperfection.*" In the application under consideration, chattering is the ripple caused by the non-infinite switching frequency.

where R_o is the equivalent average load. [4.14] outlines the setting procedure of τ_C. By setting τ_C equal to τ_L, the inner loop becomes a first-order loop whose transfer function is:

$$\frac{\overline{V}_{out}(s)}{\overline{V}_{comp(s)}} = \frac{R_o}{R_o + R_L} \frac{1}{1 + \frac{R_o R_L}{R_o + R_L} C_o s} \quad [4.15]$$

Equation [4.15] shows that the inner-loop behaves like a first-order system. The output voltage tracks the compensation voltage with a static error and a first-order-type response. When noise immunity considerations force the designer to consider a higher ripple value on the negative comparator input, the single pole approximation is no longer valid and the full equation without simplification becomes:

$$\frac{\overline{V}_{out}(s)}{\overline{V}_{comp}(s)} = \frac{R_o}{R_o + R_l} \frac{1 + \tau_c s}{1 + \frac{R_o \tau_c + L + R_l R_o C_o}{R_o + R_l} s + \frac{R_o C_o L}{R_o + R_l} s^2} \quad [4.16]$$

This simple continuous-time model for the discontinuous inner loop is extensively used to set the components' values. It remains valid while the converter evolves in the boundaries of the sliding function, i.e. the comparator input error is small.

4.2.2. *System analysis: loop delay control*

The switching frequency of the voltage regulation part is intrinsically uncontrolled. Moreover, it is mainly dependent on input and output voltage and delays. For the sliding-mode converter under consideration, the instantaneous switching frequency is derived from [1.35]:

$$f_{sw} = \left[\frac{V_{in}}{V_{in} - V_{out}} T_{don} + \frac{V_{in}}{V_{out}} T_{doff} \right]^{-1} \quad [4.17]$$

The delay-control solution for this design is to lower the speed of a fast comparator by reducing its differential input-pair bias current. The lower the bias current, the lower the overall comparison speed of the comparator. This delay can be modeled as:

$$t_d = \sqrt{\frac{2 C_l V_{in}}{n \xi V_T \alpha_r} \frac{1}{\sqrt{I_b}}} \quad [4.18]$$

where C_l and n are a design-related parameter, ξ and V_T are process-dependent parameters, I_b is the bias current of the comparator and α_t is the differential voltage input slope. [4.18] shows that the time delay decreases nonlinearly with respect to the bias current. As expected, the higher the bias current, the lower the comparator delay.

The actual propagation delay slightly differs from the single propagation delay of the comparator or equation [4.18]. The delay through the logic blocs and the switch drivers should also be taken into account. However, this simple model outlines that a single current can control the switching frequency of the converter. The DC/DC converter is not only a voltage converter, but also a current-controlled oscillator.

4.2.3. *System analysis: switching frequency control*

The DC/DC converter is inserted as a voltage-controlled oscillator (VCO) in a PLL circuit. The proposed solution is schematically represented in Figure 4.2. The inner current-emulated loop is circled as path (1) and the outer voltage loop as path (2). These two loops form the voltage regulation part as described previously. The phase regulation loop is outlined as path (3).

The voltage-controlled switching-frequency of the converter is sensed at the power bridge output and fed-back as an input of a phase frequency detector (PFD) built with two standard logic D-flops and a NAND gate. The PFD three-state output drives a charge-pump circuit, as shown in Figure 4.2, by two controlled current sources. When the reference-clock rising edge is in advance with respect to the rising edge of the power stage of the converter, the high-side current source is activated and a positive current, i_{pll}, flows through the PLL filter (R_{pll}, C_{pll_p}, C_{pll_i}) until a converter V_{LX} rising edge occurs. Therefore, the PLL-filter voltage, V_{pll}, rises and the voltage-controlled current source increases the comparator bias current. The increased speed of the comparator yields a higher switching frequency. Contrarily, when the converter power stage exhibits a rising edge at its output voltage, V_{LX}, before the reference-clock rising edge, the low-side current source is activated until the clock rising edge occurs. Therefore, the PLL-filter voltage, V_{pll}, decreases and the voltage-controlled current source decreases the comparator bias current. The resulting lower speed of the comparator yields a lower switching frequency. When the converter operates in phase with the reference clock, the two rising edges arise synchronously, none of the flip-flop is set and the two current sources are deactivated. The filter remains in high impedance and holds its voltage value, V_{pll}.

The PFD average output is linear in phase. The larger the phase shift, $\Delta\Phi$, between the reference clock and the switching converter switching phase, the wider the output pulse, α_{PFD}. The maximum value of α_{PFD} is reached when the phase shift is $\Delta\Phi = 2\pi$, i.e. the output is kept in a '1' logic state. The minimum value of α_{PFD} is reached when the phase shift is $\Delta\Phi = -2\pi$, i.e. the output is kept in a '-1' logic state. In between when there is no phase error, the PFD is never triggered and the average output, α_{PFD}, is zero. Consequently, the average transfer function of the PFD block is:

$$\frac{\alpha_{PFD}}{\Delta\Phi} = \frac{1}{2\pi} \qquad [4.19]$$

The charge-pump transfer function is then a single gain that converts the average PFD output into a current, i_{cp}, such that:

$$\frac{i_{cp}}{\alpha_{PFD}} = i_{pll} \qquad [4.20]$$

The phase-loop filter filters the charge-pump current with its impedance:

$$\frac{V_{pll}(s)}{i_{cp}(s)} = \frac{1}{\left(C_{pll_p} + C_{pll_i}\right)s} \cdot \frac{1 + R_{pll}C_{pll_i}s}{1 + \frac{C_{pll_p}C_{pll_i}}{C_{pll_p}+C_{pll_i}}R_{pll}s} \qquad [4.21]$$

The filter provides an integral action to reject steady-state phase error. A phase-advance corrector is introduced with the addition of R_{pll}. This is used to stabilize the loop, thanks to a transfer-function zero, without introducing spikes on the phase-loop filter voltage, thanks to the high-frequency pole.

4.2.4. *Design*

The proposed converter is designed using the following steps:

1) power-part specification and design: this is not detailed as this process has been extensively described previously;

2) voltage regulation part design:

a) inner-loop setting: the current observer is set according to the power inductor. The comparator delay range is specified,

b) outer-loop (compensation function) setting: the integration time constant is set according to the inner-loop specifications. The amplifier is specified;

3) frequency-regulation part design: based on transistor-level simulations, the VCO characteristic is determined and the phase-loop filter is set.

The components chosen are 470 nH and 240 nH multilayer ceramic inductors in a small package (2 mm length, 1.6 mm width and 1 mm thickness), which can handle 1.7 Amps and 2 Amps, respectively, with a 30% drop in inductance value. The inductance remains constant up to 100 MHz, which makes a perfect choice for the 1 to 10 MHz switching frequency range. The output capacitor is set between 4.7 μF and 22 μF depending on the targeted transient performances. A 3.2 MHz switching frequency comes from the power stage design trade-off. The reference clock is externally provided.

The controller design starts with the choice of the current-observer time constant. The inductors' time constants chosen are 8.5 μs and 8 μs, respectively. A 10 μs observer time constant, $R_f C_f$, is chosen with the combination of a 1000 kΩ resistor and a 10 pF capacitor. In a first approximation, the typical condition voltage ripple, ΔV_{C_f}, across the observer capacitor, C_f, is:

$$\Delta V_{C_f} \approx \frac{V_{in} - V_o}{R_f C_f} \cdot \frac{V_{out}}{V_{in}} T_r = \frac{V_{out}}{R_f C_f} \left(1 - \frac{V_{out}}{V_{in}}\right) T_r \qquad [4.22]$$

With the proposed values, the typical observer voltage ripple is 25 mV. To increase the ramp amplitude, the observer capacitor can be divided by 4; therefore, the observer time-constant value can be lowered to 2.5 μs to increase the noise immunity. Nevertheless, the effective inductor AC resistance is frequency-dependent and higher than the DC one. Therefore, using the DC resistance is a worst-case condition that results in the higher observer time constant.

The compensation-to-output voltage model with the minimal load resistance gives the maximal integration time constant that meets the phase margin criterion. The 3 dB bandwidth of the inner-loop system is between

150 kHz ($C_f = 2.5$ pF) and 350 kHz ($C_f = 10$ pF). Hence, the optimal integrator time-constant values are 1 μs and 0.5 μs. Values in the range of 2 μs to 0.13 μs are implemented.

Once the voltage part is set, the current-to-frequency characteristic has to be determined. This is the main difficulty in the design process since proper time characterization depends on various parasitic elements. First, the comparator current-to-delay function depends on the lumped parasitic capacitors that can only be estimated using post-layout extraction. The power stage delay is approximately known and assumed to be fixed. The output voltage ripple value influences the delay as well as the addition of a step in the comparator input ramp voltage. The latter step amplitude depends on the ratio of output capacitor equivalent serial inductance, R_{esl}, to inductance, L. Thus, the voltage-loop propagation delay is estimated using post-layout simulation of the comparator for which a constant compensation voltage is applied on the positive input and a representative triangle, plus a small square ripple, is applied on the negative input. A small fixed delay is added to model the power stage and driving logic delays. Furthermore, the delay simulations strongly depend on the simulator accuracy. Thus, simulating the whole converter is time consuming and gives no additional accuracy as with the proposed method.

According to the current-to-frequency model, the phase filter is determined using a standard PLL design methodology. The charge pump is designed to provide the minimum possible gain. This reduces the filter capacitor size. The main filter capacitor, C_{pll_i}, is set to cut the bandwidth one-tenth lower than the voltage regulation part bandwidth. The zero resistor, R_{pll}, is set to provide a phase boost at the open-loop 0-dB gain frequency. The high-frequency filter capacitor, C_{pll_p}, is set 10 times lower than the main capacitor, C_{pll_i}, in order to filter the charge-pump current pulses but preserving the zero resistor, R_{pll}, phase boost. This setting follows the so-called non-interaction principle and is intended to stabilize the loop at the cost of a small phase noise.

The VCO evaluated gain ranges from 0.1 MHz/μA to 1 MHz/μA with a 470 nH inductor and from 0.2 MHz/μA to 2 MHz/μA with a 240 nH inductor. Simulation spread is wider depending on the integration algorithm and accuracy. The voltage-to-current converter is designed to have a gain ranging from 10 μA/V to 1 μA/V. This compensates the gain variation of the VCO even if it is not primarily intended for. The PLL-filter capacitor is set between

10 pF and 50 pF, which sets the uncompensated open-loop 0-dB crossing frequency between 90 kHz and 600 kHz. The zero resistor, R_{pll}, is set between 300 kΩ and 6.3 MΩ, which offers multiple compensation flavors in which the converter is stable, and in some cases, the converter is unstable. The light-gray shaded area plot in Figure 4.3 shows the resulting uncompensated open-loop domain (i.e. without the phase boost provided by R_{pll}). The uncompensated phase-shift follows the -180 degree line that makes the loop unstable without compensation. The compensation resistor adds the phase boost, represented by dark gray, which increases the gain after the compensation zero. It ranges from 1 kHz up to 200 kHz, and some sets are expected to be unstable. The typical open loop is represented by a solid line. It presents a sufficient phase margin to ensure loop stability despite a low expected bandwidth (50 kHz).

4.2.5. *Results*

The voltage loop regulates the output voltage within an acceptable range. The typical output voltage versus the load current is shown in Figure 4.4. The converter achieves a 2.12mΩ DC output impedance and a no-load accuracy close to 0.1%, either at a 1.2 V output voltage or a 0.6 V output voltage. This measurement validates the choice to add an outer-loop compensation function. DCM accuracy is not as good as the CCM one. This is because the output voltage ripple is higher when the converter operates in DCM than in CCM, and the output voltage is practically higher than the reference voltage.

The transient dispersion effect for PWM converters does not occur, thanks to the asynchronous response of the controller. This diminishes the transient margin voltage because a specific margin is not required to cope with transient dispersion. A large transient response of the prototype (2 A step) with a near optimal setting is shown in Figure 4.5. On the contrary, a small transient response, shown in Figure 4.5, remains in the unsaturated operation region and the widely covered output impedance can be used to determine the transient response. Therefore, the load transient response can be predicted as intermediate between the linear small-signal output impedance response and the time-optimal control response.

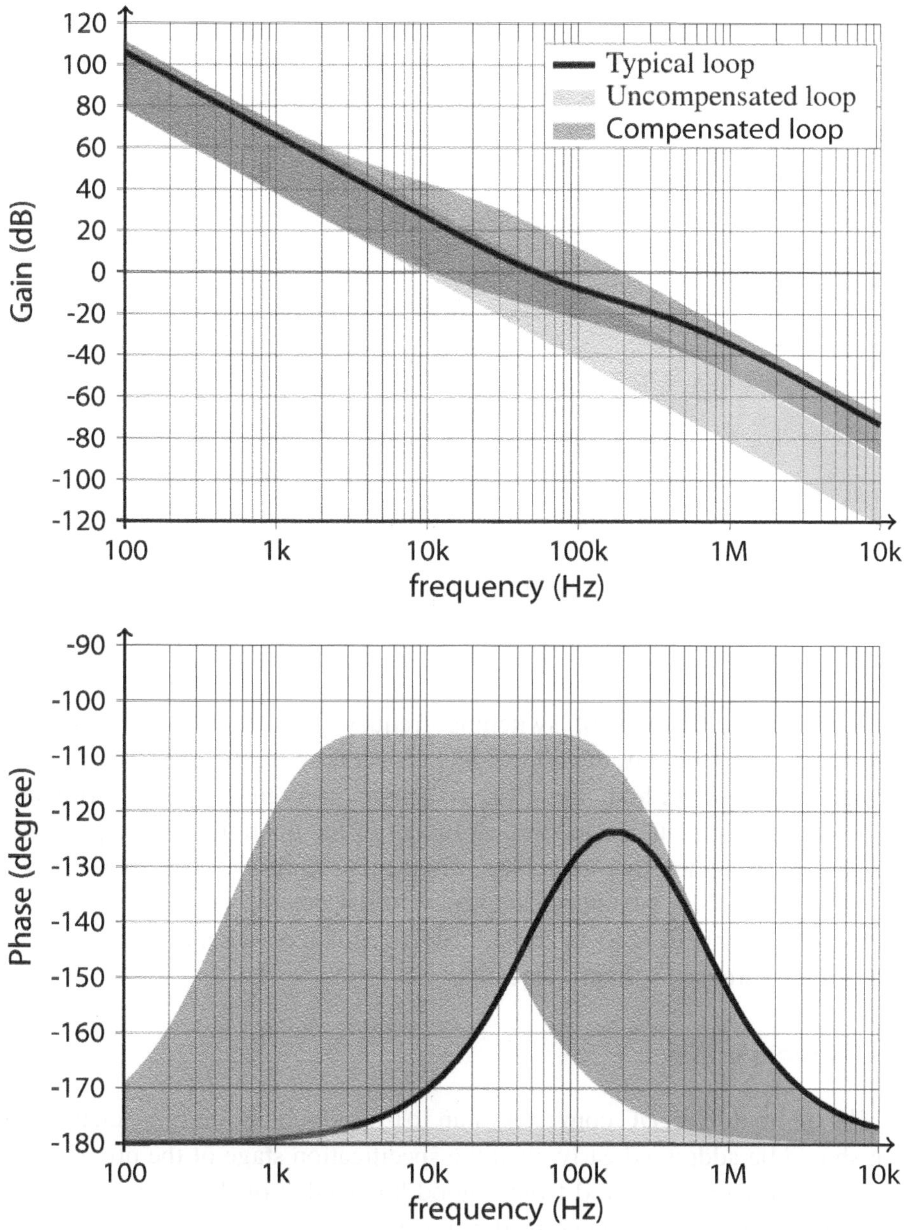

Figure 4.3. *Open-loop transfer function of the phase locked loop*

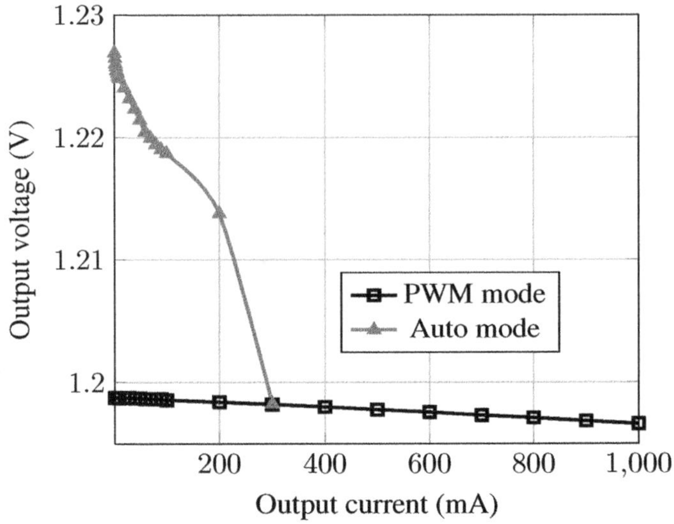

Figure 4.4. *DC output voltage vs load current ($L = 470$ nH, $C_o = 10\mu F$, $V_{bat} = 3.6$ V, $V_o = 1.2$ V)*

The main parameters of switching frequency variation are the battery voltage, V_{in}, and the output voltage, V_{out}. Switching frequency versus battery voltage with and without the PLL is shown in Figure 4.6. Experimental measurements validate the theoretical model of the operation of the proposed PLL-based scheme for switching frequency control. The phase loop effectively locks the switching frequency to the desired target frequency. In Figure 4.6, the small shift in the 3.2 MHz regulated switching frequency is due to the reference oscillator clock.

4.3. Conclusions

The on-board DC/DC converter of an IoT system faces multiple challenges that should be addressed early on at the specification stage of the project. The best solution can be derived from a good knowledge of the load, the source and the environment of the converter. For most IoT-type projects, one of the three non-isolated converters can be used. As the topology choice is rather straightforward, the optimization of the power stage is a more iterative process. At this stage, most of the overall system requirements must be known: typical currents and voltages of the source and the load(s) as well as the mechanical constraints and specific rules such as allowed switching frequencies.

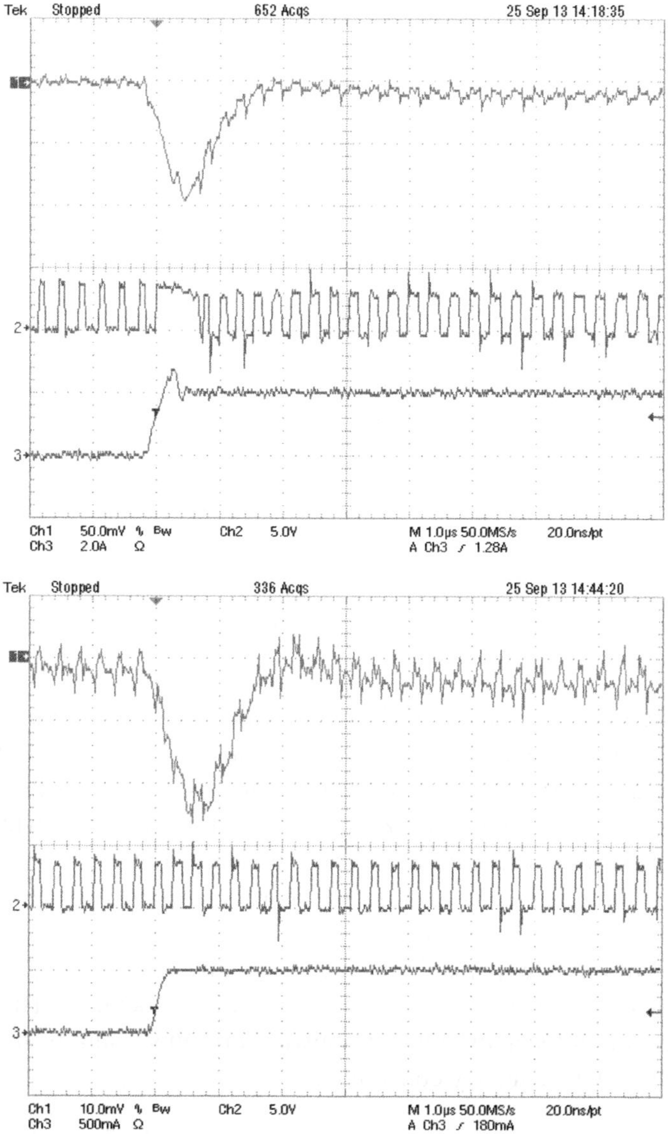

Figure 4.5. *Load transient with near optimal settings: output voltage (AC coupled) (top), power stage output voltage (middle) and load current (bottom)*

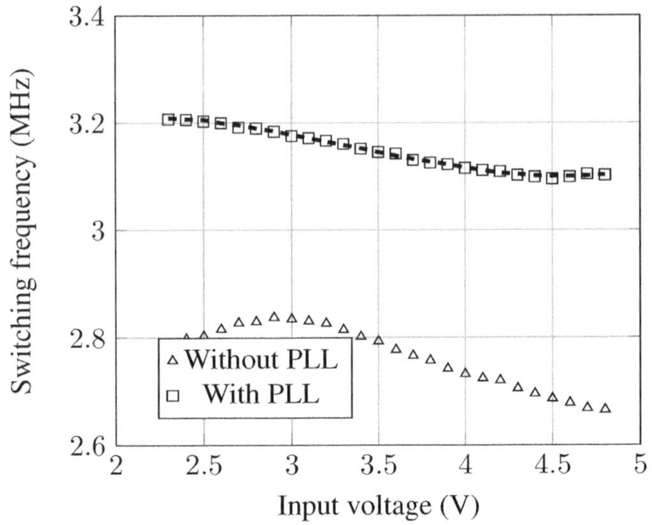

Figure 4.6. *Switching frequency (f_{sw}) versus input voltage (V_{in}) of the two prototypes with and without the phase loop*

The design and optimization of an integrated power stage differs from that of a discrete power stage; however, the fundamental equations and process remain the same. The design process of a power stage consisting of discrete components computes the optimization process for different devices in order to pick the best trade-off. On the contrary, the design process of an integrated power stage optimizes the width of a particular transistor. However, the packaging and board parasitic elements play a key role in the efficiency of the converter. Minimizing parasitic elements on the switching paths reduces switching losses as well as crossover losses of the devices. Therefore, the tight integration of the active elements always presents a significant advantage for high switching frequency converters.

Selecting the right control strategy for an on-board switched-mode power supply is one of the most difficult challenges for the power supply designer. It gathers almost every possible constraint of the system apart from the mechanical constraints. However, once the system requirements are clearly defined, we can select the easiest way to meet the specifications. As a switched-mode power supply is a nonlinear system, nonlinear control and transient simulations are the best allies of the designer. Nevertheless, small-signal linearized models and linear control theory are often useful to set the compensation functions. From these analyses, it is clear that the PDN and the passive components ultimately limit the achievable performances. Most IoT systems require integrated controllers. Digital control is not considered in this chapter; however, we can quickly derive a digital version of the proposed controls. Furthermore, there is no rule for the design of the elements themselves, and depending on the designer's experience, a common architecture can present different silicon implementations.

5

On-Chip Power Management DC/DC Switched-Capacitor Converter

"Power-on-chip" and "granular power distribution" described in the previous chapter lead to the key issue of how to integrate the power supply in the same chip as the powered circuit. For example, the fine and fast voltage modulation in multi-core processors needs numerous fully integrated power supplies on the same chip as the digital cores. Therefore, the challenge is to propose a power supply compatible with most recent CMOS technologies that embed dense digital logic. Generally, these technologies offer i) a low-voltage rating transistor, ii) a small gap between conductive layers (a few nm), iii) a relatively high dielectric constant ($\epsilon_r = 3 - 5$), iv) thin metallization layers (hundreds of nm) and v) no magnetic material ($\mu_r = 1$). The second and third issues imply considering a relatively dense capacitor (a few nF/mm^2). The fourth and fifth points reduce the capability of having a dense inductor with a low series resistance and a low EM emission: it is possible to achieve a few nH/mm^2 with hundreds of $m\Omega$ in series and in the presence of the magnetic field. Fortunately, the switched-capacitor converter (SCC) described in this chapter is an inductorless topology by nature. A SCC has *a priori* the ability to be integrated in more advanced CMOS technologies without an external passive component. The *a priori* term is a key point in this context. For example, some published results reveal the capabilities of the inductive-based converter, such as full integration of the buck structure in CMOS technology with decent power efficiency and density. As will be described later, the 3D

Chapter written by Gael PILLONNET, Thomas SOUVIGNET and Bruno ALLARD.

assembly techniques will greatly change the constraints on the technology side for the switched-capacitor converter, maybe pushing the limit where inductive-based converter were the preferable choice. In summary, the recent interest in SCCs is mainly due to the high constraint on passive components, which can be integrated in existing CMOS technologies without extra (and very costly) process steps. Some researchers predict a bright future for high-power density integrated DC/DC converters. This chapter discusses the pros and cons of SCC topology and gives some key trade-offs in order to produce an optimal design based on various constraints.

5.1. Topology description

The following analysis assumes a two-phase equally spaced operation at 50% duty cycle. If the duty cycle is outside this value, the power efficiency is reduced. The notations used in this chapter are C_i for the ith flying capacitor, V_i for the input voltage, V_0 for the output voltage and f_{sw} for the switching frequency.

There are various SCC topologies: ladder, series-parallel, Fibonacci, doubler and Cockcroft-Walton. Faced with such a variety of SCC configurations, Makowski's [MAK 95] followed by Sanders' [SEE 09] research groups studied the differences between them in depth. When the capacitor energy transfer loss dominates (see later), i.e. the intermediate energy storage element is the limiting factor, the series-parallel topology is the best compromise between the energy storage requirement and the switch voltage-current rating [SEE 10]. For an infinite capacitor value, the ladder topology is the best. In the application range given in section 5.2, the silicon size constraint implies a low capacitor value. The series-parallel topology will be described in the following sections.

5.1.1. *Ratio calculation*

A simple method is given in this section to find the ideal conversion ratio of an unloaded SCC, which is also called the unloaded ratio. Non-ideal characteristics such as on-state resistance and finite capacitor value are removed from this calculation. Ideal switches, steady-state condition, no load and constant output voltage are considered. However, they are reduced to a pure topological point of view. It should be noted that the unloaded ratio is always the ratio n = M/N, where M and N are integers.

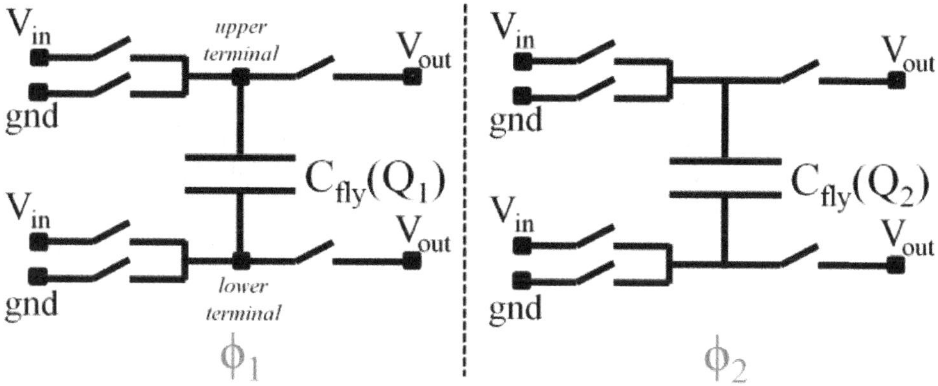

Figure 5.1. *Possible connections for the flying capacitor in a two-phase SCC*

Figure 5.1 shows all possible connections in a two-phase SCC using one flying capacitor. For example, the upper terminal of the capacitor can be electrically connected to the input voltage V_{in}, ground *gnd* or output voltage V_{out} in the first phase. The electrical charge stored in the ith phase is denoted by Q_i. At each phase, there are nine connection possibilities for the flying capacitor. The capacitor is possibly connected to two voltage nodes in the first phase and the same applies in the other phase. In total 81 combinations are possible (9×9), i.e. the capacitor connected to two voltages in the first phase followed by another phase where the capacitor is connected to two other voltages. Each possibility generates a maximal efficiency at a specific unloaded ratio.

The following example presents how to calculate the 2:1 SCC configuration (voltage gain = $\frac{1}{2}$). In this configuration, the upper terminal is connected to V_{in} in ϕ_1 and to V_{out} in ϕ_2. The lower terminal is connected to V_{out} in ϕ_1 and to ground in ϕ_2. The energies $E_{in,i}$, $E_{out,i}$ given by the input and output nodes during one period T (two phases) are defined in [5.1] and [5.2], where $\Delta Q = Q_1 - Q_2$, Q_i is the charge received by the flying capacitor C_{fly} during the ith phase and $E_{x,i}$ is the energy given or received during the ith phase by the x node:

$$E_{in} = E_{in,\Phi 1} = \Delta Q V_{in} \qquad [5.1]$$

$$E_{out} = E_{in,\Phi 1} + E_{in,\Phi 2} = \Delta Q V_{out} + \Delta Q V_{out} \qquad [5.2]$$

To obtain lossless operation, the energy provided by the input E_{in} must be equal to the delivered energy E_{out} as in [5.3]. Again, it should be noted that the ideal efficiency is obtained for a unique ratio. This is the key characteristic of a SCC compared with an inductive-based converter where the ratio has no impact on efficiency under the same condition (ideal switches and passive components):

$$\eta = 1 \longrightarrow E_{in} = E_{out} \longrightarrow V_{out} = \frac{V_{in}}{2} \qquad [5.3]$$

By using the same method for each of the 81 possible configurations in a two-phase SCC with only one flying capacitor, only four unloaded ratios can be obtained [5.4]:

$$n = \left[-1; \frac{1}{2}; 1; 2\right] \qquad [5.4]$$

Therefore, this SCC topology achieves only one optimal ratio in the "step-down" condition ($0 < n < 1$). Outside this ratio, it acts as a linear regulator, leading to low efficiency (see later).

5.1.2. *Basic scheme*

Figure 5.2 shows the basic 2:1 SCC in the CMOS technology. The converter has four MOSFET switches and one flying capacitor. The output node is connected to an output capacitor that averages the output voltage. There are two phases (charge and discharge of the flying capacitor) that control the switch states. Some transistors can be replaced by a passive switch, i.e. a diode, as in an asynchronous buck converter. The four transistors act in their ohmic and sub-threshold regions, except during the switching operations. The transistor type (N or PMOS) is chosen to maximize the source-drain voltage and thus minimize the on-state resistor in the ohmic region. In Figure 5.2 switch T1 is a p-type MOSFET as the highest voltage is connected to the source. In the 2:1 SCC configuration as a higher voltage is connected to the source. An N-type is preferred for T4 as the source is always below $V_{in}/2$ due to the loose mechanism described below. Other choices could be made if the transistor is driven by a no full-swing voltage (e.g. between V_{in} and $V_{in}/2$), or if the topology is combined with other N:M ratios. A deadtime T_{dt} between the two phases is introduced to avoid a cross-conduction path between the input voltage, output voltage and ground.

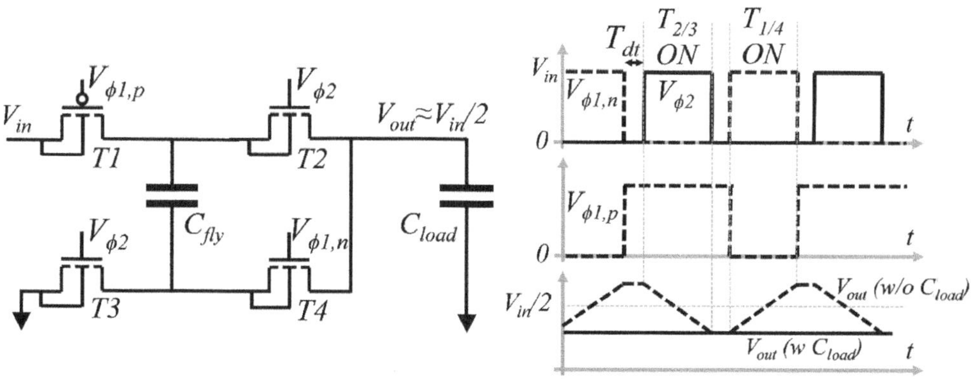

Figure 5.2. *Schematic of the 2:1 SCC configuration and main waveforms*

5.1.3. *Steady-state modeling*

The well-known steady-state schematic of a SCC is shown in Figure 5.3. In this model, the turn ratio N:M represents the voltage conversion ratio in open-circuit conditions, i.e. unloaded ratio determined by the given topology. The output impedance R_{out} links the voltage drop to the output current when a load is applied. R_{out} is due to the conductive losses from the on-state resistance and to the parasitic resistors from the flying capacitors. The impedances R_{si} and R_{so} represent the static power consumption due to the switching operation.

This model evaluates the losses inside the SCC and calculates the output voltage using [5.5] and [5.6]:

$$V_{out} = \frac{M}{N} V_{in} - R_{out} I_{out} \tag{5.5}$$

$$\eta = \frac{V_{in} I_{out}}{V_{in} I_{out} + R_{out} I_{out}^2 + \frac{V_{out}^2}{R_{so}} + \frac{V_{in}^2}{R_{si}}} \tag{5.6}$$

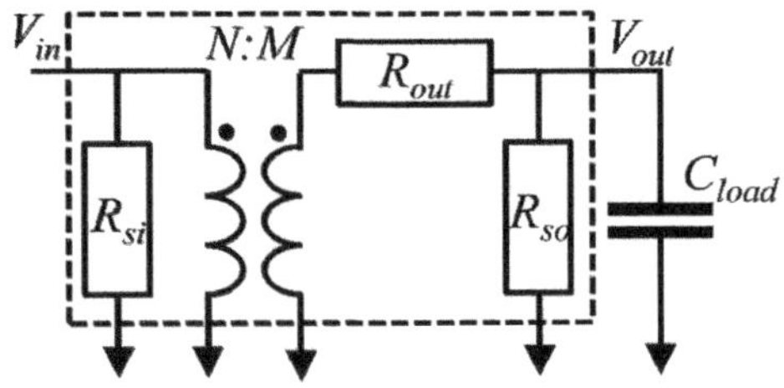

Figure 5.3. *Steady-state model of a SCC*

The impedance calculation could be made using the three methods described below.

5.1.3.1. *Method based on charge transfer*

The first method is based on a charge transfer analysis to obtain the output impedance. It is only validated in the SSL region [SEE 09] where the switch resistance has no impact on the charge and discharge process of the flying capacitor, i.e. the C_{fly} value limits the efficiency and the current is considered as constant. The calculation example for the 2:1 topology is given below [5.7, 5.8], where $Q_{\Phi i,x}$ is the electrical charge of the flying capacitor at instant x in phase i, x is the initial (x=i) or final (x=f) instant, and $\Delta Q_{\Phi i}$ is the charge given or received by the flying capacitor in phase i:

$$\Delta Q_{\Phi 1} = Q_{\Phi 1,f} - Q_{\Phi 1,i} \quad [5.7]$$

$$\Delta Q_{\Phi 2} = Q_{\Phi 2,f} - Q_{\Phi 2,i} = C_{fly} \cdot (V_{fly,\Phi 2,f} - V_{fly,\Phi 2,i}) \quad [5.8]$$

In the steady-state ($V_c(\Phi 1, i) = V_c(\Phi 2, f)$), the charge received in phase 1 and given in phase 2 by the flying capacitor are equal, thus meaning that $\Delta Q_{\Phi 1}$= $\Delta Q_{\Phi 2}$. The output node receives the same charge transfer as the flying capacitor and thus ΔQ_{out}= $\Delta Q_{\Phi 1}$-$\Delta Q_{\Phi 2}$. By following the basic energy continuity principle, the voltage across the flying capacitor is the same at the end of one phase and at the beginning of the next, for example $V_{fly,\Phi 1,f} = V_{fly,\Phi 2,i}$. Assuming that the input and output act as ideal voltage sources, the end of the flying capacitor voltage is equal to $V_{in} - V_{out}$ and V_{out}

in phases 1 and 2, respectively, for example $V_{fly,\Phi1,f} = V_{in} - V_{out}$. By rearranging the previous equations, the output current is given by [5.9]:

$$I_{out} = \frac{\Delta Q_{out}}{T_{sw}} = \frac{2C_{fly} \cdot (V_{in} - V_{out})}{T_{sw}} \longrightarrow V_{out} = \frac{V_{in}}{2} - \frac{T_{sw}}{4C_{fly}} \cdot I_{out} \quad [5.9]$$

Associating the above equation with the small-signal model given by [5.5], the output resistance R_{out} is equal to $\frac{1}{4C_{fly}F_{sw}}$. The on-state resistance does not affect the output impedance as the switches are considered ideal in this method. The main benefit of this method is to underline the inherent SCC limitation: the charge-sharing loss. Even if all components are supposedly ideal, the SCC has a finite output impedance. In addition, it has a lossless operation, i.e. R_{out} is equal to zero if the flying capacitor value and/or switching frequency are infinite, leading to an infinite SCC volume and/or switching loss, respectively. The method could be applied to other SCC topologies, but leads to the same conclusion concerning the basic limitation of this type of converter.

5.1.3.2. *Method based on an RC differential equation*

The second method for obtaining the output impedance R_{out} for any condition (SSL, FSL and regions between them) has been described in [BEN 09]. It is based on solving the two basic RC differential equations formed by the resistive paths and the flying capacitor. The hypothesis is still that there is a perfect output voltage decoupling, i.e. the infinite bypass capacitor $C - load$. The current flowing into the flying capacitor can be expressed as [5.10]:

$$i_{\Phi i}(t) = \frac{\Delta V_{\Phi i}}{2R_{on}} \cdot \exp(-\frac{t}{\tau}) \quad [5.10]$$

where $\tau = RC_{fly}$ is the time constant, $V_{\Phi 1} = V_{in} - V_{out} - V_{fly}(t = 0)$ is the initial voltage at the beginning of the phase Φ_1 , $\Delta V_{\Phi 2} = V_{out} - V_{fly}(0)$ is the initial voltage at the beginning of the phase Φ_2, Ron is the equivalent on-state resistance of the switch and t=0 is the initial value at the beginning of each phase.

The energy dissipated in the equivalent resistance R during one phase is given by [5.11], where $T_{\Phi i} = \frac{T}{2}$ is the end value of the phase and $\beta = \frac{T}{4RC_{fly}}$:

$$E_{R,\Phi i} = \int_{t=0}^{T_{\Phi i}} V_r I_{\Phi i} dt = \frac{1}{2} C_{fly} \Delta V_{\Phi i}^2 (1 - \exp(2\beta)) \quad [5.11]$$

The charge $Q_{\Phi i}$ flowing into the flying capacitor is calculated from [5.12]:

$$Q_{\Phi i} = \int_{t=0}^{T_{\Phi i}} I_{\Phi i} dt = C_{fly} \Delta V_{\Phi i} (1 - \exp(-\frac{T}{2\tau})) \qquad [5.12]$$

By combining [5.11] and [5.12], the initial voltage across the equivalent resistance R can be evaluated as:

$$\Delta V_{\Phi i} = \frac{\bar{i_{\Phi i}}}{C_{fly} F_{sw} (1 - \exp(-\frac{T}{2\tau}))} \qquad [5.13]$$

where $\bar{i_{\Phi i}}$ is the average current during one period, which is equal to $Q_{\Phi i} T$, and F_{sw} is the switching frequency.

In the steady state, the charges received $Q_{\Phi 1}$ and provided $Q_{\Phi 2}$ by the flying capacitor are equal for a 2:1 configuration, so $i_{\Phi 1}$=$i_{\Phi 2}$=i_R . The total energy dissipated by R over one switching period is given by [5.14]:

$$E_R = E_{R,\Phi 1} + E_{R,\Phi 2} = \frac{1}{C_{fly} F_{sw}} \cdot \frac{1 + \exp(-\beta)}{1 - \exp(-\beta)} \cdot \bar{i_R}^2 \qquad [5.14]$$

The power dissipated in R is [5.15] and the average output current i_{out} is given in [5.16]:

$$P_R = \frac{E_R}{T_{sw}} = \frac{1}{C_{fly} F_{sw}} \cdot \coth(\frac{\beta}{2}) \bar{i_R}^2 \qquad [5.15]$$

$$\bar{i_{out}} = \frac{Q_{\Phi 1} + Q_{\Phi 2}}{T_{sw}} = 2\bar{i_R} \qquad [5.16]$$

By combining the last two equations, the equivalent output resistance R_{out} of the small-signal model can be expressed as in [5.17] as follows:

$$R_{out} = \frac{P_R}{\bar{i_{out}}^2} = \frac{\beta}{2} \coth(\frac{\beta}{2}) \cdot 2R_{on} \qquad [5.17]$$

This analytical result is reliable in FSL and SSL, as mentioned earlier. For high frequencies and thus low β values (FSL region), the on-state resistance is mainly responsible for the losses as in [5.18]. For the low switching frequency and thus high β values (SSL region), the charge-sharing loss is predominant as in [5.19]:

$$\lim_{\beta \to \infty} R_{out} = 2R_{on} \quad [5.18]$$

$$\lim_{\beta \to 0+} R_{out} = \frac{1}{4C_{fly}F_{sw}} \quad [5.19]$$

5.1.3.3. *Method based on an Mx coefficient calculation*

The third method introduced by [SEE 09] determines the steady-state performance of SCCs using only three parameters from the topology (M_{sw}, M_{cap}, M_{bot}) and few others from the targeted technology (λ_r, λ_c, α). From [SEE 09], the total power loss can be expressed as in [5.20], where P_{sw} is the switching loss, C_{fly} is the total flying capacitor value, W is the total width of the switches, λ_r is the on-state resistance density measured in $\Omega \cdot m$, λ_c is the gate capacitance density per transistor width [F/m], I_{out} is the output current, α is the bottom-to-flying capacitor ratio, f_{sw} is the switching frequency and V_{sw} is the voltage swing to drive the switch gates (V_{sw}=V_{in}). All of these are used in the following to model the full swing transistor driving:

$$P_{losses} = P_{cond} + P_{trans} + P_{bott} + P_{sw} = M_{sw}\frac{\lambda_r}{W}I_{out}^2 + \frac{1}{M_{cap}}\frac{I_{out}^2}{C_{fly}f_{sw}} + M_{bott}\alpha C_{fly}f_{sw}V_{out}^2 + \lambda_c W f_{sw} V_{sw}^2 \quad [5.20]$$

The four power terms correspond to:

- the conduction loss P_{cond};
- the energy transfer loss P_{trans};
- the bottom plate loss P_{bott};
- the switching loss P_{sw}.

The last equation is only valid if $P_{cond} >> P_{trans}$ (FSL region) or $P_{cond} << P_{trans}$ (SSL region). These two regions, SSL and FSL defined in [SEE 09], correspond to the nature of the efficiency limitation due either to the switches or to the flying capacitor. The second method explained earlier by merging the two terms P_{trans} and P_{cond} covers all the cases, especially when $P_{cond} \approx P_{trans}$.

Reorganizing the terms of [5.20], the impedances of the steady-state model appear as in [5.21]:

$$P_{losses} = R_{out}.I_{out}^2 + \frac{V_O^2}{R_{so}} + \frac{V_i^2}{R_{si}} = (M_{sw}.\frac{\lambda_r}{W} + \frac{1}{M_{cap}}\frac{1}{C_{fly}f_{sw}}) + M_{bott}\alpha C_{fly}f_{sw}V_{out}^2 + \lambda_c W f_{sw} V_{in}^2 \quad [5.21]$$

To obtain the coefficients M_x, the targeted SCC topology has to be detailed. First, the definition is given in [5.22]:

$$M_{sw} = 2\{\sum_{i=1}^{N} |a_{r,i}|\}^2 \cdots M_{cap} = \frac{1}{\{\sum_{i=1}^{M} |a_{c,i}|\}^2} \cdots M_{bott} = \{\sum_{i=1}^{M} \frac{\Delta V_{cm,i}}{V_{out}}\} \quad [5.22]$$

where N and M are the number of switches and flying capacitors, respectively, $a_{r,i}$ represents how much charge flows through the ith switch, $a_{c,i}$ represents how much charge flows through the ith flying capacitor and ΔV_{cm} is the swing voltage of the node of the bottom plate of the flying capacitor.

The calculation for the 3:2 ratio unloaded structure is given below. Figure 5.4 shows the switching cell topology and the switching sequence in two phases. The two flying capacitors receive a charge Q/3 to provide a total charge Q to the output node during the two phases. Therefore, the coefficients $a_{c,1}$ and $a_{c,2}$ are equal to $Q/3$ divided by the full charge Q received by the output: $a_{c,1} = a_{c,2} = 1/3$. The seven active switches suffer a flow of Q/3 and thus the coefficient $a_{r,i}$ for i from 1 to 7 is equal to $Q/3$ divided by the total charge received by the output, i.e. Q. The bottom plate of the first flying capacitor $C_{fly,1}$ is connected to Vout in the first phase and $V_{out}/2$ in the second phase, then $\Delta V_{cm,1}$=V_{out} $-$ $V_{out}/2$ $=$ $1/3V_{in}$. The second flying capacitor is connected to V_{out}. The calculation of the coefficients is given in [5.23]:

$$M_{sw,3:2} = 2\{\frac{1}{3} + \frac{1}{3} + \frac{1}{3} + \frac{1}{3} + \frac{1}{3} + \frac{1}{3} + \frac{1}{3}\}^2 = 10.89$$

$$M_{cap,3:2} = \frac{1}{\{\frac{1}{3} + \frac{1}{3}\}^2} = 2.25 \qquad M_{bott,3:2} = \frac{1}{2} + 1 = 1.5 \quad [5.23]$$

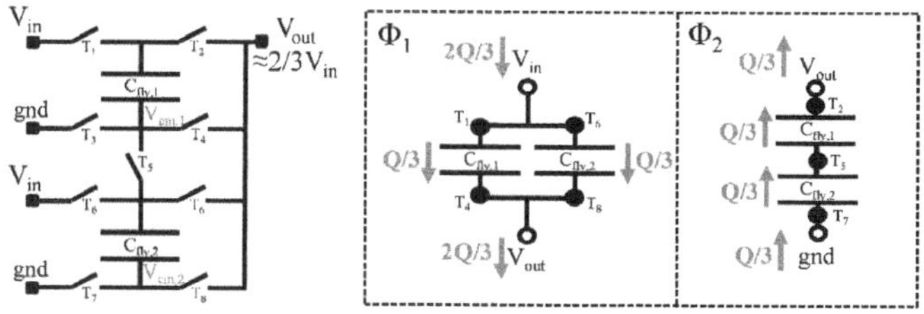

Figure 5.4. *Sequence of the 3:2 SCC topology*

Ratio	M_{sw}	M_{cap}	M_{sw}
		1 input, 1 flying	
2:1	8	4	1
		1 input, 2 flying	
3:1	10.9	2.3	2.5
3:2	10.9	2.3	0.6

Table 5.1. *Coefficient for 2:1, 3:1 and 3:2 topologies*

Table 5.1 gives the coefficients for other ratios for two flying capacitors in the series-parallel configuration.

5.1.3.4. *Method for extracting the technology parameters*

To obtain the technological coefficients $\{\lambda_r, \lambda_c,\}$ used in the third method, some electrical simulations have to be done. First, λ_r is characterized by performing a DC analysis of the transistor in the on-state condition for a given transistor width. The drain-to-source and gate-to-source voltages have to be set to a few tens of mV and to V_{in}, respectively. The transistor length is also minimized to improve the conductivity of the channel. λ_r is inversely proportional to the transistor width, i.e. a 10 times larger transistor gives 10 times less on-state resistance: $R_{on} = \lambda_r/W$. Second, λ_c is extracted from a transient simulation with a bench, which mimics a switching. By calculating the energy given to the gate during an entire switching transition for a given transistor sizing, the equivalent capacitance of the gate is extracted. λ_c is proportional to the transistor width, i.e. 10 times larger transistor gives 10 times higher gate capacitance: $C_g = \lambda_c.W$. The coefficient α is deduced from the compact model of the capacitor or extracted from a post-layout simulation. This coefficient varies with the capacitor type. For example, a poly-well capacitor has a higher value than a MIM capacitor as the bottom plate is closer to the substrate (see later for on-chip capacitor choice). In contrast, the poly-well capacitor has a better capacitance density.

The value of the flying capacitor C_{fly} is determined by the available silicon surface. If no surface is given at the beginning of sizing, the designer has to choose one and see whether the efficiency is sufficient for their specification. However, the trend is that the better the available surface, the better the peak efficiency.

5.2. Pros and cons

5.2.1. *Key advantages*

The SCC has several interesting properties. The first property is that the SCC is an inductorless topology that enables its monolithic integration in the CMOS technology without any additional process steps. As has been already stated, CMOS technologies do not offer dense inductors (a few nH/Ω), which reduces the interest in an inductive-based converter.

The second property is the possibility of miniaturizing the capacitor. Table 1 in [SEE 10] shows that the energy density stored in a discrete capacitor ($0.5CV^2$) is a few decades higher than in an inductor ($0.5LI^2$) for discrete components on a millimeter scale. For larger devices, this relation may not hold true. If this comparison could also be applied to integrated silicon technology, then the SCC is a promising candidate for small and widely distributed on-die converters. However, this hypothesis may not be as good as it seems because the passive component utilization is quite different for the SCC and the SIC. The SCC only uses a capacitor ripple voltage ΔV to transfer energy from input to output. The real transfer energy is $0.5C[V^2 - (V - \Delta V)^2]$, where ΔV is related to the charge-sharing loss. In PFM mode for a buck (one topology for the SIC), the inductor is fully magnetized and demagnetized, i.e. ΔI=I at each switching period. The energy stored in the inductor is fully used. Finally, the passive component performance can be changed if additional process steps are considered such as thick metal or vertical trench.

The third property is the relative insensitivity of the capacitor to its energy transfer frequency. First, the electrical permittivity of the materials compared with the magnetic permeability seems to be more constant, giving a key advantage when the switching frequency increases. Second, the parasitic series resistor of the capacitor is not a bottleneck at high frequency when low resistivity capacitor plates associated with regular plate connection techniques are used.

The fourth interesting property of the SCC is the switch utilization. In some contexts, the on-chip SCC is directly powered by a higher input voltage as the voltage-rating of the transistor. In some configurations, i.e. the 2:1 voltage ratio, the switches see half the input power supply instead of the full voltage swing in bulk topology. For example, a 1.8 V compatible input voltage converter can be designed with only 0.9 V voltage-rating transistors in [LE 11]. The comparable inductive structure, i.e. bulk, suffers from lower power device utilization because 1.8 V transistors are needed under the same conditions. This fundamental advantage allows the use of native low-voltage digital-core transistors (e.g. thin oxide transistor) in the SCC, suppressing the extra thick oxide process option. [SEE 10] shows that the ladder structure minimizes the V-A constraint across the switches.

The fifth property is that a SCC can remain efficient over a wide load range. Operation at zero load is possible without moving into a particular modulation scheme. In contrast, the inductor current ripple in the buck converter is directly linked to the load. At weak load, the pulse frequency modulation is activated to reduce this ripple. However, the capacitor voltage ripple in the SCC is inversely linked to the load. At weak load, the ripple is smaller at the same switching frequency. The overall efficiency can be maintained by reducing this frequency and then a modulation transition between PFM and PWM is not needed compared with SIC. This key advantage explains its wide use in low power applications.

The sixth and final interesting property is the possibility of dividing a SCC into numerous elementary units without the need for extra silicon surface. The capacitor density is almost constant with the area, which is not the case for an inductor. Cutting one capacitor into N capacitors only requires a little extra routing and a few gaps between each unit. This topology is a good candidate when designing distributed and small power supplies in order to reduce the supply routing constraints of a centralized and bulky power supply.

5.2.2. *Main disadvantages*

There are several well-known issues associated with SCCs. Before designing a SCC, we invite the reader to read the following paragraphs carefully to determine any possible inconsistencies with their own constraints.

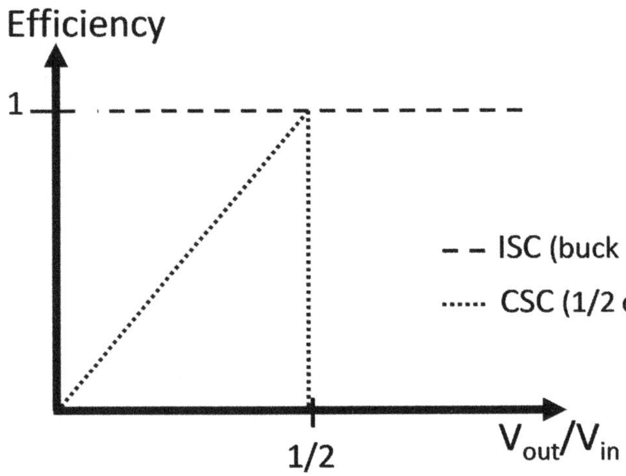

Figure 5.5. *Theoretical efficiency limit vs. voltage gain (V_{in}/V_{out}) for step-down inductive and capacitive-based converters*

For a fixed SCC topology, the 100% theoretical efficiency is only reached for a particular conversion ratio [MAK 95]. The converter acts as a linear regulator outside this ratio. If we consider an ideal switch, then Figure 5.5 shows the inherent efficiency achieved by a buck and a step-down SCC in a 2:1 configuration over the voltage gain. The SCC suffers from a fundamental efficiency limit outside its own ratio (e.g. 0.5 here). In other words, the zero load condition is relatively easy to manage in the SCC compared with the SIC. In contrast, the bulk maintains lossless conversion. It is important to point out that the conversion ratio control is done by modulating the switching frequency in the SCC and by the duty cycle in bulk. In some applications such as the processor supply in a battery-powered application, a wide range of voltage gain is needed. For this, the input converter voltage varies greatly depending on the battery discharge status. On the output side, wide voltage is required to perform DVFS (dynamic voltage and frequency scaling). Thus, the converter experiences a large voltage ratio range and the peak efficiency of the SCC at its particular ratios is rarely achieved. To address this limitation, some papers propose reconfigurable topologies [LE 11], [RAM 10b] or double outputs [JEO 12] to maintain a constant efficiency for multiple ratios. Another solution is to design a hybrid structure using an SCC in series with an inductive-based converter [LIM 15]. [PIL 15] proposes a MISO structure for an SCC by having multi-input power supplies to achieve a more constant efficiency over a large conversion ratio.

The second limitation is the dependency of the power efficiency on the SCC voltage gain even at its nominal ratio. In most published works, the SCC has been designed for low conversion ($< 1/4$) or moderate ($< 1/10$) ratios. [NG 09] shows that the SCC can attain up to 90% efficiency in an 8:1 ratio but with discrete capacitors.

The third limitation is linked to the first. A lossless regulation is difficult to achieve. Fine control is possible but at the expense of degrading the power efficiency.

The final limitation is the potential conducted EMC issues. Some research works on SCC are motivated by the large coupling noise generated by the SIC topologies. A major benefit of the SC structure is a low generated magnetic field as compared to inductive-based conterpart. However, the designer has to deal with a high input peak current demand at each SCC cycle to charge the flying capacitor. The current is only limited by the on-state resistance of the switch and the parasitic access inductor. This behavior could create undesired high-frequency conducted noises on the input power rails, which could be coupled to other circuits on the board. This noise could be mitigated by adding an input decoupling capacitor, but this adds extra area.

5.3. State-of-the-art

SCCs are gaining acceptance as alternatives to traditional switching voltage regulators. Designing this topology requires an understanding of all limitations and their impacts on global operation. As the specification greatly depends on the application target, the items below outline some general design guidelines, physical insights and trade-offs based on the literature to help the designers in their own choice.

5.3.1. *Research scope and main focus*

The inductor-based converter has addressed power tree distribution for embedded systems for a decade. A partial switching from this traditional and well-adopted structure to a capacitive-based converter needs strong arguments to convince the power community to adopt this change such as a major improvement in power density, faster regulation and significant reduction in PCB footprint or more granular power domains. In the past, the SCC has been relegated to low power applications (< 0.1 W) or voltage biasing. In parallel, new industrial technologies offer suitable power devices for inductor-based converters to address this huge market. In recent decades, the SCC has been thoroughly studied for moderate power applications (up to 10 W) by both academic and industrial research laboratories. Numerous papers have focused on minimizing the SCC silicon surface occupation on the compatible CMOS technology. Other features such as the nanosecond control response and wide-range voltage conversion have been demonstrated. However, few papers have addressed the high-voltage step-down or step-up conversion ratio (> 10) [SEE 10].

The on-chip term combines two integration levels: in-package and on-die. The latter requires integration of all SCC components (passive and switches) on the same die. In this case, the division of the flying capacitors into multiple smaller capacitors has a limited influence on the capacitance density and has no impact on the IO number. This feature allows many parallel, interleaved and reconfigurable topologies. The in-package converter is not a truly single-die converter as there are some internal connections to discrete components in the package that increasing the assembly complexity. In contrast to on-chip, the off-chip converter needs components outside its own package.

5.3.2. *Integration level*

The use of an off-chip converter is currently the most efficient way to provide multiple power domains, but it is a bulky solution mainly due to the extra components needed around the powered circuit on the motherboard footprint. Nowadays, on-chip power management seems a more suitable solution, providing a clean, fine, high-speed and granular power supply without bulky external components or additional circuits. There are two levels in which the on-chip power supplies can be integrated: in-package and on-die solutions. In-package converters combine different technologies to integrate the switching cell, passive and control in the same package as the powered circuit [ABE 07]. Using non-standard processes such as deep-trench capacitors [AND 13a], magnetics on silicon [MAT 13] or wide band-gap technology [PAL 12] leads to the converter being potentially smaller and having a higher power efficiency than an on-die solution, though the 3D heterogeneous assembly of the converter and its load could be a cost issue. The second integration level, i.e. on-die, consists of the integration of the converter on the same technology as the powered circuit [WAN 13, LE 11]. The key issue is the integration of the passive for intermediate energy storage (Figure 1). Some researchers have proposed numerous design refinements to overcome the low energy capability of the integrated passive components. For example, high switching frequency [HAZ 05, ARN 98], series-parallel switching cell connection [RAM 10a, UZU 14] or coupling of passives [WIB 08] help improve the converter performance. However, the recently published on-die power converters still achieve a performance far from the industrial targets [SAN 13], in terms of power density, voltage regulation, efficiency versus conversion ratio or direct battery connection compatibility.

5.3.3. *The point-of-load (POL) application*

5.3.3.1. *Key performance and common metrics*

The principal figure of merit (FoM) of published SCCs is the power density (W/mm^2). This characterizes the integration capability relative to its alternatives. Nowadays, the order of magnitude is less than 1 W/mm^2 for the classical CMOS technology with no extra options (e.g. the deep-trench capacitance). Compared with the approximately 10 W/mm^2 power consumption density of a digital core, the current published SCC needs 10 times the surface to power it. In other words, the surface of the converter is 10 times greater than the digital one. This is not acceptable for industrialists who focus on merging the power supply into a digital core. Furthermore, the power density is not linearly scalable; e.g. 0.1 mm^2 providing 0.1 W cannot be directly scaled up to 1 mm^2 to deliver 1W. To conclude, this FoM gives a partial view and is more often used for convincing reviewers than making a fair comparison. The main benefit of this FoM is to give a reasonable range for a designer working on an initial dimensioning.

5.3.3.2. *Superiority of the 2:1 ratio*

In the literature, most studies have focused on the 2:1 step-down ratio. The main reason for this is that this topology achieves the best efficiency at this particular point. The output impedance calculation emphasizes this fact. By using the third method explained earlier, the impedance is given by [5.24], where the equation is valid for an M:1 ratio and in the SSL or FSL regions:

$$\{M_{cap}, M_{sw}\} = \{(\frac{M}{M-1})^2; 2(\frac{3M-1}{M})^2\}; R_{out} = \frac{1}{M_{cap}C_{fly}} + M_{sw}R_{out} \quad [5.24]$$

The above equation shows that the output impedance is minimal for M = 2. If M is high, the output impedance limit is given by [5.25]:

$$\lim_{M \to \infty} R_{out} = \frac{1}{C_{fly}F_{sw}} + 18 \cdot R_{on} \quad [5.25]$$

As publications often focus on performance comparison (e.g. power density), it is easy to understand the 2:1 focus in the literature. However, the designer has to be aware that the efficiency and power density given in the published works are often achieved in the best topology case.

5.3.3.3. *Ratio selection*

The SCC is rarely in a 2:1 configuration, thus leading to a performance lower than expected compared with published ones. For example, a DC/DC step-down connected directly to a chemical battery and providing supply voltages to ICs experiences a large conversion ratio range. In the worst case, the battery is charging (5 V) and the IC needs low voltage rails (0.5 V), leading to a 10:1 step-down ratio. In the best case, the battery is in a low charging state (2.3 V) and the IC needs a maximal voltage rail (1.8 V), leading to about a 4:3 ratio. Under these constraints, the SCC topology need to have a reconfigurable ratio in order to maintain sufficient efficiency. This relative complexity has an impact on the overall efficiency and needs to dynamically manage the ratio reconfiguration. Therefore, the SCC is probably less suited for this application than an inductor-based regulator.

In some applications, the input voltage is provided by a pre-regulator and so the input voltage has a constant value. If the output voltage is also constant (no dynamic voltage scaling needed), the ratio can be constant. However, the 2:1 ratio is not necessarily the right choice to use in this fixed configuration. For example, a ratio of 2:3 has to be chosen in the case of 1.8 V input voltage and 1.0 V output voltage, which will increase the inherent loss compared to the 2:1 best case topology.

To reduce the power consumption of a digital circuit, the dynamic voltage scaling strategy requires at least two output voltage levels to achieve this power saving. This application constraint also implies maintaining power efficiency at different voltage ratios [SOM 06].

5.3.3.4. *Output power range*

In addition to the large ratio required in some applications, the designer is faced with a large output load range condition. The delivered power can vary by a few decades. In this case, the converter must have a power-down mode in which the quiescent current consumption is reduced to a negligible value in a weak load condition.

5.3.3.5. *Interleave technique*

In most works in the literature [SOM 06], the interleave technique is used to reduce the output ripple and smooth the input current peak demand. To avoid using a large input and output capacitor (see later), the designer has a very cheap solution: cutting the total SCC into N smaller sub-units that operate at the same frequency but with equally shifted N-phases between them. The "elementary" parallel SCCs perform the same function. The solution does not affect the power efficiency and maintains the total sizing because the flying capacitor and the transistor's width are just divided by N. As the efficiency of the SCC is inversely related to the flying capacitor's ripple, a quick analysis could suggest a better efficiency. However, the number of unit cells is multiplied by N, hence compensating for the efficiency gain of a smaller ripple. This interleaving effect does not change the average small-signal model of the entire SCC. The load transient is also largely improved without compromising other design parameters. The main drawback is the unbalanced energy-sharing between each interleave stage due to the mismatching and management of the many connections between each sub-unit.

5.3.3.6. *Mismatching effect of switch, capacitance and phase*

The matching effect is not often covered in the literature. This is because in order to maximize the power efficiency, the duty cycle between two phases has to be equal to 0.5, and for any other value, the efficiency is reduced. The impact has not been clearly quantified in research papers; however, to my knowledge, it has a second-order effect on efficiency. Concerning the flying capacitor mismatch, the output voltage ripple is impacted and the charge-sharing loss between interleaved SCCs is increased. The on-state resistance mismatch of the switch also has an effect on the current sharing between each interleaved SCC, creating an unbalanced power delivery for each unit cell.

5.3.3.7. *Settling time*

Some applications target a maximal settling time to increase/decrease the output voltage. To predict this time, the time constant given by the calculated SCC output impedance and output capacitor is a good starting point. To increase the slew rate of the SCC, the switching frequency could be increased greatly during a short period in order to transfer the electrical charge faster to the load capacitor through the flying capacitor.

5.3.3.8. *Effect of series resistance of the flying capacitor*

The equivalent series resistance (ESR) of the flying capacitor has a first-order effect on the performance of the SCC. Using some types of capacitor, especially semiconductor plates (n-well poly-capacitor), the total access impedance to the planes could be larger than the resistivity of the switches [VIL 03]. Therefore, the flying capacitor has to be divided into some elementary flying units to reduce the ESR. The switching frequency also has to be taken into account in the dividing process. The time constant of the distributed RC formed by the unit capacitor in series with the equivalent plate resistor could have a non-negligible effect on the charge distribution over the capacitor bank. The latter reduces the effective capacitor density (F/m^2) by introducing extra routing and additional electrical insulation between each unit cell. The design rule manual also introduces some layout constraints, for example polysilicon maximal density which reduces the effective capacity density. The latter, including these constraints and layout strategies [VIL 03], is approximately 60 to 80% of the capacitance density given to the technology. Sometimes, the best capacitance density of all capacitor types offered by the technology does not lead to the best effective capacitance density. To conclude, a lumped model consisting of a single flying capacitor with a resistor connected in series has to be avoided if the plane conductivity is expected to be negligible. A distributed element model throughout the capacitor islands has to be made in order to take into account the dynamic behavior of the ESR [RID 03]. Concerning the quality factor (Q) of the capacitor, i.e. a key parameter in RF circuits to quantify the capability of being in the capacitance region, this parameter is not a bottleneck in SCC design because the switching frequency is rarely greater than the self-resonant frequency of the flying capacitor.

5.3.3.9. *Choice of flying capacitor or switch type*

Careful design requires an understanding of the trade-offs offered by different technologies. An appropriate choice of capacitor technology is essential, as with switching element selection. The SCC switches are often made with MOSFET. The FOM $Q_G \times R_{ON}$ is essential in choosing the appropriate transistor type for low conduction and switching losses in the SCC switches. This FOM represents the inherit performance of the transistor used as a switch device and is independent of transistor width. It depends on the voltage overdrive (V_{GS} applied in the on-state), the transistor length (minimized to reduce R_{ON}) and the V_{DS} voltage of second-order magnitude (if the transistor is not completely in the linear region). In most silicon technologies, the transistor exhibits low-, medium- and high-threshold options. The transistor also has different oxide thicknesses or other flavors, for example drift topology to experience different voltage stresses. To find the λ_R parameters (defined earlier, expressed in $\Omega.m$) of each transistor type, the on-state resistance is evaluated by applying the appropriate voltage in gate/drain/source voltages in DC simulation to mimic the on-state. Then, a transient simulation of a single on-to-off transition is required to measure the energy needed for the gate to switch. This result gives the λ_C parameters (defined earlier, expressed in F/m). The FOM $Q_G \times R_{ON}$ is expressed as $\lambda_R \cdot \lambda_C \cdot V_{drive}$, where V_{drive} is the voltage swing of the switches' drivers.

The capacitor value is key in combining high power density and power efficiency. An appropriate choice of capacitor design and technology is therefore essential as are appropriate switches. A designer may choose from a number of capacitor technologies (electrolytic, ceramic, polyester, tantalum) to implement an external one. For on-chip, there are three main capacitor types: poly-oxide-semiconductor (POS), metal-oxide-metal (MOM) and metal-insulator-metal (MIM). The POS often offers the highest capacitor density, thanks to the thin oxide, and also has the highest ESR. The MIM needs an extra process step to deposit a thin insulator layer, in most technologies, leading to additional costs. As mentioned earlier, the time constant calculated from the capacitor value and the ESR is compared with the SCC's operating frequency. The self-resonant frequency needs to be checked to ensure good operation of the flying capacitor at the switching frequency, i.e. outside resonant behavior. The combination of lateral and vertical field capacitors in some layout patterns also demonstrates a larger density capability and lower series resistance [APA 02].

5.3.3.10. *Input/output decoupling*

As mentioned before, the introduction of the interleave technique reduces the output ripple without increasing the silicon-die area. By increasing the interleaving, the requirement of the output filter capacitor is reduced linearly. In contrast, the load transient dropout is affected by the output capacitance value as less energy is stored at the output (to supply a large and fast energy demand). Thus, faced with this load variation, the size of the output capacitance compensates for the relatively low feedback bandwidth. Outside (i.e. above) the feedback bandwidth, the output capacitor is the only element able to smooth the high-frequency current demand. In some applications, the output capacitance can be added by the load itself. For a digital glue power supply, the load and active decap [MAK 15] introduces a non-negligible extra capacitor on the output node, helping to reduce the undesired voltage oscillation.

The SCC input current differs from the inductive-based converter. There is more peak demand over a short period compared with a buck converter. The current could generate an input voltage drop due to the input access impedance formed by the wire, bond and PCB trace. In the 2:1 topology, the input current peak at the beginning of phase 1 (the charging phase) is only limited by the on-state resistance of the switch and is directly proportional to the voltage drop across the capacitor due to its discharge. In fact, the higher the switching frequency and interleave number, the less problematic the input current peak. In some cases, the input current demand is challenging as this current flows into a power plane that has a non-negligible inductive part. This large di/dt creates a non-negligible voltage drop, which potentially impacts the output voltage oscillation. To overcome this problem, the designer has to include a parasitic inductor from the input trace in their electrical simulation and add input capacitance to smooth the input current peak, if needed. [AND 16] is an interesting paper that underlines the effect of the input voltage drop on the transient response.

5.3.3.11. *Bottom plate loss*

The bottom plate loss, already described in this chapter, is specific to the on-chip capacitor. The loss occurs at each cycle by charging and discharging the bottom plate without transferring the stored energy to the output node. The effect of this loss is proportional to the flying value and the switching frequency. As a high power density is needed, the switching frequency is rarely low; hence, the bottom plate loss is often a significant contributor in the total loss. In topologies with multiple flying capacitors, each bottom plate contributes differently as M_{bott} coefficient values vary.

5.3.3.12. *Cooling requirement*

The cooling requirement could be difficult especially when the power density increases. For example, an 80% power-efficient SCC with 1 W/mm^2 power density dissipates 0.2 W/mm^2. The packaging has no constraints concerning heat removal requirements but sufficient heat draining must be in the layout through sufficiently large metal tracks up to the pads and pins of the package.

5.3.3.13. *Experimental characterization of an on-chip converter*

The on-chip SCC is fully integrated on a silicon die and is associated closely with the load. Therefore, it is sometimes hard to define a non-invasive way to monitor the performance and functionalities. A replica circuit or a bypass converter mode could be a solution to determine the power consumption of the powered circuit itself. The measured efficiency is calculated as the ratio of the power delivered with the SCC disabled (and the voltage delivered off-chip) to the power delivered with the SCC enabled.

5.3.3.14. *Output voltage regulation*

A well-regulated power supply is critical for the stability and performance of an IC system. One pressing issue is to provide a clean power rail within a tight tolerance for the loading circuit. The converter also needs to provide a clean output voltage, regardless of line and load transients. The abrupt change from light to heavy load occurs in a fraction of a microsecond. For example, Intel requires a voltage regulator with $+/-1\%$ steady-state voltage accuracy and less than 10% voltage spike at a few Amps per microsecond. The first method is to control the SCC itself by using different feedback methods. The regulation can be achieved by modulating the output impedance R_o and the unloaded conversion ratio, n (see Figure 5.3). The two feedback techniques can be combined to achieve a large output voltage range modulation while maintaining efficiency.

For a slight voltage gain variation, the output impedance is controlled by changing the on-state resistance, the flying capacitor value or the switching frequency. Then, the additional dropout across R_o is directly lost by the resistive element in the SCC. Outside its nominal ratio, the SCC acts as a linear regulator. The popular method is to modulate the switching frequency as proposed by [BAY 00 and SEE 08a].Only voltage-controlled oscillators such as a ring oscillator and a proportional-integrator controller are then needed. By reducing the switching frequency at lower output voltage, the lower output power delivery limits the switching loss, but increases the output voltage ripple. Then, a large output bypass is needed to maintain a quasi-constant output voltage. The other drawback is the limited control bandwidth to ensure feedback stability. In contrast, [GRE 06 and ZHU 96]suggest varying the switch conductance to modulate the SCC output impedance. The duty cycle could also be varied to modulate the voltage gain [KIM 05a]. In all solutions, the quiescent power consumption required to maintain a high control bandwidth results in further efficiency degradation at low output power delivery.

When a large voltage gain transient occurs, the unloaded conversion ratio is preferably changed to have a lossless regulation. However, n has a limited series of values (n = N:M, where N and M are integers), reducing the fine output voltage regulation. A high n increases the flying capacitor number and the switch bank, reducing nominal efficiencies at each N:M ratio. The ratio transition leads to an output voltage overshoot due to the charge and discharge of the flying capacitor bank.

An active clamp circuit can be an alternative solution by avoiding a high bandwidth SCC control. Designing low-area SCCs relies on a low capacitor value and a high switching frequency to achieve a sufficiently fast load transient response [WU 01]. The active clamp acts as a linear regulator in parallel with the SCC, which is capable of sourcing or sinking the full load current for a short period. In the steady state, the active clamps do not operate. The voltage drop is limited by the finite bandwidth of this circuit.

Other researchers outline a conservative strategy. As the linear regulator is already made to provide clean power rails in the existing digital cores, then why not use it as a post-regulation stage by connecting the linear regulator in series to the SCC. This allows high bandwidth control, but the major drawback is that it reduces the power efficiency. If the voltage drop between the LDO input and output voltages is low (e.g. a few hundreds of mV), then the linear regulator presents a relatively high efficiency as it is directly linked to this voltage ratio. Furthermore, the linear regulator does not significantly increase the surface required for the SCC.

5.3.3.15. *Step-up SCC topology*

Some applications require an on-chip voltage higher than the input voltage source such as an audio amplifier, electrostatic actuators or non-volatile memories. The Dickson architecture is well known to boost the voltage higher than the input voltage but the threshold voltage of diode or diode-connected transistors reduces the voltage gain at each stage [DIC 76]. Therefore, using a cross-coupled voltage doubler is an attractive solution at low input voltage [STA 01].

5.3.3.16. *Alternative topology: resonant SCC*

Even though research work on SCCs is progressing at a phenomenal rate, other topologies suitable for on-chip power management have been introduced. To avoid the intrinsic limitations of the SCC, a resonant SCC was introduced in 2000 and is based on the combination of the resonant converter and SCC [YEU 04, KIM 05b]. By including a very small inductor in the charging and discharging paths, the transfer loss is greatly reduced. The quality factor of the resonant tank is a key parameter in these topologies as it is directly linked to the transfer loss. A soft-switching behavior is also the main benefit of the resonant topology because it allows the switching devices to operate at zero current or voltage switching, thus reducing the switching loss. This feature allows the switching frequency to be increased and hence reduces the passive component values that form the resonator tank. Furthermore, the input current spike is reduced compared with the SCC, thus lowering the EMI issues at the PCB level. As the SCC is intrinsically exposed to the difference between the input and output voltages, the current spike is only limited by the on-state resistance of the switch. In a resonant converter, the switching mostly operates at zero current and hence smooths the current demand on the input source. This is one of the reasons that the SCC is limited to a small power rating, typically lower than a few Watts. [PIL 16] show the ability of the resonant topology to maintain a constant power efficiency for a large voltage gain range by different control methods. Modulating the phase duration, increasing the waiting phase or reconfiguring the switch topology [YEU 04] are lossless methods to change the voltage gain between the input source and output. These control approaches also allow step-up and inverting operations. However, the resonance creates additional stress on the flying component and the design of the low ESR inductor is critical. Nevertheless, recent results are promising for this method to be used for on-chip power management [SCH 15].

Combination with other converter topologies. To alleviate the major drawbacks of the SCC, the hybridization of this topology with inductive one has been proposed. This is the focus of Chapter 8.

5.4. Design example

As introduced earlier, complex system-on-chips require local, highly dynamic, dedicated power supply solutions. The number of independent power domains is very close to that of intellectual properties embedded in the SoC. A huge number of input–output resources would be spoiled only to deliver the power to the chip. It is now safe and more economically reasonable to consider a limited number of independent power lines tightened to the chip while the required power domains are generated internally.

This change of paradigm regarding power supplying calls for integrated converters with components compatible with the technology of the SoC (Figure 5.6). Switched-capacitor DC/DC converter is a good candidate.

The following sections give some details about the implementation of a SCC for dynamic voltage scaling application. The choice of the SCC is first argued. The principle of the SCC is described and the control-loop strategy is discussed. Silicon results are presented.

5.4.1. *Landscape of demonstrated solutions*

Reduction in the power consumption in advanced SoC is challenging. Power management techniques have significantly changed the design of

SoCs. The most promising strategy is massive DVFS to deliver power according to performance requirements. DVFS is widely used for the most demanding blocks in a SoC, such as CPU, but cannot be further extended to other blocks for various reasons. Traditional architectures are based on on-chip power supplies. Each domain is independently powered by a voltage regulator with low impedance connection. Increasing the number of power domains will increase the number of regulators and the board area. It becomes difficult to optimize routing on constrained board. The power delivery network is also impacted. The resources required for the power supply rails in the package and in the die are limited and have to be shared.

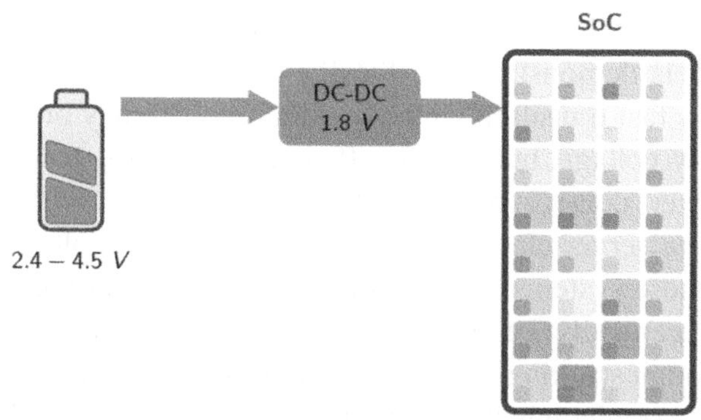

Figure 5.6. *New paradigm for SoC embedded power supply. For the color version of this figure, see www.iste.co.uk/allard/systems.zip*

This causes impedance degradation of the PDN. The performances and current consumption of SoCs continue to increase while the voltage tends to decrease to mitigate the power consumption. The impedance of the PDN must also decrease to keep the same performances [ERI 01]. Moreover, complex power delivery and bus speed limit the timing transition between different operating points.

Recent advances in magnetics [SUL 13] and packaging technologies offer new platforms with high integration levels. Power supply manufacturers integrate passive devices on the die (homogeneous architectures) or using SiP (heterogeneous architectures). A comparative analysis of commercial PMICs published recently is presented in [FOL 10]. Relative sizes of inductors tend to decrease while the switching frequency of DC/DC converters increases. Since the switching losses increase, a lot of efforts are made to improve or maintain high efficiency.

The TPS8267 series by Texas Instruments is an example of a commercial product where a 600 mA step-down DC/DC is integrated along with the inductor and input/output capacitors. A maximum efficiency of 90% has been reported as well as a quiescent current of 17 A during the PFM operation. Those PMICs significantly reduced the BOM and board area.

Fully or partially integrated voltage regulation appears as a suitable approach to increase the DVFS use. A recent growth of literature demonstrates the interest of such methods [AME 09, ZEN 10, GJA 11, KOS 12, VAI 13]. The linear voltage regulator is certainly the simplest approach since no passive devices are required. Such an approach is presented in [GJA 11, VAI 13] and a practical implementation for the POWER microprocessor is demonstrated in [TOP 14]. The comparator-based voltage regulator allows fast transition with a maximum efficiency of 90%. Efficiency decreases at low voltage, but is still sufficient to enhance power consumption saving. A power density of 34.5 W/mm^2 has been reported, which is appropriate for a server-class microprocessor. Per-core voltage generation is interesting in these systems in terms of cost and area, but thermal limitations and efficiency can only be improved by switching voltage regulators. 3D integration through stacking voltage regulator die above the IC is a possibility [ONI 07], but semiconductor manufacturers have also developed and integrated magnetic devices and high density capacitors in their process to enable inductive and capacitive voltage regulation on the die. Efficiency versus power density is one of the most frequently used performance metrics to compare DC/DC converters. Other figures of merit offer the possibility for a more detailed analysis [NEV 14, VIL 03, SAN 13].

A landscape of the most recent and relevant works on integrated conversion is depicted in Figure 5.7. The challenge is to achieve both high efficiency and high power density at an acceptable cost or in a manner compatible with the synthesis of the circuit to be powered. The usual common value for the SoC power density is 1 W/mm^2, but a small silicon area penalty would in fact require a density of 10 W/mm^2 for the DC/DC converter. The quality of the passive devices is essential. As expected, the highest performances are achieved by Intel Haswell architecture with magnetic core inductance, but the technology cost is unaffordable for many applications [BUR 14]. Some DC/DC converters using air-core inductors on package exhibit acceptable efficiencies of about 1 W/mm^2 [SCH 07, HAZ 05, STU 13, HUA 13, ISH 10, BER 09, SUN 07b], while those using on-die air-core inductors still suffer from lower power densities [KUD 11, WEN 11, WIB 08, ABE 06]. Switched-capacitor (SC) converters demonstrate a high efficiency at a density mainly limited by the density of the capacitors [LE 11, LE 10, LE 13, MEY 11, BRE 09, VAN 10, ELD 13, SOM 09, VIL 12, SEE 08b]. The limit in power density featured in [SOM 09] clearly calls for high density capacitors. Deep trench capacitors have been introduced and enable a peak efficiency of 85–90% up to 1 W/mm^2 [AND 14, AND 13b, CHA 10]. Architectures distributed across digital cores also bring a relevant point [JAI 14]. Flying capacitors are distributed over the digital core and the area penalty is limited to one of the power switches and the controller. Generally, SC converters present the best trade-off for low-cost monolithic integrated solution, which are being investigated in depth [SAN 13]. Resonant SC converters or cascaded SC converters with inductive converters are also explored to take advantage of both capacitors and inductors. Resonant SC converters use an inductor in series with the flying capacitor to produce a resonant energy transfer [STA 13, KES 14]. Efficiency can be improved for a given power density, but inductors remain costly. SC converters cascaded with inductive converters seem promising [PIL 08]. An unregulated SC converter is used to convert the high input voltage to a lower one. Then, an inductive converter achieves a fine regulation. This combination allows large voltage step-down and high bandwidth regulation.

5.4.2. *Selected architecture*

The starting point of the reflection is the consideration that a large digital IP is at least managed through power switches able to interconnect different voltage levels to the digital load or to cut it from power. A rule of thumb is that 10% of the IP silicon area is consumed by such power switches.

The SCC is located in this captured area (Figure 5.8 (top)). The distribution of power switches also calls for a distribution of elementary SCC cells. As stated earlier in this chapter, multi-phase switched-capacitor DC/DC enables the operating frequency to increase and to limit the requirement in capacitor total value for a given specification of output voltage range and peak output current.

The regulated output voltage is 0.2 V to 1.2 V from 1.8 V input. A density of 0.5 A/mm^2 is expected when using MIM and MOM capacitors. The landscape shown in Figure 5.7 has established the limitations introduced by such a technology choice. However, MIM and MOM capacitors associated with standard digital MOSFETs guarantee a compatibility of the SCC with the SoC.

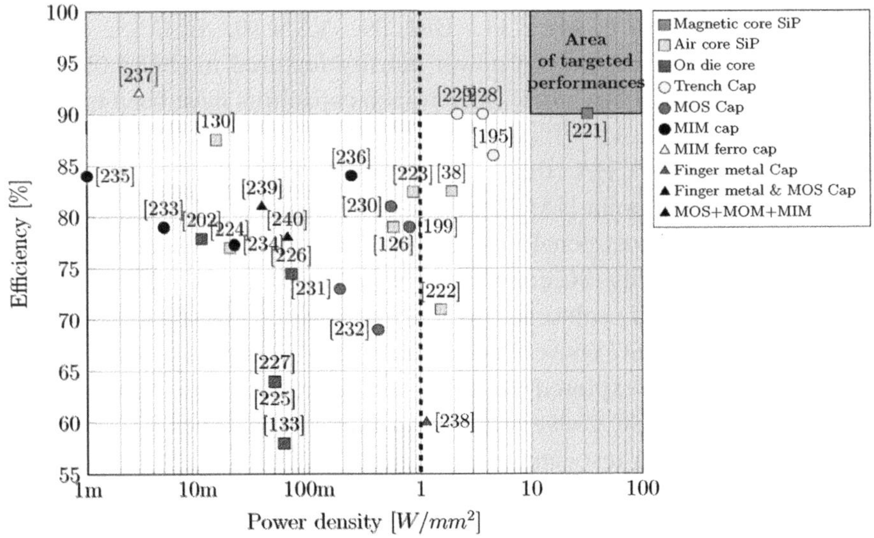

Figure 5.7. *Peak efficiency vs. power density of recent published integrated inductive and capacitive DC-DC. For the color version of this figure, see www.iste.co.uk/allard/systems.zip*

The elementary SCC cells will be connected in a ring fashion to ensure a natural multi-phase operation (Figure 5.8 (bottom)). A control IP is necessary, but will not be inserted in the reserved area as this control block may be shared or distributed in the digital IP.

As shown in Figure 5.3 and evaluated in [5.5], the output voltage is regulated on the SCC cell conversion ratio ($\frac{M}{N}$) and on the effective output resistance [5.26]:

$$R_{out} = f(C_{fly}; f_{sw}; R_{on;D}) \qquad [5.26]$$

A switched-capacitor converter, independent of the conversion ratio, presents a peak efficiency of typical small-arc form [5.6]. The desired output voltage range $[V_{out,min}; V_{out,max}]$, at peak efficiency or near this, requires a specific ratio of conversion for each value of the desired output voltage [5.5], i.e. reference voltage, V_{ref}, and taking into account the actual value of the input voltage, i.e. $M(V_{ref}, V_{in})$ and $N(V_{ref}, V_{in})$ in [5.5].

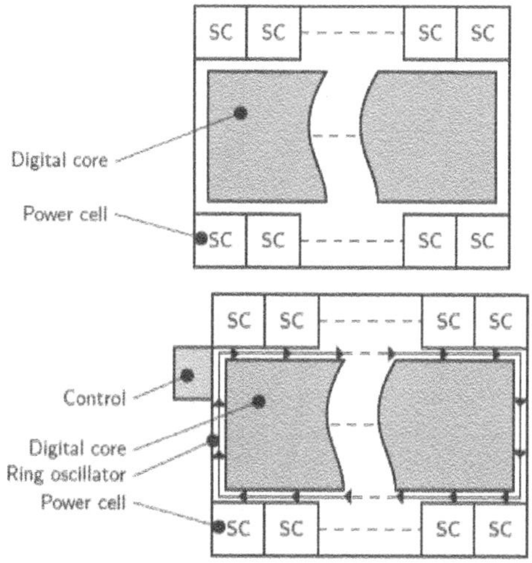

Figure 5.8. *Proposed architecture for the switched-capacitor converter. For the color version of this figure, see www.iste.co.uk/allard/systems.zip*

5.4.2.1. *Multi-ratio cell*

First, a multi-ratio cell is to be selected to cover the combined range in M and N. The literature is abundant on different methods to synthesize such a multi-ratio cell [JAI 14]. A gear-box scheme is then necessary to adapt the effective conversion ratio depending on the reference voltage tuning. The gear box is essentially a digital block, which is out of the scope of this chapter. As seen earlier in this chapter, the conversion ratio depends on a set of switches that are activated to control the flow of charges between the different capacitors. For a given reference voltage of the converter – hence a targeted regulated output voltage – an ideal ratio of conversion exists closest to the actual ratio of output-to-input voltages. The gear box is dynamic as the input voltage may vary as well as the reference voltage. Transient performances are involved, which mainly define the line-transient performances of the converter.

The multi-ratio cell shown in Figure 5.9 is introduced, which offers the set of topologies listed in Table 5.2. There is definitely a trade-off between the cell complexity and the number of intermediate output voltage values offered for a peak efficiency.

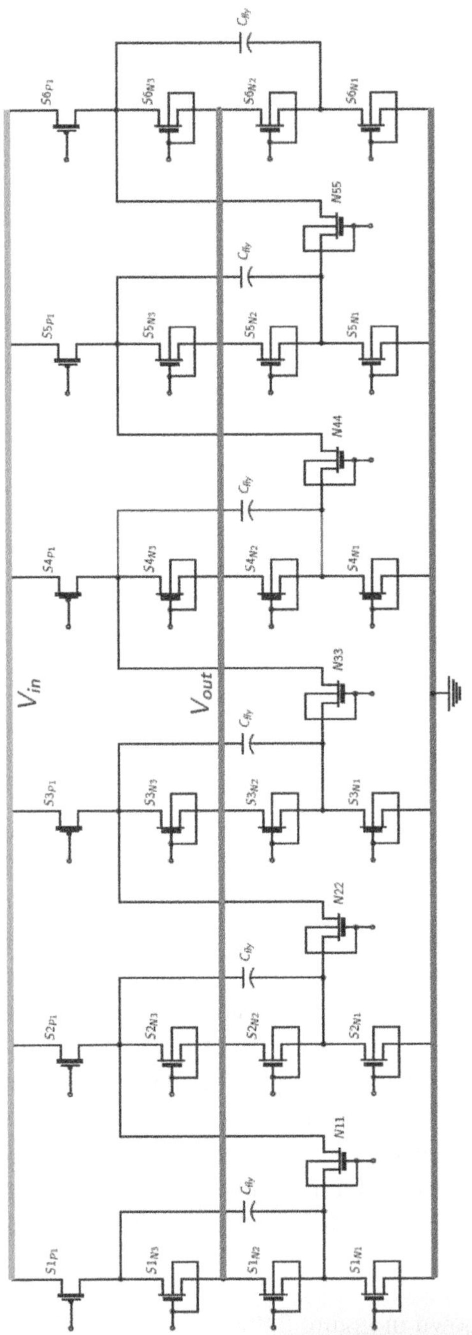

Figure 5.9. *Duplication of the multi-ratio cell in the ring arrangement. For the color version of this figure, see www.iste.co.uk/allard/systems.zip*

Ratio	3:1 $V_{out} < 0.6\ V$	2:1 $V_{out} < 0.9\ V$	3:2 $V_{out} < 1.2\ V$	4:3 $V_{out} < 1.35\ V$
Phase 1	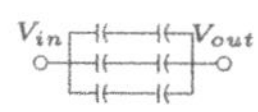	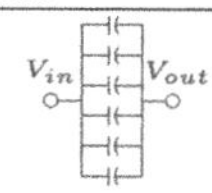	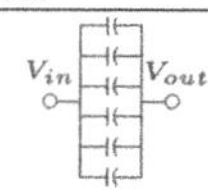	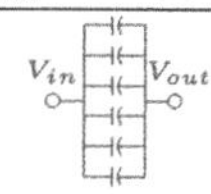
Phase 2	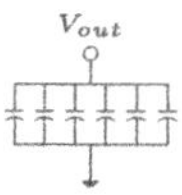	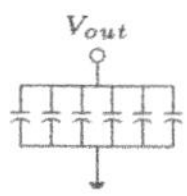	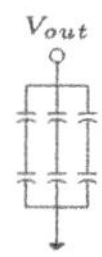	

Table 5.2. *Composition of topologies offered by the proposed multi-ratio cell*

5.4.2.2. *Frequency modulation*

When the ideal conversion ratio is selected, the output voltage requires a feedback control with respect to other impacting parameters such as the variability of manufacturing technology, the temperature and, in particular, the output current. Fine regulation is achieved by adequately varying the output resistance [5.26]. Obviously, the switching frequency is the most convenient electrical parameter as varying the value of a capacitor involves complex strategies, the switch on-resistance requires to control the gate voltage but the relation to R_{on} is highly nonlinear and the duty ratio, D, impacts electromagnetic conducted pollution.

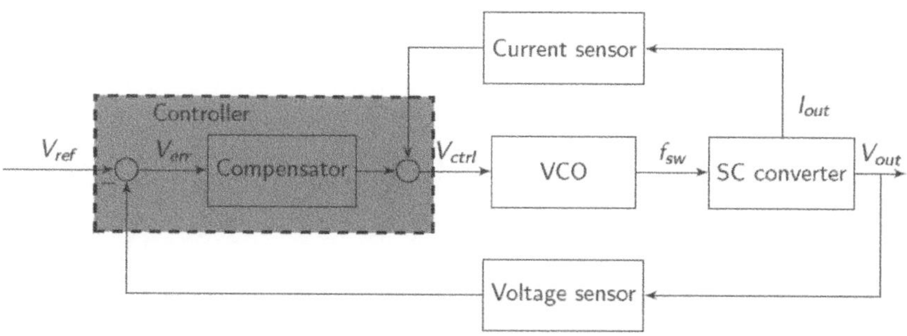

Figure 5.10. *Feedback loop schematic based on frequency modulation*

In Figure 5.10, the SCC is controlled by a voltage-controlled oscillator, the input voltage of which is set by a voltage controller. The voltage controller order depends on the system objective: second or third order in a closed loop.

6

High-Switching Frequency Inductive DC/DC Converters

An exercise prior to deciding on a topology of the DC/DC converter is to compare the capabilities of well-known architectures with respect to given specifications. A first step is then to define metrics that can cover the majority of converters without losing too much information. Transient aspects have been excluded from this analysis as their performances are mainly dependent on the test conditions. Transient considerations will be discussed later in this chapter. Thermal aspects are also not considered as the DC/DC converters studied present losses well below 10 W/cm^2, so thermal drain approaches are generally sufficient.

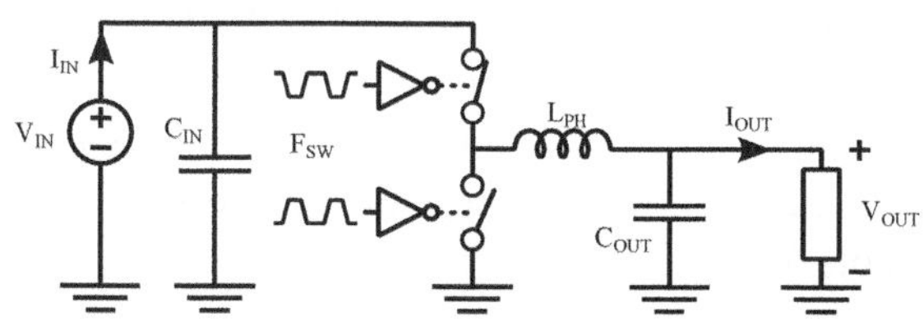

Figure 6.1. *Basic metrics for steady-state evaluation of a buck-type converter*

Chapter written by Christian MARTIN, Florian NEVEU and Bruno ALLARD.

Name	Symbol	Unit	Definition
Efficiency	η	%	P_{OUT}/P_{IN}
Switching frequency	F_{SW}	MHz	–
Technology node	–	nm	–
Input voltage	V_{IN}	V	–
Output voltage	V_{OUT}	V	–
Output current	I_{OUT}	mA	–
Conversion ratio	α	–	V_{OUT}/V_{IN}
Output power	P_{OUT}	mW	$V_{OUT}\times I_{OUT}$
Output capacitance	C_{OUT}	nF	–
Number of phases	N_{PH}	–	–
Phase inductance	L_{PH}	nH	–
Filter corner frequency	F_{LC}	GHz	$1/2\pi\sqrt{L_{PH}C_{OUT}}$

Table 6.1. *Metrics for the review*

6.1. Context and topologies

The fundamental metrics discussed are as follows: input and output voltages, input and output currents, output inductance and capacitance, and switching frequency. In addition, the technology node and the total converter area (when available) are also discussed. Additional metrics can be derived from basic metrics such as conversion ratio, output power and power conversion efficiency, or from even more intricate indicators such as the efficiency enhancement factor (EEF) defined in [WEN 08]. These metrics can be divided into three groups: functional specifications (input and output voltage, output current and technology node), performances (efficiency and area) and design parameters for the rest of the metrics. All the metrics are summarized in Table 7.1.

Once the metrics are identified and extracted from selected publications, they are plotted against each other, allowing us to highlight the possible trends and trade-offs. Figure 6.2 plots for each studied converter its output filter natural frequency (y-axis) versus its switching frequency (x-axis). Dots are parameterized with the number of phases of the converter. The plot clearly shows that increasing the switching frequency leads to an increase in the output filter natural frequency, thus reducing the component values of the output filter (output inductance and/or capacitance). Reducing the component values helps make them smaller ones, thus it is a step towards more integration (either in package or monolithically).

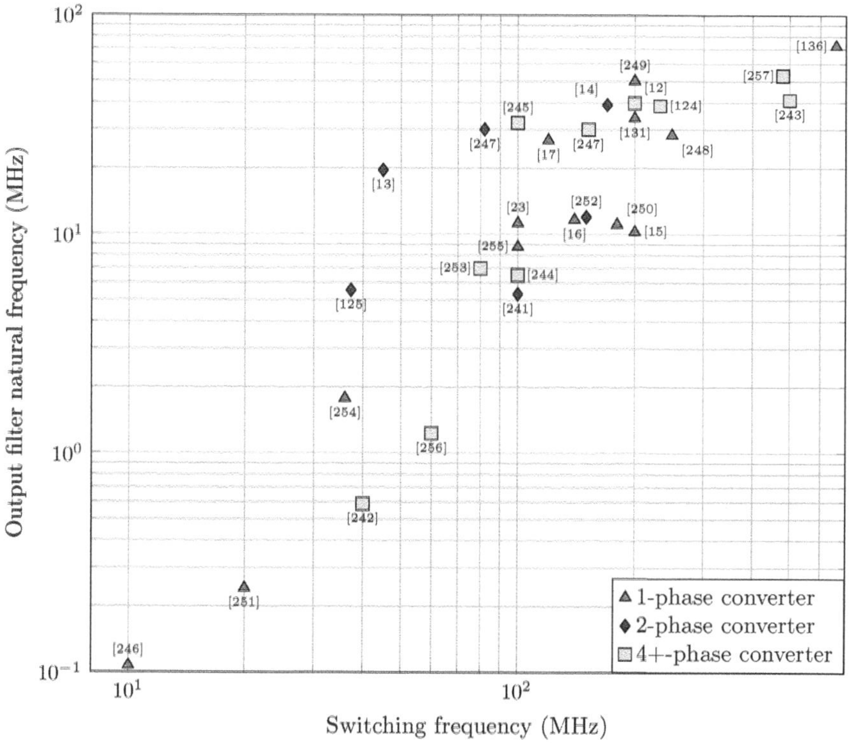

Figure 6.2. *Output filter natural frequency versus switching frequency. For the color version of this figure, see www.iste.co.uk/allard/systems.zip*

However, a higher switching frequency increases the switching losses, reducing the converter's efficiency. This impact is deduced from Figure 6.3, where the efficiency is plotted against the switching frequency. The dots are parameterized with the conversion ratio. For a given conversion ratio, the efficiency tends to drop when the switching frequency increases.

Another aspect of this plot is the impact of the conversion ratio on the efficiency at a given switching frequency: converters with a lower conversion ratio tend to have a lower efficiency. This trend can be explained by evaluating the efficiency gap between the converter and a hypothetical linear converter operating in the same conditions. In a first approximation, the efficiency of a linear converter is equal to the conversion ratio. However, the high efficiency of a given converter will make it appear further from the linear case as the conversion ratio is small.

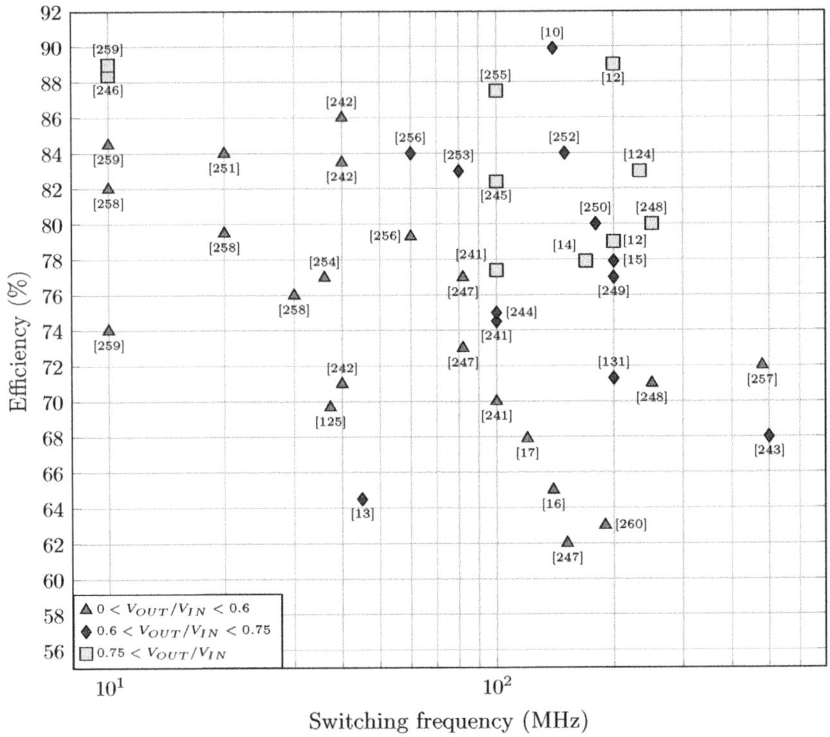

Figure 6.3. *Efficiency versus switching frequency, parameterized with conversion ratio. For the color version of this figure, see www.iste.co.uk/allard/systems.zip*

To compare between various converters, it becomes necessary to use a figure of merit (FoM), taking into account both the efficiency and the conversion ratio. The EEF calculation (%) is detailed in section 7.1.1.

Figure 6.4 plots the EEF of each converter (calculated according to Figure 6.2) against the switching frequency. The maximum EEF achieved tends to reduce as the switching frequency increases, confirming the negative impact of the switching frequency on conversion performance.

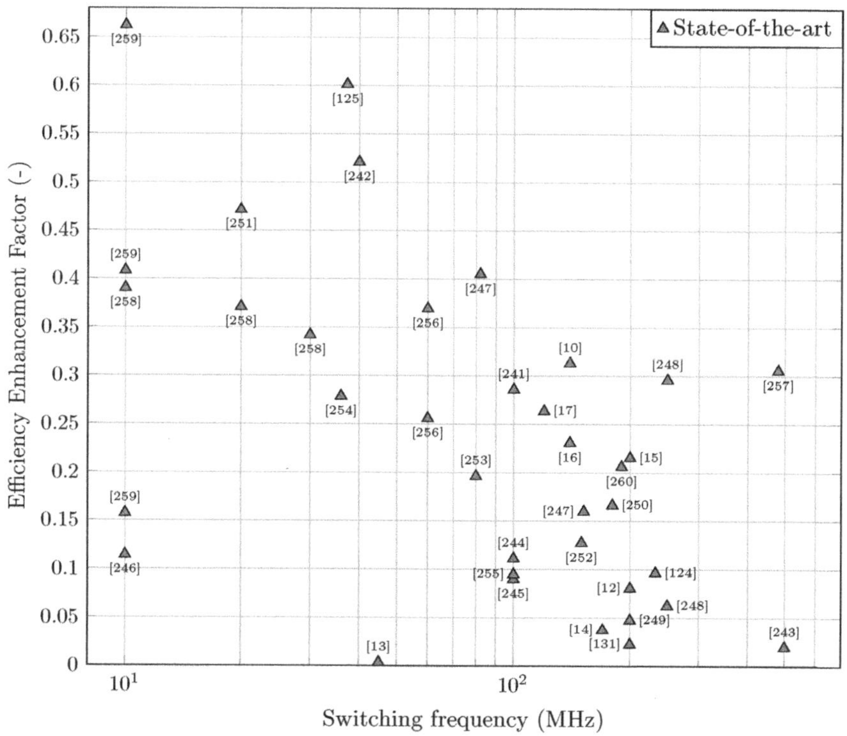

Figure 6.4. *EEF versus switching frequency*

The landscape in Figure 6.5 shows the efficiency of the converter with respect to the output power. Most of the converters target the power range of 100 to 1000 mW. When considering the impact of the output power on conversion efficiency, no trend can be identified. This means that the output power is not a decisive metric for power efficiency.

For integrated DC/DC converters, a crucial parameter is the manufacturing technology of active components.

Figures 6.6 and 6.7 illustrate how the technology has an impact on the converters. Figure 6.6 plots the efficiency of each converter against its manufacturing technology (of active components). The points are plotted with respect to the switching frequency as it impacts the efficiency. The global trend is that thinner technologies enable higher efficiencies. Furthermore, the use of more advanced technologies also allows for higher switching frequencies.

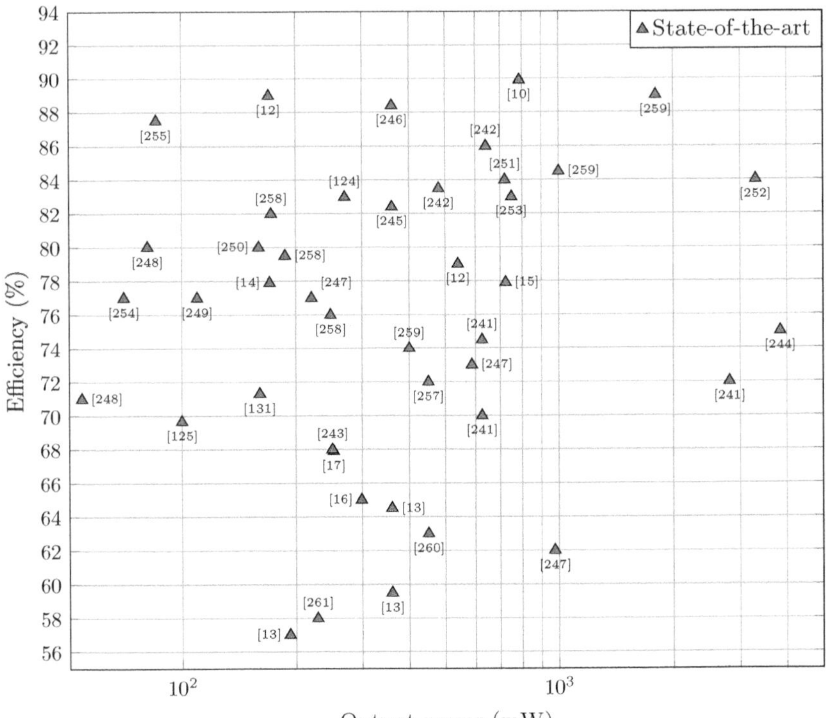

Figure 6.5. *Efficiency versus output power*

However, as the technology shrinks, converters tend to operate with a lower input voltage. Figure 6.7 plots the input voltage of the converters with respect to their manufacturing technology. With shrinking of the silicon technology, a notable trend is the decrease in the maximum value of input voltage. Transistors in advanced technologies have a shorter gate length and thinner gate oxide, thus the maximum operating voltage is reduced.

The major design trade-offs are as follows: for a given set of specifications (input and output voltages, output power), a high switching frequency is required to reduce the output filter (in terms of component values). If the step-down conversion ratio is small, it is hard to achieve high efficiency, especially if the switching frequency is really high. However, an advanced technology helps reduce the losses and achieve high efficiency, but challenges arise when the input voltage of the converter is higher than the nominal voltage of the technology.

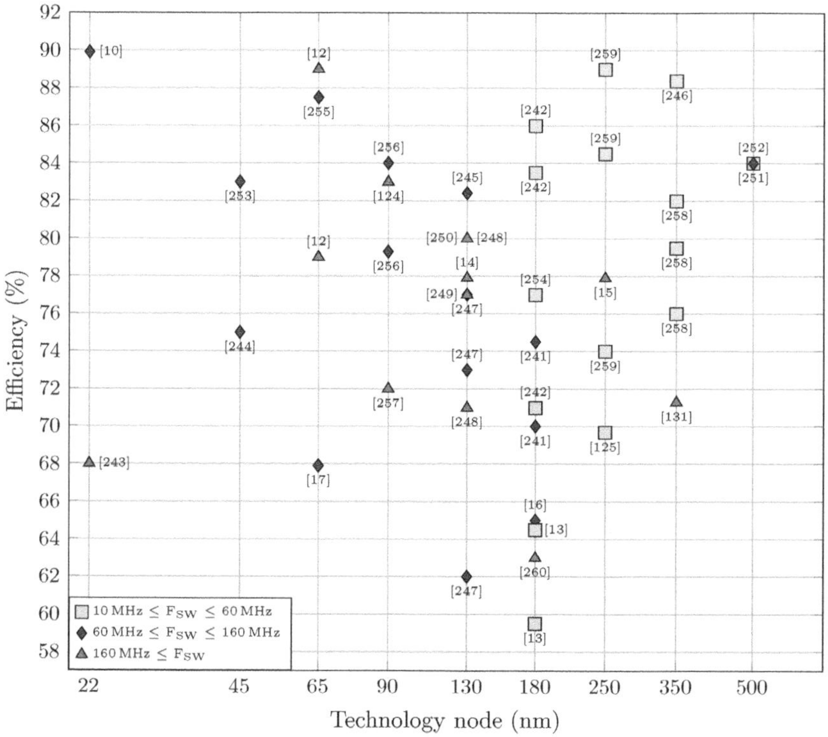

Figure 6.6. *Efficiency versus technology node, parameterized with switching frequency. For the color version of this figure, see www.iste.co.uk/allard/systems.zip*

6.1.1. *Discussion on figures of merit*

As mentioned previously, efficiency, while being widely used and easy to understand, may not be the most suitable indicator for comparison between various converters. In fact, efficiency is impacted by many parameters, but the most problematic one is the conversion ratio. Comparison of the efficiency between two converters that do not have the same conversion ratio becomes tricky. It then becomes necessary to use an indicator that combines both the efficiency and the conversion ratio.

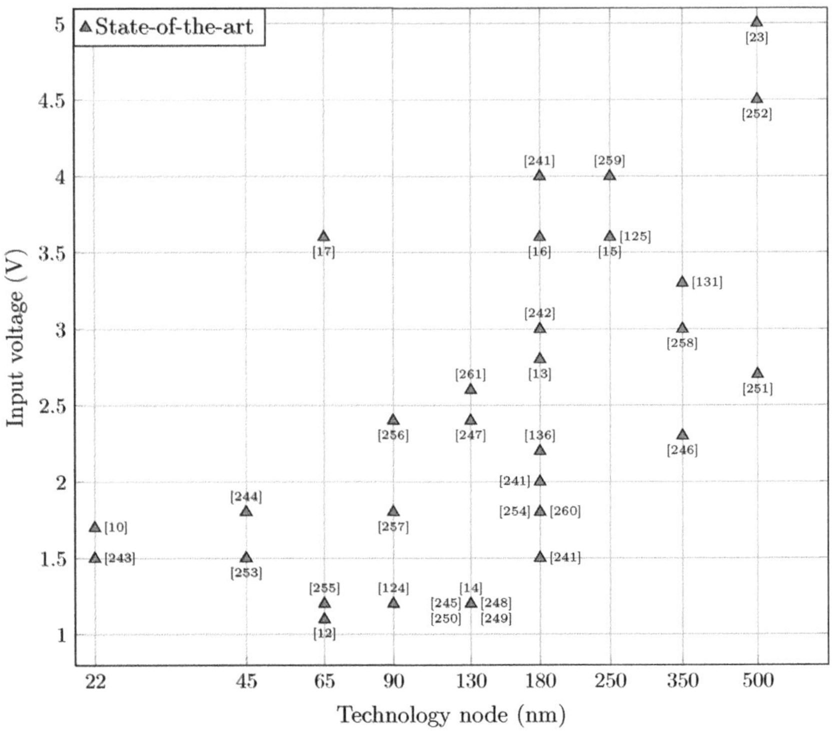

Figure 6.7. *Input voltage versus technology node*

6.1.1.1. *Efficiency enhancement factor*

The EEF has been proposed as a performance metric in [WEN 08]. The idea behind this metric is to compare the performance of the actual converter with that of a hypothetical linear converter operating in the same conditions, i.e. at the same input and output voltages, output current and conversion ratio. The conversion ratios of the switching converter and the linear converter are denoted by k_{SW} and k_{lin}, respectively, which are equal. The efficiency of the switching converter is denoted η_{SW}; For the hypothetical linear converter, it is denoted η_{lin}, which is equal to k_{lin}. The EEF is defined as the input power difference between the two converters (ΔP_{IN}), divided by the input power of the hypothetical linear converter (P_{IN_lin}). The equation is given by:

$$EEF = \frac{\Delta P_{IN}}{P_{IN\_lin}} = \frac{P_{IN\_lin} - P_{IN\_SW}}{P_{IN\_lin}} = 1 - \frac{\frac{P_{OUT\_SW}}{\eta_{SW}}}{\frac{P_{OUT\_lin}}{\eta_{lin}}}, \quad [6.1]$$

and as $\eta_{lin} = k_{lin} = k_{SW}$ and $P_{OUT\_lin} = P_{OUT\_SW}$, we get

$$EEF = 1 - \frac{k_{SW}}{\eta_{SW}}. \quad [6.2]$$

The value of the EEF is always less than 1 and has no minimal low value (the EEF tends to $-\infty$ when the efficiency of the DC/DC converter tends to 0). A positive EEF means that the switching converter is more efficient than its linear counterpart, and a negative EEF means that it is not.

However, this FoM is limited as it tends to favor the converter with a small conversion ratio. When considering [6.2], we see that the maximum value depends on the conversion ratio. For instance, a converter with a conversion ratio of 0.8 cannot have an EEF greater than 0.2. The EEF is calculated for efficiency results presented in [CHE 14a] in Figure 6.2. For the same converter, the EEF can vary greatly, depending on which point is studied. Furthermore EEF is not a straightfoward FoM to select a converter.

6.1.1.2. *Losses reduction factor*

The losses reduction factor (LRF) is proposed here as a new FoM that is used to address the shortcomings of the EEF. It is defined as the ratio of the losses of the hypothetical linear converter (as defined in section 6.1.1.1) to the losses of the converter. The details of its calculation are given in [6.4]:

$$LRF = \frac{P_{IN,lin} - P_{OUT}}{P_{IN,SW} - P_{OUT}} = \frac{\dfrac{P_{OUT}}{\eta_{lin}} - P_{OUT}}{\dfrac{P_{OUT}}{\eta_{SW}} - P_{OUT}}$$

$$= \frac{\dfrac{1}{\eta_{lin}} - 1}{\dfrac{1}{\eta_{SW}} - 1} = \frac{\dfrac{1 - \eta_{lin}}{\eta_{lin}}}{\dfrac{1 - \eta_{SW}}{\eta_{SW}}} = \frac{\eta_{SW}}{\eta_{lin}} \cdot \frac{1 - \eta_{lin}}{1 - \eta_{SW}} \quad [6.3]$$

$$LRF = \frac{\eta_{SW}}{k_{lin}} \cdot \frac{1 - k_{lin}}{1 - \eta_{SW}} \quad [6.4]$$

The LRF can range from 0 (not included) to $+\infty$, and is not limited by the conversion ratio. It tends to 0 when the efficiency of the actual DC/DC converter is close to 0. When the LRF is greater than 1, it means that the efficiency of the switching converter is better than its linear counterpart. LRF directly indicates the factor by which the losses will be cut when the DC/DC converter studied is used instead of the linear one. As it is a factor, it needs to be seen in a logarithmic scale. LRF is calculated for efficiency results presented in [CHE 14a] in Figure 6.2. LRF exhibits the low conversion ratio values, but as this factor is better seen in a logarithmic scale, it is better than the EEF for the converter with a high conversion ratio. Nevertheless, it is still hard to compare between various converters using the LRF as it still favors converters with a small conversion ratio.

6.1.1.3. *Predicted efficiency*

It is easy to compare two converters already working in the same conditions. Given the efficiency at an operating point, it would be good if it were possible to predict the efficiency at another operating point (for instance, at another conversion ratio). It is necessary to consider loss mechanisms in a DC/DC converter to determine the impact of a different conversion ratio. We assume that the conversion ratio is modified by varying only the output voltage, i.e. the duty cycle of the converter. This means that all the other relevant values (i.e. input voltage, switching frequency and load current) are considered constant.

In a switching DC/DC converter, there are switching losses and conduction losses. Switching losses depend on the switching frequency and the input voltage, while conduction losses depend on the root mean square (RMS) current [GIL 02]. The RMS value contains two contributing parameters: the load current and the current ripple. Of these parameters, only the current ripple depends on the duty cycle. If we suppose that the contribution of the current ripple is relatively small, we can deduce that the losses are not going to vary greatly when changing the output voltage, i.e. the conversion ratio. It is then possible to calculate a "predicted" efficiency (η_{pr}) at any given output voltage if the efficiency at one operating point (η_{op}) is available. The "predicted" efficiency estimates the efficiency of the converter if it operates at an interpolated output voltage ($V_{OUT,pr}$), delivering the associated output power ($P_{OUT,pr} = V_{OUT,pr} \times I_{OUT}$). The "predicted" efficiency η_{pr} is defined in [6.5]:

$$\eta_{pr} = \frac{P_{OUT,pr}}{P_{loss} + P_{OUT,pr}} \qquad [6.5]$$

Power and losses are then defined as follows, using the reported operating conversion ratio (k_{op}) and the "predicted" output voltage ($V_{OUT,pr}$) and its associated "predicted" conversion ratio (k_{pr}), assuming a constant input voltage:

$$P_{OUT,pr} = \frac{P_{OUT,op}}{V_{OUT,op}} \cdot V_{OUT,pr} \qquad [6.6]$$

$$P_{loss} = \frac{P_{OUT,op}}{\eta_{op}} - P_{OUT,op} \qquad [6.7]$$

$$\frac{V_{OUT,pr}}{V_{OUT,op}} = \frac{k_{pr}}{k_{op}} \qquad [6.8]$$

We can then calculate the value of η_{pr} as follows:

$$\eta_{pr} = \frac{\dfrac{P_{OUT,op} \cdot V_{OUT,pr}}{V_{OUT,op}}}{\dfrac{P_{OUT,op}}{\eta_{op}} - P_{OUT,op} + \dfrac{P_{OUT,op} \cdot V_{OUT,pr}}{V_{OUT,op}}} \qquad [6.9]$$

$$\eta_{pr} = \frac{\dfrac{V_{OUT,pr}}{V_{OUT,op}}}{\dfrac{1-\eta_{op}}{\eta_{op}} + \dfrac{V_{OUT,pr}}{V_{OUT,op}}} \qquad [6.10]$$

$$\eta_{pr} = \frac{k_{pr}}{k_{op}} \cdot \frac{1}{\dfrac{1-\eta_{op}}{\eta_{op}} + \dfrac{k_{pr}}{k_{op}}} \qquad [6.11]$$

$$\eta_{pr} = \frac{\eta_{op} \cdot k_{pr}}{k_{op} - \eta_{op}\left(k_{op} - k_{pr}\right)} \qquad [6.12]$$

Table 6.2 shows the predicted efficiency values calculated for each reported point. The calculation does show a rather good prediction (less than 1% error). The use of the predicted efficiency will allow us to obtain a consistent comparison between converters at the same conversion ratio.

V_{IN}	3.3V	3.3V	3.3V
V_{OUT}	2.4V	1.8V	1.2V
Conversion ratio (k_{op})	0.72727	0.54545	0.36364
Reported efficiency (η_{op})	91.8%	88.6%	84.3%
EEF	0.2078	0.3844	0.5702
LRF	9.614	6.477	4.199
Predicted efficiency (η_{pr}) at 2.4V output	91.8%	91.20%	91.48%
Predicted efficiency (η_{pr}) at 1.8V output	89.36%	88.6%	88.96%
Predicted efficiency (η_{pr}) at 1.2V output	84.84%	83.82%	84.3%

Table 6.2. *EEF, LRF and predicted efficiency calculations on the results published in [CHE 14a]*

6.1.2. *Outstanding state-of-the-art performances*

Based on results presented above, we can see that some converters follow the major trends. The underlying factors of this fact are discussed below.

When considering the filter frequency, some converters use very small output filters with respect to their switching frequency. For instance, the converter presented in [ABE 07] has a 20 mHz filter corner frequency for a 45 mHz switching frequency. An on-chip inductor (with low inductance density) and hysteretic control are used to maintain a relatively low current ripple. However, the efficiency achieved is not very high, which is only 64% when converting 2.8 V to 1.8 V. The low efficiency value can be explained by the small inductor, leading to a high current ripple and thus resulting in an important RMS current through the switches.

In terms of efficiency, the converter presented in [BUR 14] shows very good performances. It operates at a high frequency (140 mHz) with a significant conversion ratio (0.62) and achieves about 90% conversion efficiency. It achieves these performances by taking the benefit from the performances of a 22 nm FinFET CMOS technology. The power stage is designed with two MOSFETs in series, allowing the use of a very short length transistor with thin oxide. In terms of assembly, it is highly integrated. Power inductors are air-core devices (with no magnetic losses) that are embedded in package, and capacitors are on-chip MIM capacitors.

In terms of voltage rating, depicted in Figure 6.7, the converter depicted in [OST 14] emerges. It withstands a 3.6 V input voltage with a 65 nm bulk complementary metaloxidesemiconductor (CMOS) technology. To achieve this, it uses two MOSFETs in a cascode configuration as a single switch. The transistors are 2.5 V MOSFETs with thick oxide.

The state-of-the-art has already addressed some challenges, but further improvements are still expected, for instance pushing further the trade-off between efficiency and the switching frequency, and reducing the footprint of the passive components, which is still a big challenge for these converters.

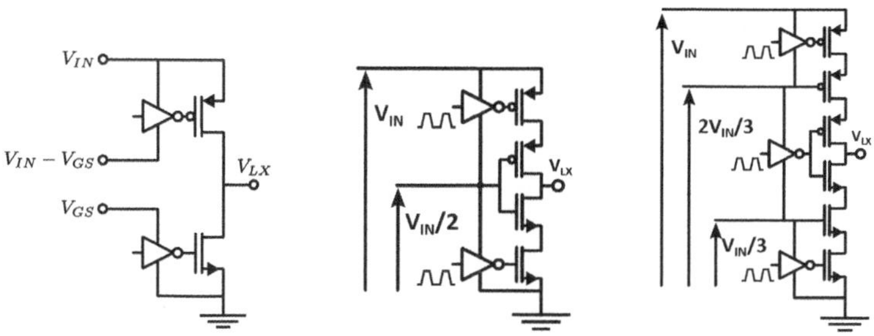

Figure 6.8. *Typical power stage configurations: standard (left), two-stage cascode (center), three-stage cascode (right)*

6.2. Cascode power stage

The choice for a cascode configuration power stage stands as an alternative to the standard power stage, either with complementary transistors or with dual N-transistors, when the value of the input voltage is higher than the nominal drain-to-source voltage of the transistor. Figure 6.8 shows three options. The details of the limitations of the standard power stage at high switching frequencies have been provided in [BAT 12b]. The two-transistor cascode configuration has been reported in [OST 14], [BUR 14]. Good results have been experimentally demonstrated and the conclusion is favorable to the cascode behavior. The configuration is generic and can theoretically be extended to a large number of transistors due to the systematic arrangements [SCH 04]. The three-transistor cascode power stage is shown in Figure 6.8 (right): physical implementation was first reported by Serneels *et al.* in [SER 05, SER 07] in the context of a high-voltage driver. Recently, it was demonstrated in the context of a class-D amplifier [SER 12]. In addition, the capability of the three-transistor cascode power stage was not demonstrated at high switching frequencies.

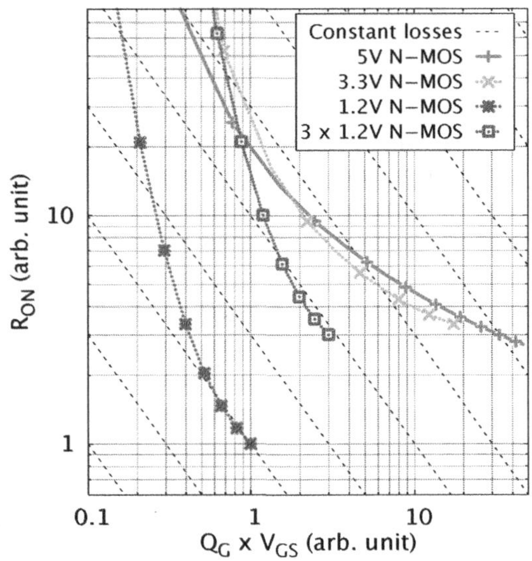

Figure 6.9. *FoM for transistor options in a standard CMOS 40 nm node. For the color version of this figure, see www.iste.co.uk/allard/systems.zip*

The selected technology for silicon implementation (Table 6.1) must be considered here as a case where the nominal technology voltage (1.2 V) is not consistent with the input voltage (3.3 V). High-voltage transistors may be designed as lateral-drain structures (LDMOS) or extended-drain structures (EDMOS). Transistors have an impact on the power stage configuration in terms of losses and silicon area. Secondary issues include safe-operating areas, sensitivity to temperature or electromagnetic interference. Losses are related to static on-state resistance and parasitic capacitances for given operating conditions (voltage, current, frequency, temperature). Depending on the channel width of a transistor, it is possible to evaluate the losses for given operating conditions. In nominal operating conditions for the targeted power stage, it is then easy to compare the impact of various transistor options depending on the channel width or the silicon area. Figure 6.9 shows iso-value curves of losses in the (R_{on}, $Q_G.V_{GS}$) plane, which are indicated by dashed lines. Lower losses are towards the bottom left corner. The FoM of several transistor options are plotted. As expected, a standard transistor features better in terms of losses and silicon area but not in terms of the targeted input voltage. The 5 V and 3.3 V transistor options are readily available to stand the input voltage, but will not help reach the efficiency target. Fortunately, stacking three standard transistors offers a better trade-off, hence the choice for the cascode power stage.

MOSFET polarization has been added to the power stage as many voltages appear inside the power stage and level shifters.

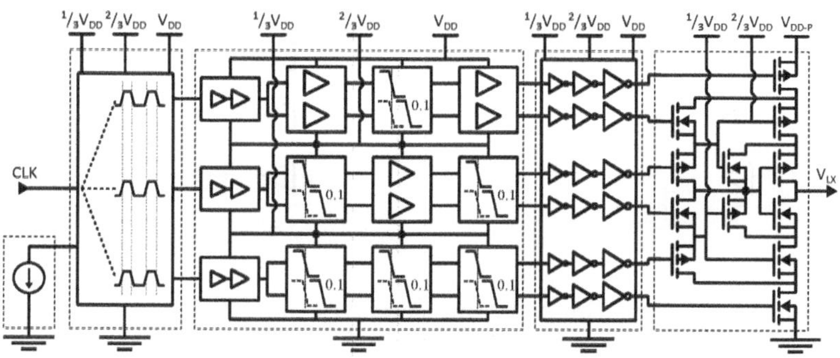

Figure 6.10. *Schematic of the power stage*

Transistors in the power stage cannot stand more than 1.2 V, as shown by a systematic arrangement Figure 6.10. The three-floating drivers in shown in Figure 6.8 (right) are distributed on two rows of local drivers, shared between the three transistors in the stack. This design is scalable to any odd number of transistors in the stack. Local intermediate voltage levels required are 1/n of the transistor number in the stack. The generation of these local voltage levels as well as the associated efficiency are discussed at the end of this chapter. The power stage requires three driving signals with perfect synchronization. Skew within these signals is the key to the robustness of the power stage. At 100 MHz (10 ns period), an accuracy of less than 100 ps has to be considered. Otherwise, a transistor in the arrangement may have to stand a voltage higher than the nominal rating, thereby inducing stress and aging. The layout of the power stage must comply with the following objectives: synchronous signals should be propagated from the global control input to the drivers' inputs through dedicated level shifters to take care of the local change in voltage swings.

The power stage is designed in a first step for continuous conduction mode (CCM) as the objective is to demonstrate the capability of a cascode configuration and the benefit of the 2.5D assembly. In Figure 6.11, the color code represents the voltage levels from input voltage to the ground, respectively, in red, blue, green and gray. In CCM operation, Figure 6.11 shows the mapping of voltages when the output voltage is, respectively, high (left) or low (right).

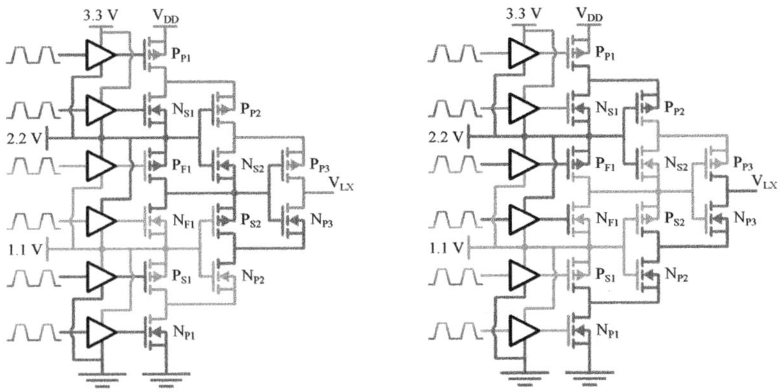

Figure 6.11. *Operation of the two-stage cascode in CCM: high output voltage (left), low output voltage (right). For the color version of this figure, see www.iste.co.uk/allard/systems.zip*

Active-well transistor is mandatory to cancel the polarization of parasitic junctions in the cascode power stage (Figure 6.12, left and center). The final layout is shown in Figure 6.12 (right) where the power stage is optimized at a 150 mA output current and 100 MHz operation. A standard power stage with 3.3 V transistors is also designed according to similar specifications and with similar care for the purpose of comparison.

The full power stage requires four decoupling capacitors for the various input voltages. These capacitors, as well as the output filter capacitor, are located off-chip to the passive interposer where the active die and the inductor will also take place. The values of decoupling capacitors and the induct value are optimized with respect to the accuracy constraints of the output current ripple and the output voltage. A proper decoupling of the main input voltage requires 33 nF, accounting for an ESR of 100 mΩ. Other decoupling capacitors are set to 11 nF, accounting for an ESR of 150 mΩ. Finally, the output filter capacitor is set to 16 nF, with a maximum ESR value of 150 mΩ followed by an inductor of 60 nH (reference PFL1005-60NMRU).

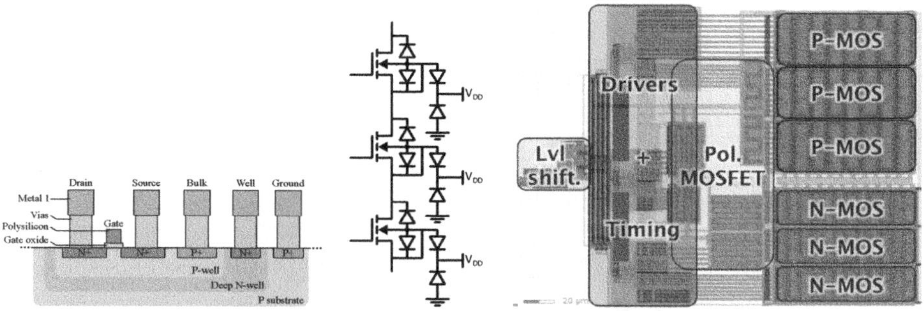

Figure 6.12. *Cross-section of the active-well transistor (left), three-transistor cascode parasitic junctions (center) and layout of the power stage (right). For the color version of this figure, see www.iste.co.uk/allard/systems.zip*

6.3. High-quality decoupling

In [BAT 12a] is demonstrated a single stage of a buck converter for a 200 MHz operation has been demonstrated. Passive components are placed off-chip. The results indicate that dumping resistors have to be implemented on-chip to kill oscillations introduced by parasitic devices (L and C) between the off-chip devices and the on-chip power stage. The authors report a penalty of at least 2 points of efficiency. It is of utmost importance to reduce the length of interconnection between decoupling capacitors and the power stage. The active die has been tested in a chip-on-board configuration with on-board passive devices. Figure 6.13 shows the lengthy connections accounting for large parasitic devices.

Here, an interposer was manufactured by IPDiA to accommodate all decoupling capacitors, the output filter capacitor and all landing pads necessary for the flip-chipping connection of the active die and the inductor [LAL 13], Figure 6.14. Table 6.3 lists the performances of the embedded capacitors.

The manufactured converter is shown in Figure 6.15 (left). For the purpose of comparison, the sample is connected to a board through wire bonding. No extra passive devices are added. The standard power stage and the cascode power stage are experimentally compared to explain the benefit of the interposer, on the one hand, and the benefit of a cascode stage, on the other hand. The details of the measurements are presented in section 6.5.3.

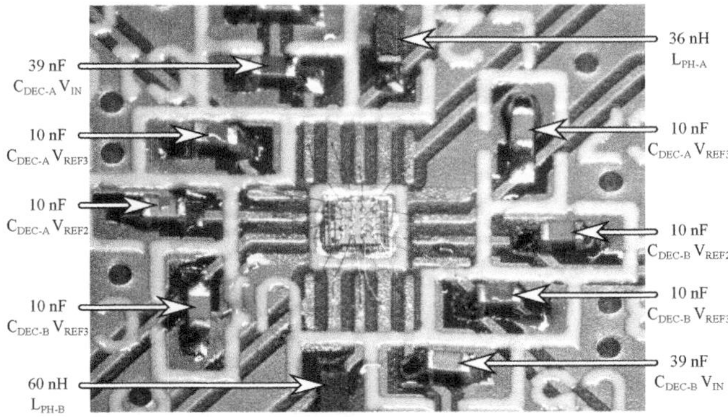

Figure 6.13. *Active die with off-chip passive devices on board. For the color version of this figure, see www.iste.co.uk/allard/systems.zip*

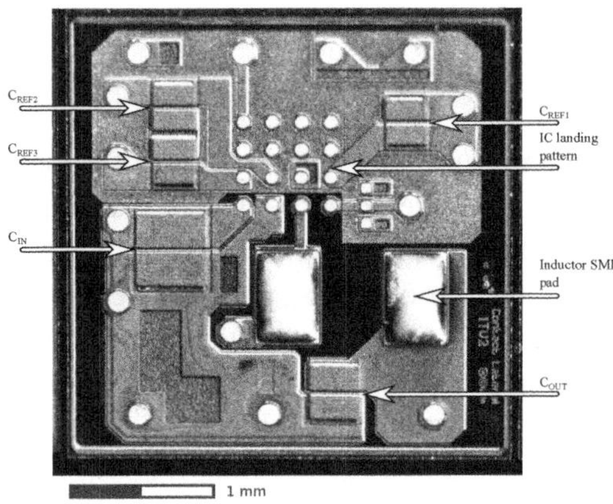

Figure 6.14. *Capacitive interposer manufactured by IPDiA [LAL 13]. For the color version of this figure, see www.iste.co.uk/allard/systems.zip*

	C_{MEAS} (nF)	ESR ($m\Omega$)	SRF (MHz)
C_{IN}	33	70	330
C_{OUT}	16	110	490
$C_{REF1,2,3}$	11	150	630

Table 6.3. *Measured parameters of designed capacitor structures*

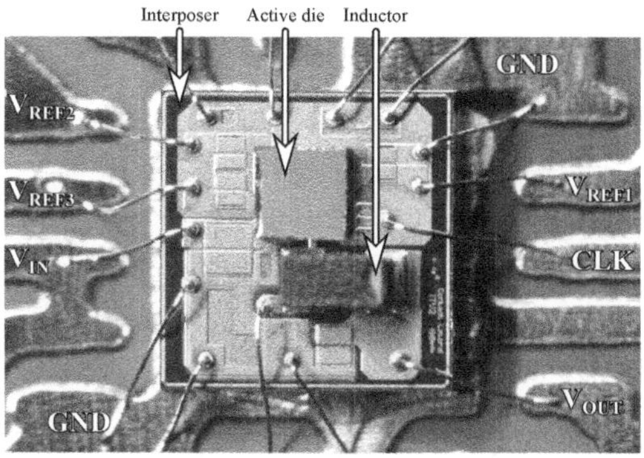

Figure 6.15. *Converter sample (left) and on-board assembly (right). For the color version of this figure, see www.iste.co.uk/allard/systems.zip*

Measurements are not easy due to the compactness of the assembly. For the sake of internal signal sensing, the interposer in shown in Figure 6.14 is not in optimal geometry. Figure 6.16 pictures a schematic of the model of the main parasitic devices used in the assembly when probing the output voltage of the power stage (V_{LX}) and the output voltage of the converter (V_{OUT}). Unfortunately, this system of parasitic devices creates ringings and interferences in several configurations. Thus, all signals could not be measured. As an example, Figure 6.17 shows the experimental measurement at V_{out} node with or without the connection of a probe to the V_{LX} node.

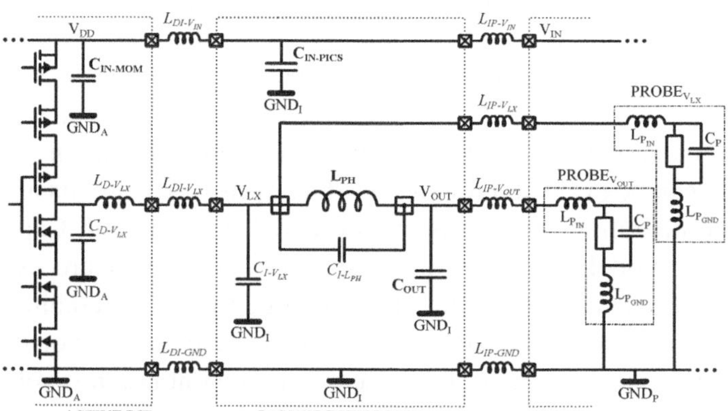

Figure 6.16. *Equivalent circuit to the 2.5D converter from the viewpoint of the parasitic device*

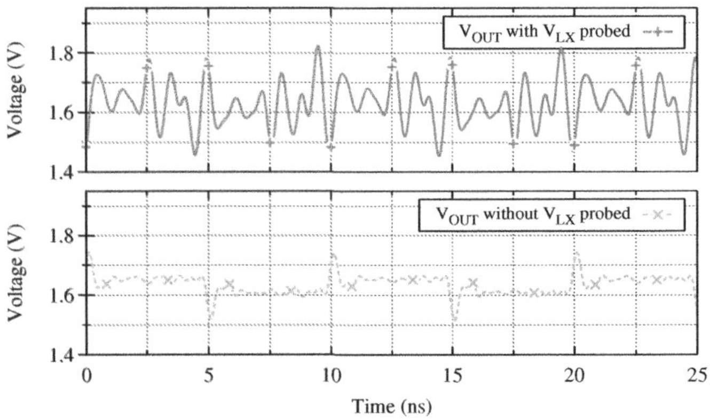

Figure 6.17. *Measurements of the output voltage in two sensing configurations*

6.4. Design considerations for passive components

This section focuses on the design considerations for passive components. This includes an in-depth study of the converter architecture, modeling of the losses of active components and dimensioning of the output filter. Different selected architectures (one-phase buck, two-phase buck, coupled two-phase buck and three-level buck) have been simulated in the conditions listed in Table 6.4. Simulation has been performed using standard L, C, R switches and source elements in a SPICE simulator. For this study, we consider an ideal switch which implies that both cascode and standard power stages have an equivalent behavior to that of a classical switching cell. RMS current values for each structure given by both analytical equations and circuit simulations can be evaluated for various duty cycles. All circuits are evaluated in the same operating conditions, as presented in Table 6.4. Capacitors and inductors are modeled by parasitic resistance values that are estimated by considering the datasheet. As the goal is to compare the architectures, this choice is not critical. Moreover, we show that the relative error between the results of analytical equations and those of circuit simulation is very small.

Figure 6.18 shows the contributions to the losses of all components in each converter architecture. Red indicates the inductor losses, green the output capacitor losses, blue the input capacitor losses and purple the flying capacitor losses (only for a three-level converter). For the coupled-inductor converter, the coupling factor is supposed to be optimal for each duty cycle. The output capacitor losses are very small compared with the other losses. Except for the one-phase buck, they are negligible. This means that the parasitic resistor of the output capacitor is not critical in terms of losses. The inductor losses are fairly constant whatever the architecture. They are also the dominant losses. This implies that the choice of the inductor plays a crucial role in the converter performance. A value of 5 $m\Omega/nH$ has been chosen for single-phase converters, and 10 $m\Omega/nH$ for two-phase converters. Coupling reduces the inductor losses to some extent by decreasing the current ripple and thus the RMS current. This is mainly evident when the duty cycle comes close to 0.5. The input capacitor losses significantly decrease in two-phase converters. This is one of the major benefits of the multi-phase approach. The three-level converter shows significantly higher losses than the other architectures because of the flying capacitor. As detailed in the last chapter, the three-level converter is a topology apt to reduce the current ripple by a significant factor through the inductor compared with a one-phase buck converter. To benefit from the reduced current ripple, we use a much smaller inductor, leading also to a significantly reduced resistance, thereby diminishing inductor losses. However, the gain would still not counterbalance the important losses due to the flying capacitor. The best potential architecture in terms of efficiency is then the two-phase converter with coupled inductors. It has the advantages of the two-phase converter with a reduced phase current ripple.

Name	Value	Unit
V_{IN}	3.3	V
I_{OUT}	280	mA
T	5	ns
L_{PH}	30	nH
$R_{L_{PH},1-phase}$	150	mΩ
$R_{L_{PH},2-phase}$	300	mΩ
$R_{C_{IN}}$	250	mΩ
$R_{C_{OUT}}$	250	mΩ
$R_{C_{FLY}}$	150	mΩ

Table 6.4. *Conditions for evaluation of architectures*

Regarding losses, the coupled two-phase converter is the structure of interest. Figure 6.19 shows the evolution of the optimum coupling factor with the duty cycle. When the factor is negative, the inductors are coupled in opposite polarities. It quickly drops to -1 when the duty cycle tends to 0.5. For a relatively low duty cycle, the optimum coupling factor is not very high.

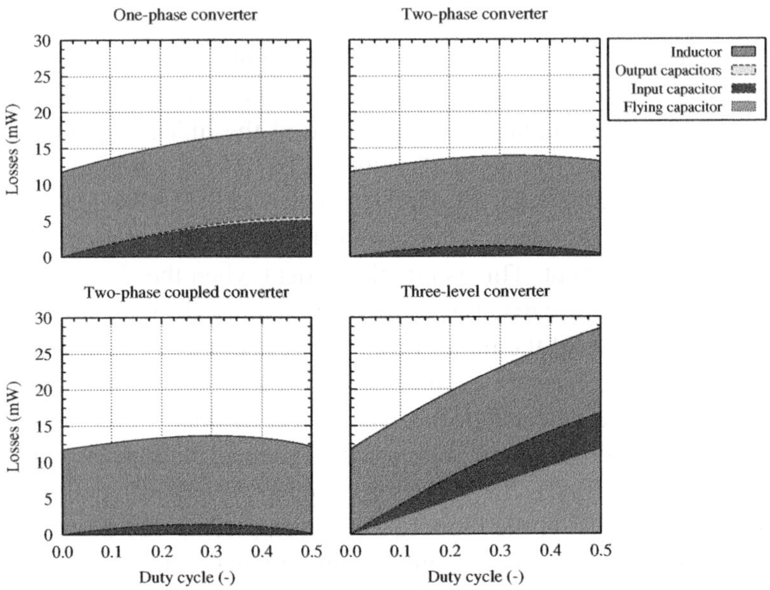

Figure 6.18. *Breakdown of losses in the architectures evaluated, for a duty cycle varying from 0 to 0.5. For the color version of this figure, see www.iste.co.uk/allard/systems.zip*

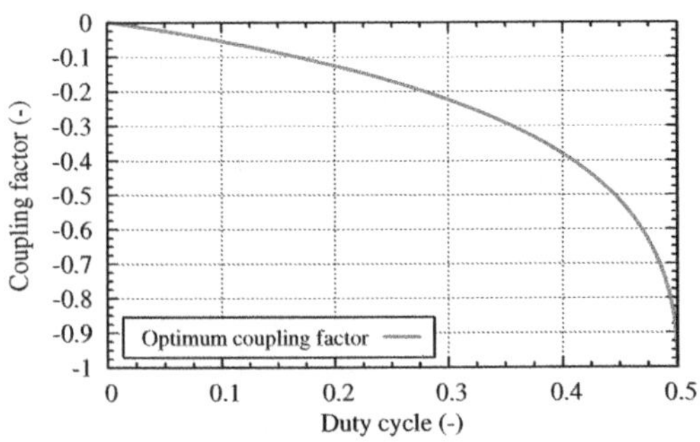

Figure 6.19. *Optimum coupling factor for a coupled two-phase buck converter, for a duty cycle varying from 0 to 0.5*

Figure 6.20 shows the evolution of the current ripple at different coupling factors. We see that the current ripple decreases until it reaches the optimum coupling factor, and then it increases to a very high value; at the $k = -1$ limit, it tends to infinity. Thus, having a coupling factor close to 1 is not desirable at all, even if it is the optimum coupling factor. The coupling factor is unstable in terms of current ripple if the duty cycle is not 0.5. In conclusion, the coupling of the inductors is beneficial for reducing the current ripple value, but the coupling factor must be chosen wisely in order to avoid instability and to considerably increase the current ripple.

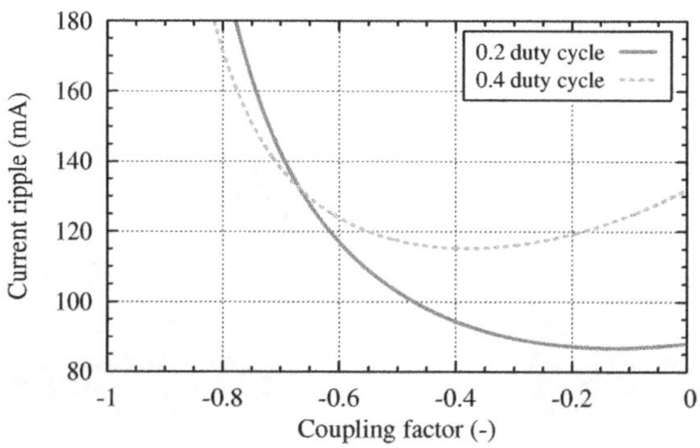

Figure 6.20. *Current ripple for various duty cycles, for a coupling factor varying from 0 to* -1

6.5. Integrated inductor characterization

This section covers the harmonic characterization of various integrated magnetic devices and time-domain characterization.

6.5.1. *Harmonic characterization*

An Agilent 4294A impedance meter is used for the stand-alone tests of passive components. It features a four-point measurement with a bandwidth up to 110 MHz. In addition to the impedance meter, probes are added to get contacts to the passive components. True-Kelvin coaxial probes are used for four-point measurements up to 110 MHz. Probes are shielded very close to contact points, allowing for high frequency measurements. Several integrated inductor structures have been designed and then characterized under probes using the impedance meter. Micro-photographs of two single inductors fabricated by Tyndall are depicted in Figure 6.21. Characterization of the single-phase structures is shown in Figures 6.22(a) and 6.22(b) for the 20 nH and 30 nH inductors, respectively. The frequency behavior for the two structures is quite similar. The inductance values of both devices are approximately –10% lower than the design values. This error can be explained by a variation in the thickness of the magnetic material. The DC resistance of the 20 nH inductor is 80 mΩ and that of the 30 nH inductor is 120 mΩ, which gives a value of 4 $m\Omega/nH$ for both structures. The inductance value starts to decrease by about 4 MHz and the resistance increases simultaneously. The inductance value is still significant at 110 MHz, about 85% of the low frequency value.

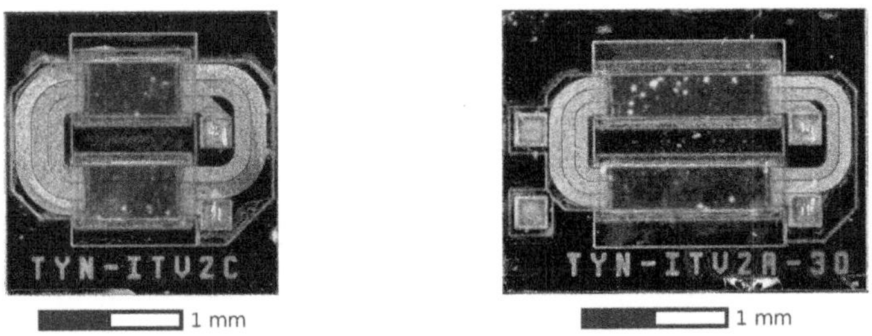

(a) 20 nH inductor

(b) 30 nH inductor

Figure 6.21. *Micro-photographs of single inductors. For the color version of this figure, see www.iste.co.uk/allard/systems.zip*

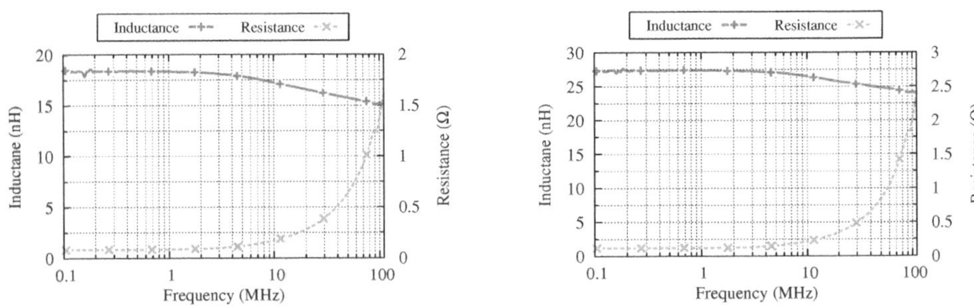

(a) 20 nH inductor

(b) 30 nH inductor

Figure 6.22. *Harmonic characterization of the inductors. For the color version of this figure, see www.iste.co.uk/allard/systems.zip*

Commercial inductor devices are also measured (Murata – PFL1005-36NMRU and PFL1005-60NMRU), as their inductance value is given for a frequency of about 8 MHz in the manufacturer datasheet and only DC resistance is available. Characterization gives more accurate data on these devices. Measurement results for the 36 nH and 60 nH inductors are shown in Figures 6.23(a) and 6.23(b), respectively. These inductors have a low DC resistance (30 mΩ for the 36 nH inductor and 44 mΩ for the 60 nH inductor). Their DC resistance (about 0.8 mΩ/nH) is much lower than the values of Tyndall devices, but the volume of the commercial inductors is higher. The inductance value of these inductors at low fequency is slightly higher than their nominal value, but their specified inductance value is about 10 MHz, as expected. The resistance rises faster with frequency than in the Tyndall devices, meaning that the magnetic material/process used in the commercial inductor is less suitable for high frequency operations.

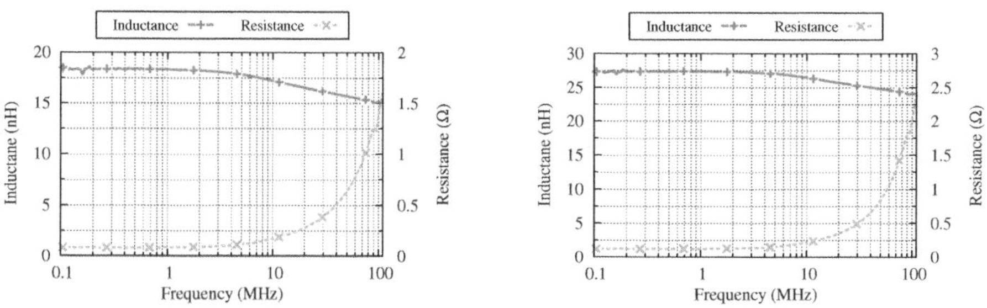

(a) 20 nH inductor

(b) 30 nH inductor

Figure 6.23. *Harmonic characterization of the commercial inductors. For the color version of this figure, see www.iste.co.uk/allard/systems.zip*

6.5.2. *Time-domain characterization*

Inductor structures are tested in time domain to evaluate how they would behave in conditions similar to those of a buck output filter. The power stage is emulated using a wideband amplifier with a high output current capability (TI THS4302 amplifier) and both a capacitor and a resistor are added at the output. The capacitor value is 20 nF and the resistance load is 10 ω. A schematic of the test circuit is shown in Figure 6.24. The current through the inductor is measured using the CT6 probe. Input (V_{LX}) and output (V_{OUT}) voltages at the inductor terminals are measured as close as possible to the inductor to obtain the true voltage across the inductor. For testing coupled inductors, the same capacitor and resistor are used, and a power stage emulator and current sense are added.

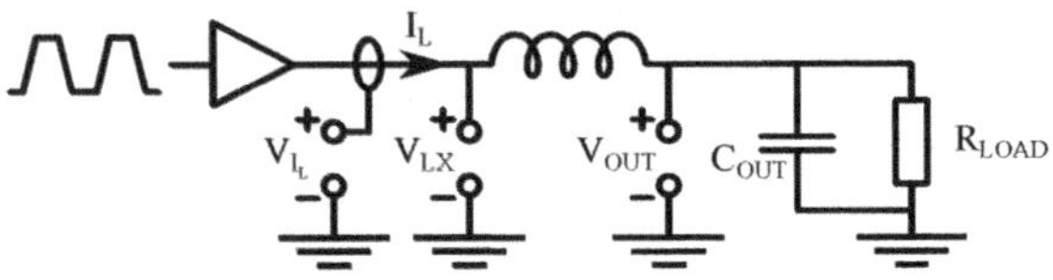

Figure 6.24. *Schematic of the test circuit for the inductors*

Despite being fairly simple in principle, voltage measurements performed on this test bench are quite difficult, especially at high frequencies. Parasitic inductances combined with capacitance at the input of voltage probes create resonant loops. As the inductor is wire-bonded to the test board, true voltage measurements across the inductor are not possible. Input and output bonding wires are included in this voltage measurement. A picture of the inductor and the probes is presented in Figure 6.25.

Figure 6.25. *Picture of the test bench for the inductors. For the color version of this figure, see www.iste.co.uk/allard/systems.zip*

From this measurement, it is possible to compute the DC output current ($\frac{V_{OUT}}{R_{OUT}}$) and the voltage across the inductor terminals ($V_{LX} - V_{OUT}$, denoted by V_L). The computation of the output current is necessary to obtain the full current through the inductor as the CT6 probe only provides AC measurements. Based on these measurements (current and voltage across the inductor), it is possible to calculate an equivalent inductance and resistance value.

This so-called "LR" identification is performed using an optimization loop. The current is calculated using L, R values by the numerical integration of the voltage waveform. It is also calculated using a recursive function, defined in [6.13], for a set of I_0, L and R parameters:

$$\begin{cases} I_L[0] = I_{L,0} \\ I_L[n] = I_L[n-1] + \Delta I_L[n] \\ \Delta I_L[n] = \dfrac{\Delta T}{L}\left(\dfrac{V_L[n] + V_L[n-1]}{2} - R \times I_L[n-1]\right) \end{cases} \quad [6.13]$$

The equation then yields a calculated current waveform, which is compared with the measured waveform using the mean square distance. This process is iterated over a range of I_0, L and R values until the mean square distance is minimized. The result of this identification process is presented in Figure 6.26. The calculated current after identification matches well the measured current. An inductance value of 33.4 nH and a resistance of 2.23 ω are obtained. The inductance value is higher (approximately +8 nH) than the value measured with the impedance meter. This is explained by the fact that in the test circuit, some copper traces and the bonding wires located between the two voltage probes are included in the inductor measurement. Copper traces and bonding wires of few millimeters are enough to create an error of about 10 nH.

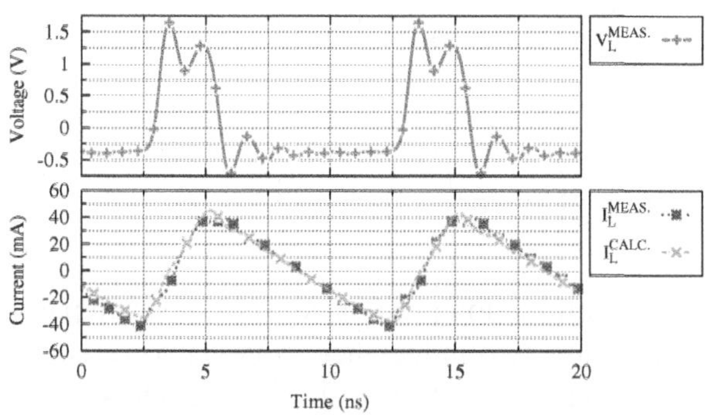

Figure 6.26. *Measured and calculated waveforms for the 30 nH inductor at 100 mHz, 25% duty cycle. For the color version of this figure, see www.iste.co.uk/allard/systems.zip*

Frequency (MHz)	L (nH)	R (ω)
5	35.5	0.7
10	34.5	0.9
20	33.9	1.5
50	33.0	1.9
100	33.1	2.2

Table 6.5. *Summary of the identification results at various frequencies for a 30 nH single inductor*

Similarly, the coupled inductors are measured and identified. In this case, two interdependent currents are measured, as well as two V_{LX} voltages. Currents are calculated using recursive formulas, presented in [6.14]:

$$\begin{cases} I_{L1}[0] = I_{L1,0} \\ I_{L2}[0] = I_{L2,0} \\ I_{L1}[n] = I_{L1}[n-1] + \Delta I_{L1}[n] \\ I_{L2}[n] = I_{L2}[n-1] + \Delta I_{L2}[n] \\ \Delta I_{L1}[n] = \dfrac{\Delta T}{(1-k^2)\,L_1}\left(\dfrac{V_{L1}[n]+V_{L1}[n-1]}{2} - R_1 \times I_{L1}[n-1]\right) \\ \qquad +k\sqrt{\dfrac{L_2}{L_1}}\left(\dfrac{V_{L2}[n]+V_{L2}[n-1]}{2} - R_2 \times I_{L2}[n-1]\right) \\ \Delta I_{L2}[n] = \dfrac{\Delta T}{(1-k^2)\,L_2}\left(\dfrac{V_{L2}[n]+V_{L2}[n-1]}{2} - R_2 \times I_{L2}[n-1]\right) \\ \qquad +k\sqrt{\dfrac{L_1}{L_2}}\left(\dfrac{V_{L1}[n]+V_{L1}[n-1]}{2} - R_1 \times I_{L1}[n-1]\right) \end{cases} \quad [6.14]$$

In this case, the identification process consists of seven parameters, namely two initial current values, two inductance values, two resistance values and a coupling factor. This makes the identification process less straightforward than in the case of a single inductor. The results of the identification of coupled inductors are presented in Figure 6.27. The comparison between the measured and calculated results are obtained.

Identification results for single and coupled inductors are presented for various frequencies in Tables 6.5 and 6.6, respectively. The results are found to be consistent with harmonic characterization, especially for the single inductor. For coupled inductors, when the frequency increases, inductance values tend to increase, which is not an expected behavior. However, identification is not very effective at high frequencies. The coefficient of determination (R^2) between the calculated and measured current waveforms is below 0.9 at 100 MHz for coupled inductors, meaning that identification may not be very accurate. This is due to the nonlinearities of magnetic inductors that are not taken into account during the identification process.

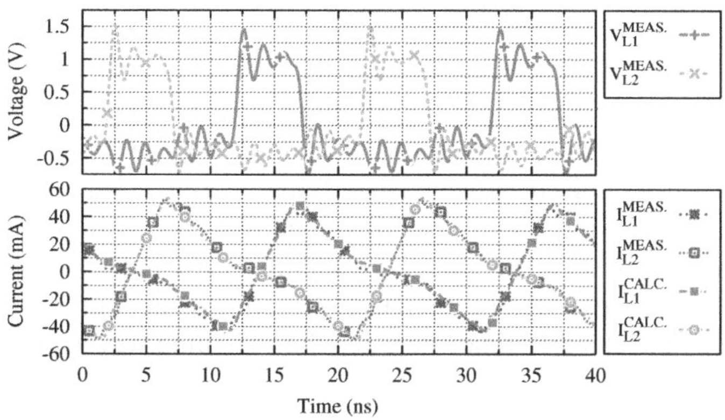

Figure 6.27. *Measured and calculated waveforms for a 45nH coupled inductor at 50MHz and 25% duty cycle. For the color version of this figure, see www.iste.co.uk/allard/systems.zip*

Freq. (MHz)	L_{BOT} (nH)	R_{BOT} (ω)	L_{TOP} (nH)	R_{TOP} (ω)	k (-)
5	47.9	1.0	44.5	0.8	0.33
10	47.5	1.1	43.7	1.0	0.33
20	47.0	2.0	46.5	1.7	0.32
50	49.2	2.4	45.0	1.5	0.31
100	53.4	3.0	48.7	1.8	0.31

Table 6.6. *Summary of the identification results at various frequencies for a 45nH coupled inductors*

The identification gives information on the coupling factor, which is about 0.33. Since parasitic inductances are added on each phase (due to copper traces and bonding wires), the true coupling factor value is somewhat higher than that found from the identification process. As it is designed to be about 0.4, measurements are consistent.

These time-domain measurements show that harmonic measurements and simple LR model of the inductor are sufficient. This validates the model used for the specification and design of the output filter.

6.5.3. *Converter experimental characterization*

The verification of the operation of the cascode power stage is an important concern. Synchronicity and skew are key parameters for the robustness of the power stage. So far, only an experimental observation is carried out on observable signals. As mentioned previously, not all of them may be sensed without detrimental interference. The voltage at V_{OUT} node has been monitored in various operating conditions for both the standard and cascode power stages on the interposer with the same inductor value (60 nH) and switching frequency (100 MHz). As shown in Figure 6.28, it is impossible to detect a difference between the switching waveforms.

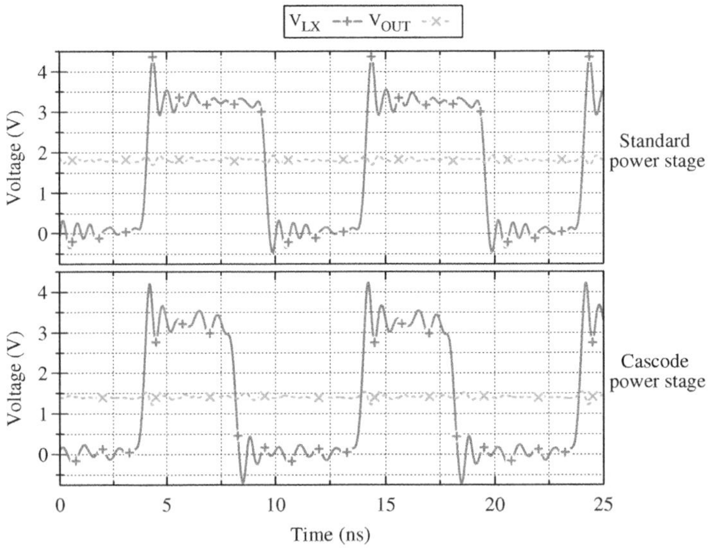

Figure 6.28. *Behavior of the standard and cascode power stages on the interposer (V_{IN} = 3.3 V, V_{OUT} = 1.8 V, I_{OUT} = 150 mA, f_{sw} = 100 MHz). For the color version of this figure, see www.iste.co.uk/allard/systems.zip*

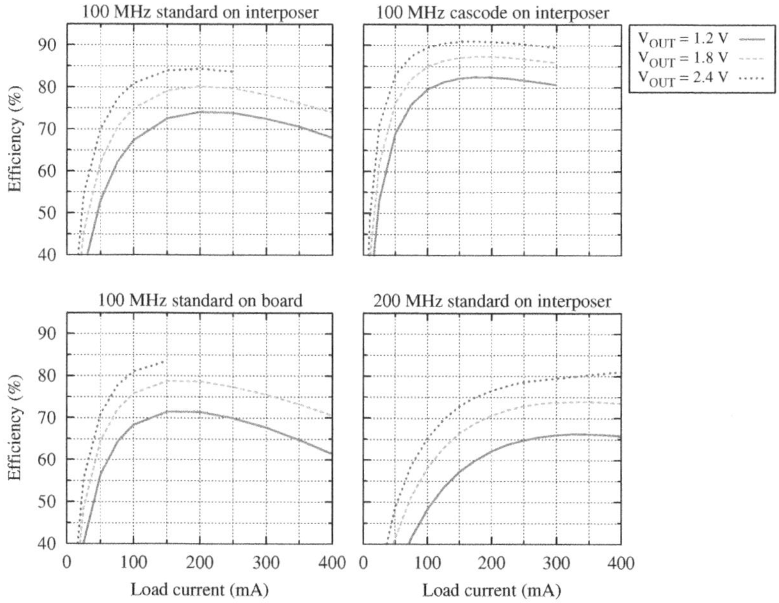

Figure 6.29. *Efficiency of standard and cascode power stages on board or on the interposer. For the color version of this figure, see www.iste.co.uk/allard/systems.zip*

Efficiency is measured from input and output power. The standard power stage is supplied under 3.3 V and its the drivers have separate input voltages. A mechanism of charge recycling between the drivers of the cascode power stage has not been explained; however, the result is that the currents drawn from the intermediate input voltages are very low. Figure 6.29 details some of the major results. The first column compares the performance of the standard power stage onboard and on the interposer. It is obvious that in similar operating conditions, the interposer offers a benefit on efficiency of at least 4 points. This is mainly due to the reduction in interconnection parasitics. Ringing on output voltage node must be very low and no particular precautions is to be taken inside the active part to dump the situation. The first row in Figure 6.29 shows that the cascode power stage offers superior performances than its standard counterpart. Approximately 10 points of efficiency have been gained. They are repaid by some complexity in the design but the silicon penalty is affordable (+0.006 mm^2, i.e. +14%). This efficiency budget can be used to operate the cascode power stage at 200 MHz with performances still comparable to that of the standard power stage at half the switching frequency. The converter footprint would be smaller if smaller passive component values were selected to keep the output current and voltage ripples equal.

The results shown in Figure 6.29 are given in terms of losses shown in Figure 6.30. Losses in the drivers and total losses from input voltages are separated for both power stages. Post-layout simulation (PLS) results are also included. Breakdown of losses in the standard and cascode power stages reflect the trend given globally by efficiency values. The PLS results are in good agreement with measurements for drivers but not for the other losses. In this case, the PLS results do not take into account the interconnection parasitics.

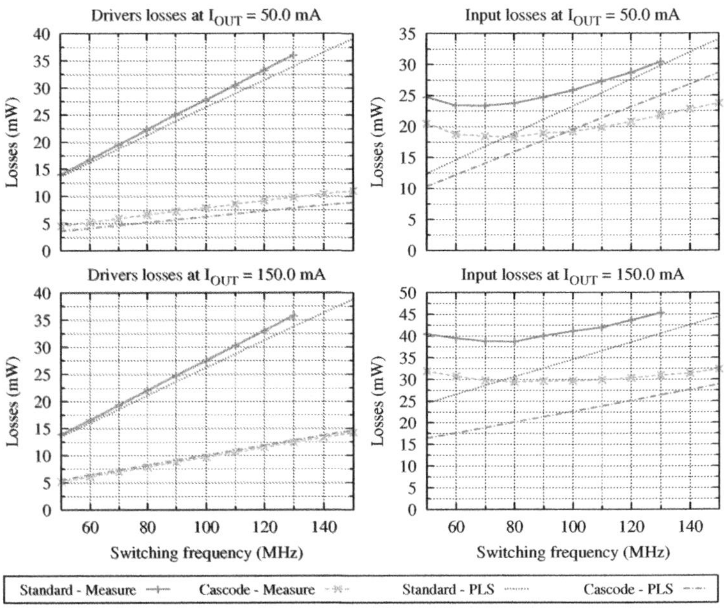

Figure 6.30. *Simulated and measured losses for standard and cascode power stages. For the color version of this figure, see www.iste.co.uk/allard/systems.zip*

Finally, it is interesting to consider the sensitivity of the cascode power stage to operating conditions, i.e. for values slightly different from those considered for optimization (nominal current, nominal output reference voltage, nominal frequency). Figure 6.31 (left-hand side) shows that the switching frequency does not significantly affect the efficiency what is not the case of the input-to-output voltage ratio. Figure 6.31 (right-hand side) indicates that the optimization of the power transistor silicon area is very sensitive. Drivers' losses remain insensitive to the output current and the efficiency drops heavily apart from the optimization point. This corroborates the interest for multi-phase architectures where the maximum output current is handled over the full number of phases and phase-shedding is considered for lower values of the output current.

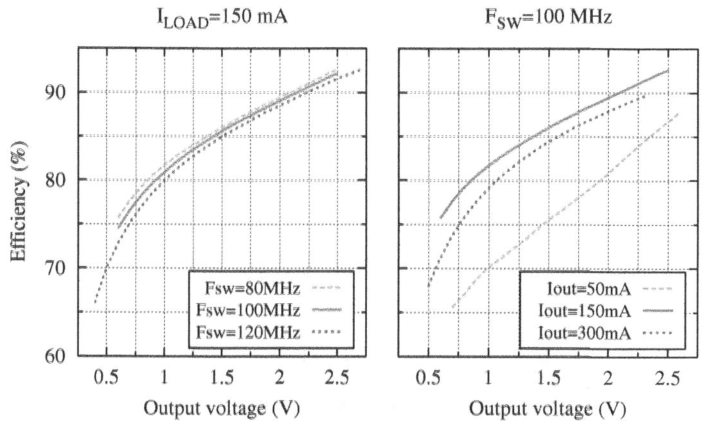

Figure 6.31. *Efficiency of the cascode power stage: sensitivity to operating conditions. For the color version of this figure, see www.iste.co.uk/allard/systems.zip*

6.5.3.1. *Perspectives*

The three-transistor cascode power stage has been demonstrated in the CCM and open-loop operation. One of the improvements is to integrate the generation of the intermediate voltage level from the main input voltage through a switched-capacitor DC/DC converter. A redesign of the power stage is necessary to introduce two global input signals in order to turn off both the top and bottom sides of the output stage. Voltage-mode control has already been demonstrated at 200 MHz for a standard power stage. This control option can be applied without any changes in the cascode power stage. More interestingly, sliding-mode control could benefit from the insensitivity of efficiency to the switching frequency if the frequency range is limited. The fabricated interposer possesses landing pads for one inductor. A new version of the interposer must be fabricated to accommodate coupled inductors. Such a magnetic device has been manufactured by Tyndall Institute based on specifications, as shown in Figure 6.32 [MAT 12]. Unfortunately, a rather discrete test board, as shown in Figure 6.33, is not sufficient as the magnetic device no longer benefits from the proximity of the capacitive interposer.

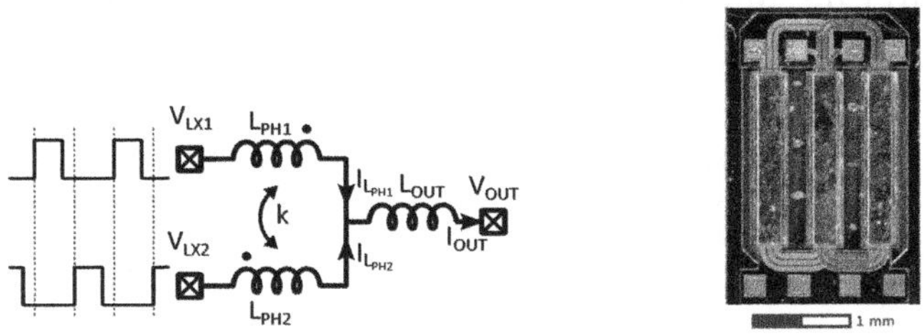

Figure 6.32. *Magnetic device suitable for a two-phase buck converter (left) and integrated sample (right). For the color version of this figure, see www.iste.co.uk/allard/systems.zip*

Figure 6.33. *Test board ready for functional verification of the two-phase buck converter. For the color version of this figure, see www.iste.co.uk/allard/systems.zip*

6.6. Conclusion

Small-footprint DC/DC converters and/or low-form-factor DC/DC converters require further integration. When considered for system-on-chip, switched-capacitor architectures are preferable but high efficiency is related to the quality of embedded capacitors, and on-chip deep-trench capacitors are quite expensive. When the targeted power density leads to the selection of an inductive architecture, the switching frequency is the degree of freedom to push inductive-based converters toward air-core inductors, compatible with on-chip manufacturing. However, the output power capability is limited. An intermediate solution to accommodate magnetic-core inductors is a 2.5D assembly using a capacitive interposer. A cost-effective industrial process must be selected for each part of the system but with perspective to meet performance expectations. This chapter highlighted the benefits of an interposer for shorter interconnections between devices thus reduced parasitic effects under high switching frequency operation. In contrast, this chapter has reported the performances of a three-transistor cascode power stage as a better trade-off between high input voltage and high efficiency in advanced CMOS technology nodes.

6.7. Acknowledgments

This work was supported by the European Commission through the Seventh Framework Programme (FP7), under the project grant PowerSWIPE No. 318529.

7

Hybrid and Multi-level Converter Topologies for On-Chip Implementation of Reduced Voltage-Swing Converters

This chapter describes emerging converter topologies that are becoming a viable, more compact alternative to predominantly used conventional inductive-based converters. These topologies use the combination of inductive energy transfer and capacitive energy processing to minimize the flux linkage of inductors and, in some cases, the total voltage stress of components, to reduce the size of power converters while improving power processing efficiency.

7.1. Introduction

High-frequency, low-power SMPS, used in consumer electronics, portable applications and computers, are required to meet stringent voltage regulation requirements [ERI 01, CHE 05, SEE 10] using a cost-effective small-volume implementation. The regulation is usually achieved with an application-specific on-chip integrated controller (controller IC) occupying a small silicon area [SAN 13, MAH 12, AHS 15a]. The controller is often required to effectively minimize voltage deviations caused by load transients, thus minimizing the requirements for a bulky output filter capacitor [MCR 15].

Chapter written by Aleksandar PRODIC, Sheikh Mohammad AHSANUZZAMAN, Behzad MAHDAVIKHAH and Timothy MCRAE.

In numerous electronic devices, such as portable applications, multiple switched-mode power converters, based on the buck, boost or buck-boost configuration, that use inductive energy transfer [ERI 01] provide well-regulated voltages or currents to various functional blocks (loads), such as processors, memory, LED panels, and other digital and analogue loads.

The active components of multiple converters (power switches and controllers) are often integrated inside a single or multiple power management integrated circuits (PMIC), while the reactive components of those converters (inductors and capacitors) are usually discrete. These discrete components often take a large portion of the overall device volume and/or printed circuit board (PCB) area and, in some cases, are among the main contributors to the overall device volume. Depending on the application, the passives often take between 12% and 80% of the overall PCB area [CHE 05]. In space-limited applications, reactive components, especially inductors, are the main obstacles to further minimization of electronics devices and, in some cases, prevent introduction of novel functional blocks that could potentially add new features to those devices. In the targeted applications, the relatively large volume of inductors compared to that of the capacitors, which is often reflected through their height, is related to a much lower energy storing capacity of the inductors. Comparisons of the energy capacity between the inductors and capacitors used in the targeted low power applications [SEE 10] shows that, inside the same volume, the capacitors can store between 100 and 1000 times more energy.

Switched-capacitor (SC) converters perform power processing without inductive components and therefore allow for much more compact implementation than their conventional counterparts [SEE 10, SAN 13]. However, the efficiency of SCs converters drastically degrades when required to operate away from the fixed (and optimal) conversion ratio. Also, compared to the conventional solutions, SCs suffer from inferior dynamic performance. As such, SC converters are not the most suitable solution when variable conversion ratios and/or fast transient responses are required. Emerging hybrid and multi-level converter topologies are becoming viable alternatives to predominantly used conventional inductive-based and SC converters. Compared with conventional inductive converters, hybrid and multi-level converters [MAH 12, AHS 15a, MCR 15] allow for implementation with smaller inductors without efficiency and dynamic performance-related penalties, characteristic of SC converters. These emerging topologies combine inductive energy transfer with capacitive energy processing to reduce the inductor voltage swing, minimizing its flux linkage and therefore energy storage requirements.

7.1.1. *Inductor volume reduction through voltage swing minimization*

The size of the inductor in a general switched-mode converter (Figure 7.1) is usually defined based on the maximum allowable inductor current ripple as in [7.1], where V_{high} and V_{low} are inductor voltages during the two portions of a switching period T_s, D is the duty ratio value and λ is the flux linkage:

$$\Delta I_L = \int_0^{DT_S} \frac{V_{high}}{L} dt = \int_{DT_S}^{T_S} \frac{V_{low}}{L} dt = \frac{\lambda}{L} \qquad [7.1]$$

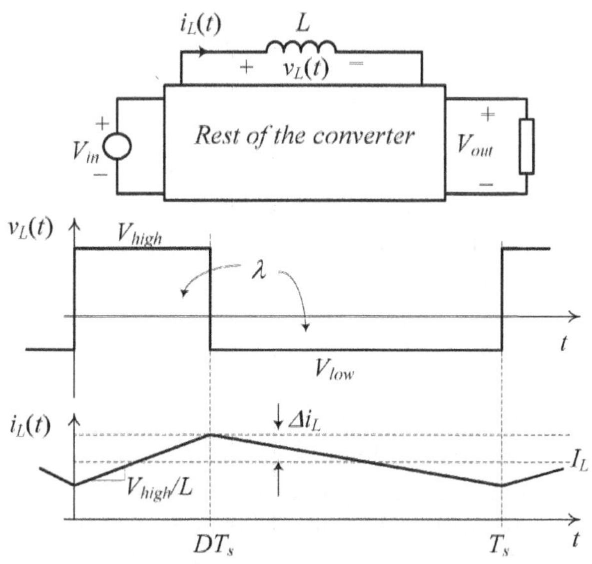

Figure 7.1. *General switched-mode converter (top) and its inductor waveforms (bottom)*

The current ripple value is determined by design trade-offs between the size of the converter and its power processing efficiency. The most commonly used approach to minimizing the inductor size is based on increasing the switching frequency fs = 1/Ts and thus reducing the flux linkage of the inductor by minimizing its energy storage requirements. However, this approach inevitably increases both the semiconductor switching losses and magnetic core losses per unit volume, negatively affecting the overall system efficiency.

In reduced voltage swing converters, an alternative approach to reducing the flux linkage is adopted. Rather than increasing the switching frequency, voltages applied across the inductor, i.e. V_{high} and V_{low} of Figure 7.1, are reduced compared with those of conventional topologies. The reduced inductor voltage swing is also often accompanied by two additional advantages, namely lower voltage stress of the switching components and improved dynamic response, through the increased slew rate of the inductor current. The lower voltage stress across switching components potentially results in the reduction of overall switching losses and implementation in a less costly IC process with semiconductor switches that have a better figure of merit (FoM) [BAL 08] than those of conventional converters.

To achieve reduced voltage swing, non-conventional power stage topologies, involving a larger number of switching components, are usually required as well as more complex control schemes. Neither of these issues represent an unsolvable obstacle. In on-chip implemented solutions, integration of a fairly large number of semiconductor components is quite common and recently emerged digital controllers for low-power applications [RAD 11, LUK 07] have created enabling technology for the realization of quite sophisticated control algorithms.

As mentioned previously, in the class of converters reviewed here, the swing of the switching node voltage, denoted by v_{sw}(t) in Figure 7.2, is also often reduced. As a result, the improvements in efficiency are achieved. The power savings can be achieved by the reduction of switching losses, which are proportional to We $= 1/2 \cdot V_{off}^2 \cdot C_{ds}$ product (Figure 7.2), and, in some cases, by a possible reduction of the overall conduction losses. The reduction of conduction losses is due to the fact that the reduced blocking voltage of the switches V_{off} allows for the design with lower voltage rating transistors having lower conduction losses. The lower on-resistance of the switches can be attributed to both the smaller required length of the transistors and a better silicon process for implementation [BAL 08], resulting in exponential improvements of transistor characteristics. So, although the number of switches in the conduction path of the reduced voltage swing converters is usually higher, their total conduction losses are not necessarily higher than those of conventional topologies.

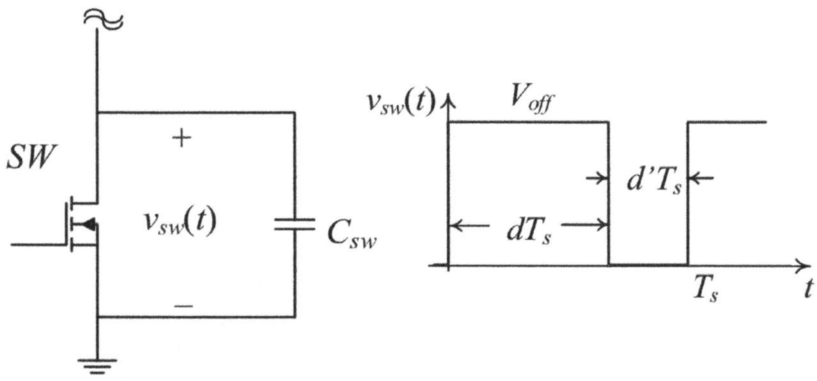

Figure 7.2. *Semiconductor switching waveforms*

Depending on how the inductor voltage swing is reduced, the converters discussed here can be roughly divided into two general groups: hybrid [AHS 15a, MCR 15] and multi-level converter topologies [MAH 12]. In sections 7.2 to 7.4, hybrid converter topologies that combine SC and inductive-based circuits are discussed. In particular, two distinctive architectures of hybrid converter topologies, namely cascaded and serial output/input converter structures, are shown, and practical implementations of converters using these two architectures are demonstrated. Furthermore, an example of an ultra-compact on-chip integrated power management system for mobile applications that uses hybrid power processing principles is shown.

Section 7.5 discusses two basic classes of multi-level converter topologies, namely capacitive divider- and flying capacitor (FC)-based converters, as well as topologies created by merging those two concepts. As their name suggests, in the capacitive divider-based multi-level topologies, the voltage swing across the components is reduced by using capacitive dividers, in which the voltage is balanced with inductive components of the converter. In FC-based multi-level converters, the voltage is often reduced with an addition of a DC voltage holding capacitor whose position during each switching period is varied such that all the components of the system are exposed to a lower voltage swing. Again, several practical design examples are given. Also, several challenges related to the practical design and control of these classes of converters are addressed.

7.2. Cascaded hybrid SC-inductive topologies

In hybrid SC-inductive topologies, the advantages of the capacitors' higher energy storage capability are used and combined with the conventional inductive-based topologies. Compared with the all-capacitive based solutions, the cascaded topologies provide a better voltage regulation and flatter efficiency curves over a wide range of conversion ratios. A block diagram of typical hybrid capacitive-inductive converter [AHS 15a] is shown in Figure 7.3. It usually consists of serial connection, of a capacitive energy processing stage, operating at a fixed conversion ratio at the peak efficiency point, followed by an inductive-based converter (most commonly a buck converter) that provides a tight voltage regulation.

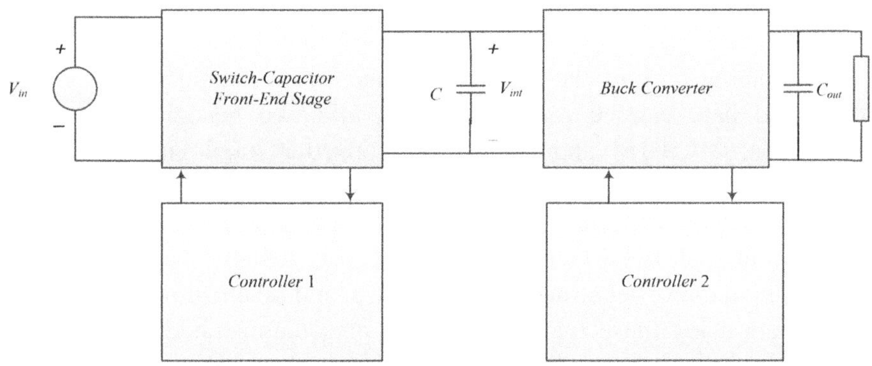

Figure 7.3. *Cascaded hybrid SC-inductive converter*

Thus, the voltage difference between the input and output of the buck stage is reduced, as well as the voltage swings across the inductor and the power switches of the downstream stage. As an example, in Figure 7.4, reduction of the inductance value for a 5 V output buck converter is shown as the intermediate voltage produced by the SC stage reduces from 10 V to 5.5 V. It can be seen that the required value of the inductance reduces by about five times. However, it can be seen that the reduction in the input voltage also results in the degradation of the inductor slew rate during light-to-heavy load transients, even though the inductance value has been reduced. As a result, the dynamic response of the converter can be negatively affected and the requirements for the output capacitance value C_{out} are increased.

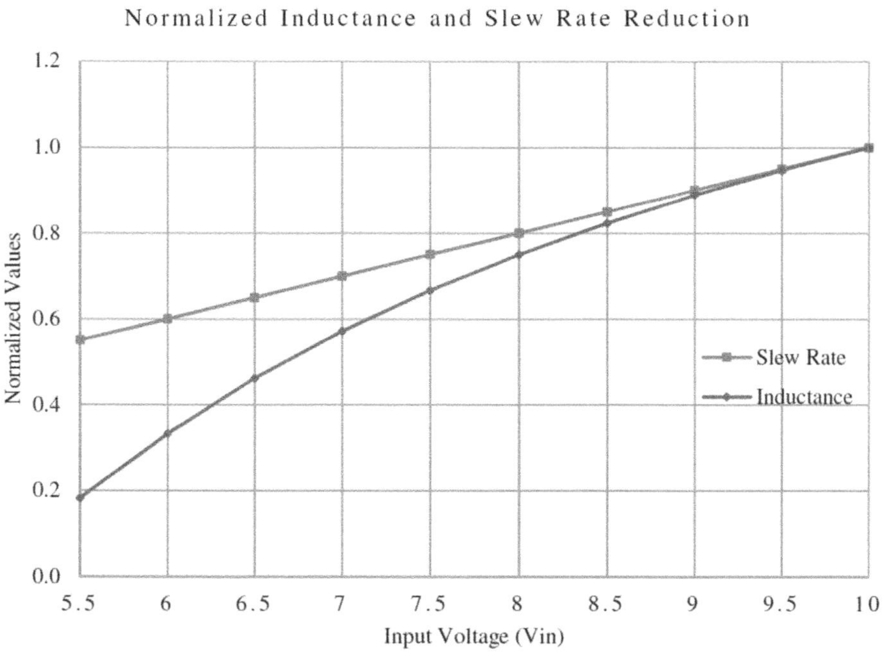

Figure 7.4. *Reduction in inductor size and inductor current slew rate for a 5 V buck converter for which input bus voltage varies between 5.5 V and 10 V. For the color version of this figure, see www.iste.co.uk/allard/systems.zip*

In the following section, an implementation of a modified cascaded hybrid converter topology that minimizes the previously mentioned trade-off is provided.

7.2.1. *Merged switched-capacitor multi-phase buck (MSCB) converter*

The MSCB converter, as shown in Figure 7.5, merges a three-level, SC front-end converter with a two-phase buck converter. This converter is well-suited for DC/DC point-of-load (PoL) applications such as tablet computers and games consoles, which need a large input-to-output step-down ratio and a relatively high current. One of the main targets in the converter design for these applications is the reduction in the DC/DC converter volume, mainly contributed by passive components [RAH 06, XIA 04, KAI 04], high efficiency over the full range of operation and fast dynamic response.

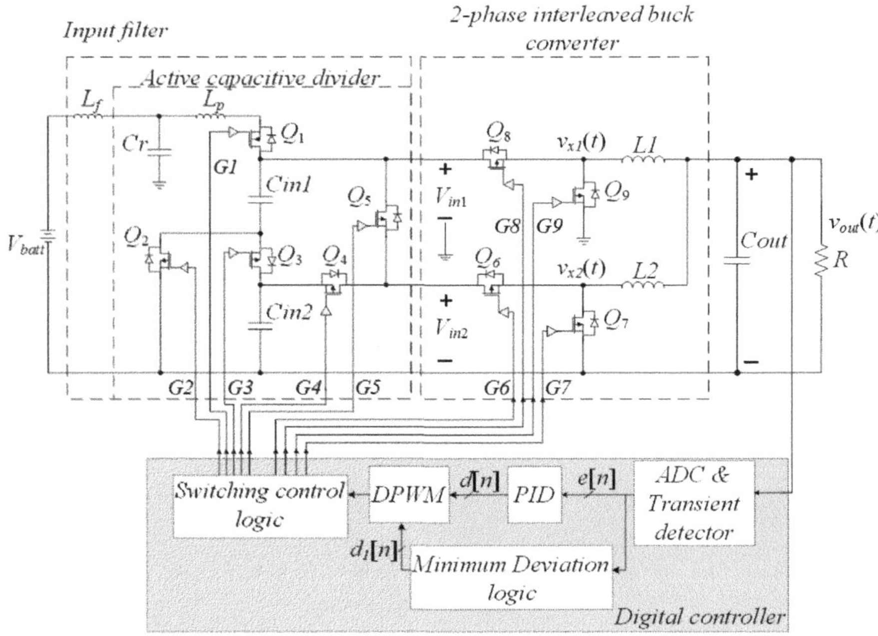

Figure 7.5. *Block diagram of a two-phase MSCB converter and its digital controller*

In the MSCB converter, a soft-charging SC circuit provides each of the buck stages with a half of the input voltage ($V_{batt}/2$) during the steady-state converter operation. Therefore, by reducing the voltage swing across the inductor and switches of the buck stage by half, the reduction in the volume and switching losses of the buck stage is achieved, as shown in [MAH 12]. The voltage divider capacitors in the MSCB converter are soft-charged through a quasi-resonant circuit and discharged by inductors of the buck stage. This allows for sharing the switches between the two stages, eliminating the bulky flying capacitor and reducing conduction losses compared with existing two-stage solutions [SUN 07a, SUN 06, PIL 08]. At heavy loads, the converter efficiency is comparable to the conventional multi-phase buck, and at light-to-medium loads, it is improved.

Compared with the existing solutions, the transient response is also improved by reducing the size of passive components (filter inductor and output capacitor) and using a full range of input voltage during light-to-heavy load transients, resulting in a larger inductor current slew rate. Also, unlike the single phase MSCB presented in [RAD 11], it provides common ground between the input and the output, allowing for multi-phase operation and, hence, a significant extension of the power rating.

In addition, by operating the buck phases of the MSCB converter with equal duty cycles, equal current sharing between buck phases and equal charge balance between capacitors of the SC stage can be achieved through an inherent feedback network that significantly simplifies the controller complexity.

7.2.1.1. *Principle of operation and system description*

In the system shown in Figure 7.5, the capacitive divider provides two input voltages for the buck stage, i.e. $V_{in1}(t)$ and $V_{in2}(t)$, the values of which are approximately equal to a half of the input battery voltage V_{batt}. The operation of the converter is controlled by using a single voltage-mode digital pulse-width modulation controller, where, as described in [MAH 12], the regulation of the attenuator tap voltages $V_{in1}(t)$ and $V_{in2}(t)$ is provided through an inherent feedback loop existing in this topology. For small values of error signal e[n], the converter operates in steady-state mode. The output voltage is regulated using a PID regulator, a digital pulse width modulator (DPWM) [PRO 02, SYE 04], a minimum-deviation controller [RAD 10] that operates during load transients and a switch selector that sets the transistor switching sequence as described below.

7.2.1.1.1. A: Steady-state operation and elimination of the output capacitor of the SC stage

The operating modes of the attenuator, i.e. SC stage, and the key converter waveforms are shown in Figures 7.6 and 7.7. The converter operates in such a way that the upper buck phase shown in Figure 7.5 (controlled by Q8 and Q9) is always supplied by the voltage across C_{in1}, i.e. $V_{in1}(t)$, and the lower buck phase (controlled by Q6 and Q7) by $V_{in2}(t)$. The SC stage operates in synchronization with the buck, and, in each switching cycle, it goes through two modes, as shown in Figure 7.6.

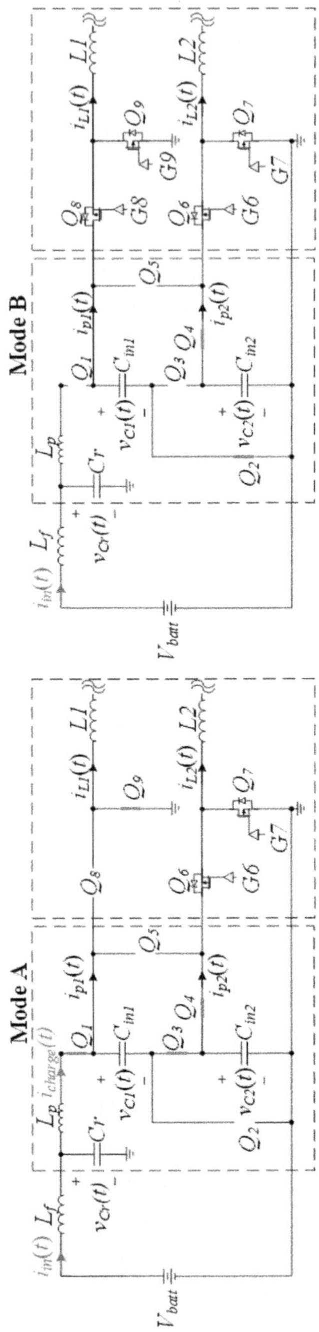

Figure 7.6. *Operating modes of the capacitive attenuator. For the color version of this figure, see www.iste.co.uk/allard/systems.zip*

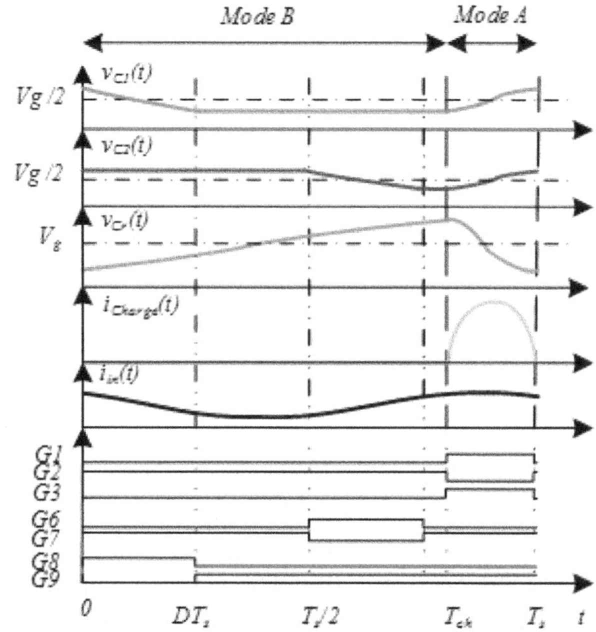

Figure 7.7. *Key waveforms of the MSCB converter. For the color version of this figure, see www.iste.co.uk/allard/systems.zip*

In mode A, it charges the cascade connection of C_{in1} and C_{in2}, through a quasi-resonant circuit formed by a small capacitor C_r, a parasitic PCB inductor L_p and switches Q1 and Q3. In this mode, transistors Q2, Q5 and Q8 are open and the upper phase of the buck converter operates in a synchronous rectification mode, i.e. transistor Q9 is turned on. During this time, the lower buck phase can be in any of the two regular switching states. The quasi-resonant switch is used to obtain zero current switching, eliminating switching losses of switches Q1 and Q3. In mode B, the switch control logic changes the circuit configuration, such that C_{in1} is only connected to the upper phase and C_{in2} to the lower phase. During this mode, the buck phases can be in any of the two switching states and discharge the capacitors when their main switches (Q6 and Q8) are turned on, through the inductors of the downstream stages.

By inspecting the conduction path during different switching periods, it can be seen that compared with a buck converter, conduction losses are slightly increased by the addition of SC stage switches. However, the switches of the SC stage (Q1 to Q5) are rated for $V_{batt}/2$ with lower R_{on} compared with those of buck stage switches. Considering the soft switching of the SC stage and switching $V_{batt}/2$ at the switching nodes of the buck stages, it can also be found that switching losses for the MSCB converter are significantly lower than that for the buck converter. Comparisons of the efficiencies between the MSCB converter and two-phase interleaved buck show better efficiency for the MSCB converter at light-to-medium loads and comparable efficiency at heavy loads [MAH 12].

7.2.1.1.2. Inherent center tap voltage regulation

The previously described operation inherently provides the regulation of SC tap voltages, eliminating the need for a relatively large charge-balancing output capacitor existing in the front stages of other SC-based two-stage solutions [SUN 07a, SUN 06, PIL 08]. The voltage regulation as well as the current sharing between the phases can be described through an analysis of the DC equivalent circuit of the converter shown in Figure 7.7. In this case, the current source I_{charge} represents the average current provided to the divider over one switching cycle, i.e. during mode A (Figure 7.5). The equivalent resistances of the phases are R_{eq1} and R_{eq2}. To simplify the explanation, it is assumed that both phases operate with the same effective duty ratio D.

Capacitor charge balance equations [RAH 06] for the input capacitors result in [7.2], where I_{L1} and I_{L2} are DC values of the phase inductor currents:

$$DI_{L1} = DI_{L2} = I_{charge} \qquad [7.2]$$

By using [7.2] and solving the circuit shown in Figure 7.3, the following expression for the difference in the tap capacitor voltages can be obtained as in [7.3]:

$$V_{c1} - V_{c2} = (R_{eq1} - R_{eq2})I_{L1/2}/D \qquad [7.3]$$

The equations show that current sharing can be achieved and that for the targeted applications, where R_{eq1}–R_{eq2} is relatively small, the tap voltages in the steady state remain approximately the same. The equations also show that the current and voltage sharing is not affected by the mismatches in the inductor and capacitor values. In other words, an inherent feedback for maintaining the same currents in both phases exists. The phase with a large equivalent resistance will also have a higher tap voltage so that its current is the same as in the other one.

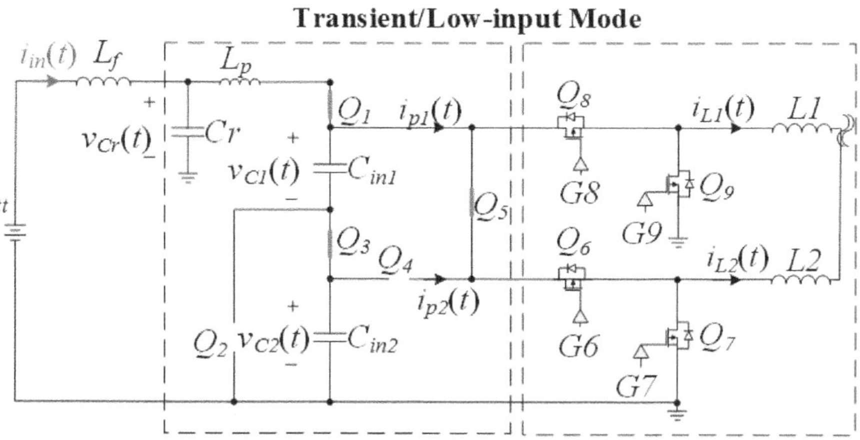

Figure 7.8. *Approximate DC equivalent circuit of the converter. For the color version of this figure, see www.iste.co.uk/allard/systems.zip*

7.2.1.2. *Soft switching*

SC converters often suffer from excessive switching losses due to direct energy transfer between the capacitors [MCR 15] . To eliminate this problem, a quasi-resonant switch is used. The switch is formed by a parasitic inductance of the PCB, L_p, and a small capacitor C_r in series with C_{in1} and C_{in2} (Figure 7.5), and transistors Q1 and Q3. The resonant circuit parameters are chosen such that the charging of C_{in1} and C_{in2} is completed over the duration of mode A, as shown in Figure 7.7. This period can be expressed as [7.4]:

$$T_s - T_{ch} \approx \pi\sqrt{L_p.C_r} \quad [7.4]$$

and should not exceed the conduction time of the synchronous rectifier, Q9, to maintain the low voltage at the input of the upper buck phase.

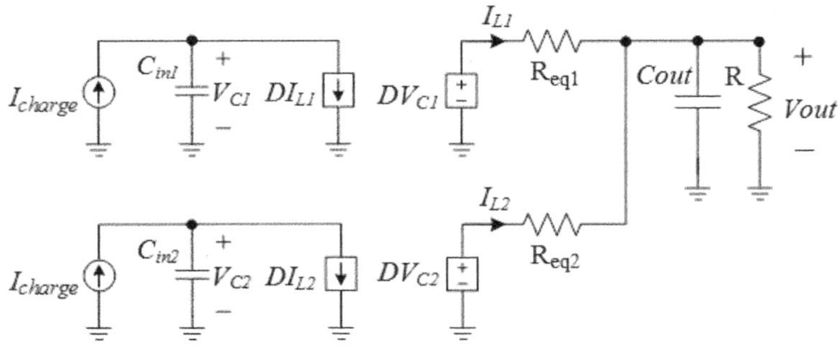

Figure 7.9. *Converter configuration during a light-to-heavy load transient and low-input voltage range* $V_{batt} < 2V_{out}$

7.2.1.3. *Transient and low-input mode*

During light-to-heavy load transients or low battery voltage conditions ($V_{batt} < 2V_{out}$), the converter uses a full range of input voltage, V_{batt}, to operate buck phases, as shown in Figure 7.9. Figure 7.10 compares a 2 A to 8 A light-to-heavy load transient for a 7 V to 1 V MSCB converter and a two-phase interleaved buck converter, where a minimum-deviation control method described in [RAD 10] is applied in both cases. This results in a larger inductor slew rate and faster transient response, and allows the reduction of the output capacitor. Heavy-to-light load transients are also improved due to the use of smaller buck phase inductors of the MSCB converter. Figure 7.10 also illustrates that, as predicted by [7.2], the current is shared equally between the buck phases during the steady state.

7.3. Hybrid serial input/output converters

Hybrid serial input/output (HSI/O) converters [MCR 15, MCR 16, YEA 14] are a particular case of partial power processing converters, as described in [LUK 07, ROD 98, SHO 14, ZHA 13], where the total output power is divided between two or more converters. HSI/O converters are different in that they use a combination of inductive SMPS and SC converters to process the output power. This type of architecture can yield several advantages compared with common SMPS. These include:

– lower passive component volume;

– lower peak voltage stress;

– modular system design.

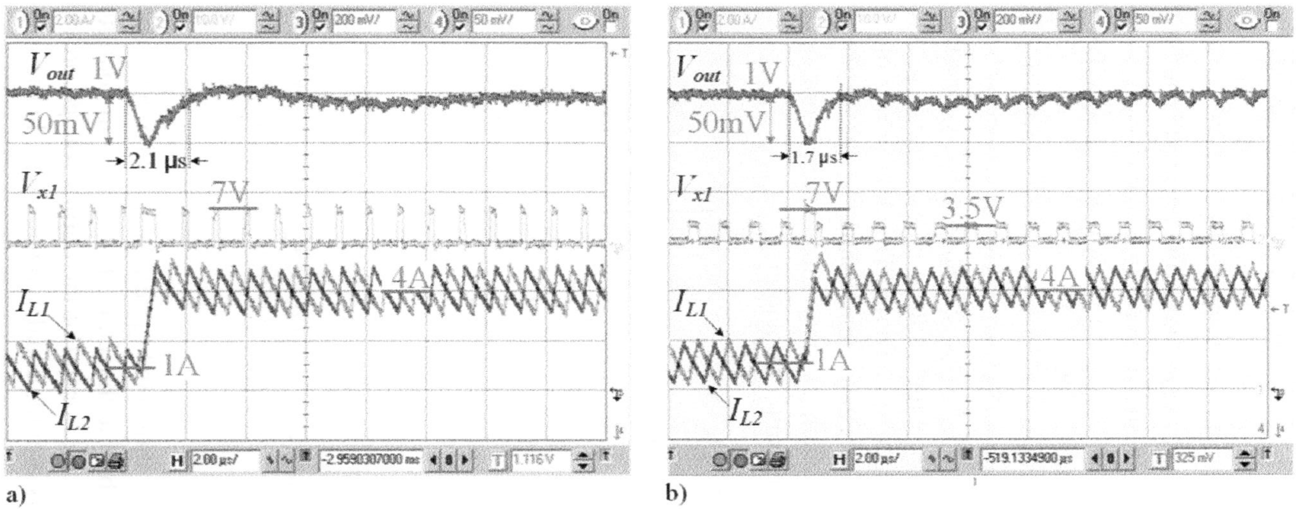

Figure 7.10. *a) Transient response for a two-phase interleaved buck converter. b) Transient response for a MSCB converter; Ch4 (top): output voltage, Vout(t) (50mV/div); Ch2 (upper middle): switching node voltage for the top phase, vx1(t) (10V/div); Ch1 (lower middle): top phase inductor current, iL1(t) (2A/div); Ch3 (bottom): bottom phase inductor current, iL2(t) (2A/div). Time scale is 2 us/div. Operating conditions: Vin = 7 V, Vout = 1 V, load current step of 2A to 8A. For the color version of this figure, see www.iste.co.uk/allard/systems.zip*

Diagrams of a step-up (serial output) and step-down (serial input) hybrid converters are shown in Figure 7.11. For the step-up configuration, the input power is drawn in parallel from the input source and the outputs are connected in series. Source-load inversion can be performed to achieve a step-down converter, as reported in [YEA 14]. Any number of converters can be stacked inside a HSI/O, but only two are shown here for simplicity. The idea is to combine the high power density of SC converters with the tight output voltage regulation of inductive converters. This is because only conventional SC converters have high power processing efficiency at a relatively small number of discrete voltage conversion ratios [SEE 10, SAN 13] and suffer from large efficiency drops at other ratios. For this reason, the SC converter is left unregulated and regulation of the output voltage is accomplished by closing the control loop around the inductive converter. Dividing power in this way leads to the reduction in passive component volume at the expense of an increased number of switches, which, as mentioned earlier, is not a particular challenge for the targeted on-chip implementation.

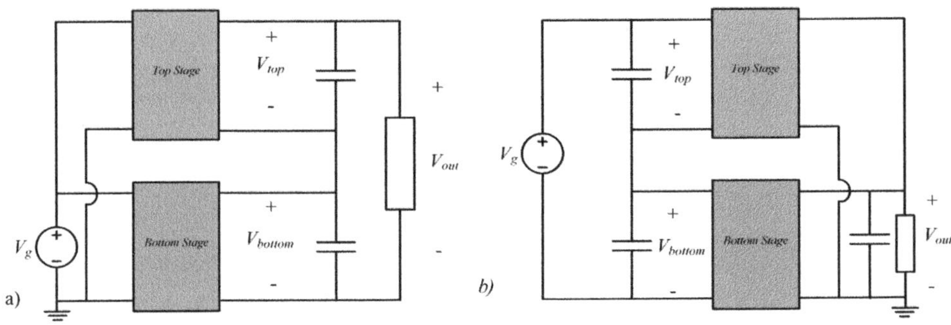

Figure 7.11. *Schematic of a step-up hybrid serial output (HSO) converter (left) and hybrid serial input (HSI) converter (right). The top stage requires isolation between its input and output voltages to avoid discharge of the bottom cap. Both converters have the same output current. For the color version of this figure, see www.iste.co.uk/allard/systems.zip*

At least one of the converters shown in Figure 7.11 should be a SC converter for the system to be a true hybrid architecture. In the step-up configuration, power sharing is controlled by the portion of the output voltage provided by each converter as all converters provide equal output current. In the step-down configuration, where the input voltage is divided among several converters and converter outputs are connected in parallel, power sharing is determined by the relative amount of current provided by each converter. Because either the inputs or outputs of two converters are stacked on top of one another in an HSO converter, one of these converters should have some kind of isolation between the input and output to prevent the discharge of the ground-referenced capacitor. This can be done either with galvanic isolation (flyback or forward converter) or the use of an isolated or differential switched-capacitor converter.

Design examples of both floating SC and grounded SC step-up HSO converters are shown in Figure 7.12. In Figure 7.12 (a), a converter designed for integrated LED lighting applications [MCR 16] is shown, where total pin and component count are expected to be as low as possible. This results in a solution where the SC voltage gain is low and power is roughly divided into half between the two converters. In Figure 7.12 (b), a solution for automotive LED lighting applications [MCR 15] is shown. Here, a flyback converter is used to isolate the top stage, allowing the SC to be non-isolated. Both of these solutions, that are taken as examples, reduce the total passive component volume by four to five times compared with conventional topologies.

7.3.1. *HSI/O power processing efficiency and power division*

Power processing efficiency can be found by using the following simplified analysis. The power delivered to the load of a step-up HSO is given by [7.5], where, as shown in Figure 7.12, V_{out} is the output voltage, I_{out} is the output current, V_{fb} and V_{sc} are the output voltages of [7.6], with P_{ind}=$V_{ind}I_{out}$ being the portion of the output power P_{out} delivered by the inductive converter and P_{sc}=$V_{sc}I_{out}$ being the remaining portion delivered by the switched-capacitor circuit:

$$P_{out} = V_{out}I_{out} = (V_{sc} + V_{ind})I_{out} = P_{sc} + P_{ind} \quad [7.5]$$

$$P_{in} = \frac{P_{sc}}{\eta_{sc}} + \frac{P_{ind}}{\eta_{ind}} \quad [7.6]$$

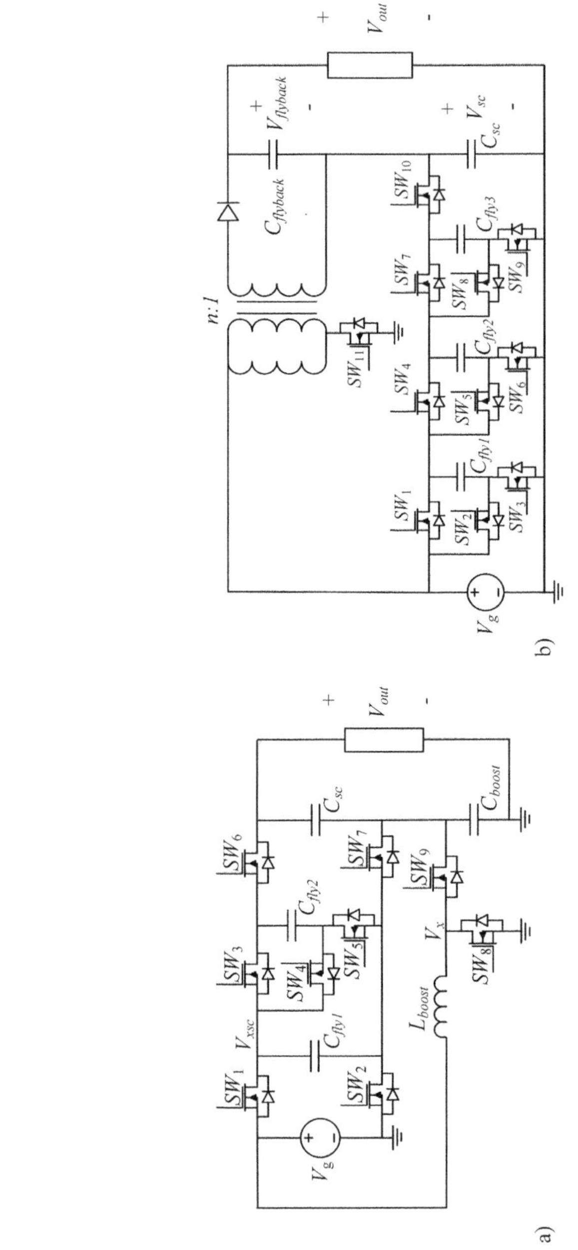

Figure 7.12. *a) SC-boost HSO converter topology [COR 15d]. b) SC-flyback HSO converter topology [MAT 14]*

The input power can be found by dividing the output power by the respective power processing efficiencies of the inductive and SC stages, i.e. η_{ind} and η_{sc}. By dividing P_{out} by P_{in} and noting that the output currents are shared, we obtain the expression for the efficiency of the entire converter [7.7]:

$$\eta = \frac{P_{out}}{P_{in}} = \frac{P_{out}}{(\frac{P_{sc}}{\eta_{sc}} + \frac{P_{fb}}{\eta_{fb}})} = \frac{V_{out}}{(\frac{V_{sc}}{\eta_{sc}} + \frac{V_{fb}}{\eta_{fb}})} \quad [7.7]$$

This equation gives a guideline not only for designing an HSO converter with targeted power processing efficiency, but also for the entire class of partial processing converters, where the output voltage is the sum of the output voltages of the two stages. It shows that a very efficient converter cannot be combined with an extremely inefficient one that processes only a small portion of the total power and expect high efficiency for the entire system. A similar equation can be derived for HSO step-down converters where the output voltage is shared and the output current is divided between the two converters [7.8]:

$$\eta = \frac{P_{out}}{P_{in}} = \frac{P_{out}}{(\frac{P_{sc}}{\eta_{sc}} + \frac{P_{fb}}{\eta_{fb}})} = \frac{I_{out}}{(\frac{I_{sc}}{\eta_{sc}} + \frac{I_{fb}}{\eta_{fb}})} \quad [7.8]$$

This total system efficiency equation leads naturally to the question of how to divide the total output power between the two converters in a reasonable way. Initial thoughts may indicate that designing the more efficient converter to process more power is the obvious choice, but depending on the application, this may not be the best method. Answering this question requires looking at the total volume of the final solution, the number of components and the number of switches. For instance, allowing a SC converter to process 95% of total power for a high step-up application could yield a solution with an unmanageable amount of switches and very large capacitors, while a simple flyback converter solution may end up being smaller.

7.3.2. *Switched-capacitor conversion ratio*

SC converters have an inherent trade-off between switching frequency, capacitor size/volume and switch on-resistance for a given desired efficiency. Furthermore, the efficiency of a SC converter can be associated with a given output voltage ripple of the converter. This implies that a smaller ripple yields a higher efficiency, and also larger capacitors and larger switches. For optimal performance, the converter should operate between the slow switching limit (SSL), where the capacitance dominates the output impedance of the converter, and the fast switching limit (FSL), where the switch on-resistance dominates the output impedance, as this will minimize the die size for a given efficiency [SAN 13]. As mentioned previously, at least one of the converters in an HSI/O converter is required to have a floating output. For low power system-on-chip (SoC) applications, galvanic isolation is usually not required. A simple solution is to design the SC converter to have a floating output. Operation of a floating 1–2 SC converter can be explained, as shown in Figure 7.13. During Φ_1, switches SW1, SW2, SW3 and SW5 are turned on, connecting the flying capacitors to the input in parallel and charging them up to the Vg. During Φ_2, switches SW4, SW6 and SW7 are turned on, disconnecting the flying capacitors from the input and connecting them in series to the output capacitor of the SC converter, which is stacked on top of the boost output. Because switches SW1 and SW2 are out of phase with switches SW6 and SW7, the negative terminal of the SC output capacitor is never connected to the ground, and thus the output capacitor of the inductive converter is never discharged.

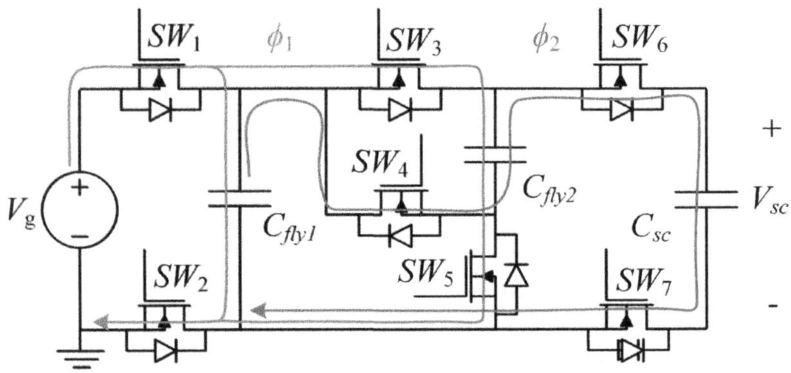

Figure 7.13. *1–2 SC converter with the two alternating switching phases indicated in red and blue. For the color version of this figure, see www.iste.co.uk/allard/systems.zip*

7.3.3. *Passive volume and switch voltage stress*

In the targeted applications, inductors have about 1,000 lower energy storage capability per unit volume than the capacitors [SAN 13]. This allows HSO converters to have a significantly smaller passive component volume because some portion of the total power is not processed by magnetics. Using stored energy as a rough approximation of the volume of passives and multiplying magnetic-based passive components by 1000 to account for the lower energy density, we can see where the benefit comes from. For example, taking into account a 3.3 V–12 V, 5 A boost converter, we can see that the average inductor current is 18 A. If an HSO converter is designed to divide the output voltage equally between a SC and a boost, the current through the HSO boost will only be about 9 A. For the same inductance, the energy storage of the conventional boost inductor is four times larger than that of the HSO boost, and because inductors are not very energy dense, this volume will tend to dominate the overall solution. A volume comparison between a SC-boost implemented with discrete components and the boost converter from [MCR 16] is shown in Figure 7.14. It should be also noted that the increase in the total volume of the switches for the hybrid structure is mainly due to the discrete implementation, not providing integration of multiple components in a single package. It can be expected that the full on-chip implementation results in significantly lower penalty in terms of the volume of the switching components.

Compared with both boost and flyback converters, HSO converters can have a lower peak switch voltage stress for a given conversion ratio. This is due to the fact that each converter provides only a portion of the total output power. For the step-up configuration, the converters are only exposed to a portion of the total output voltage depending on the power division; for the step-down configuration, the converters are only exposed to a portion of the input voltage. For example, both switches of a boost converter are exposed to V_{out}=M.V_g, where M is the conversion ratio of the converter. The primary side and secondary switches of a flyback converter are exposed to V_g+nV_{out} [CHE 05] and V_g/n+V_{out}, respectively, where n is the turns ratio of the n:1 flyback transformer. In the HSO, the maximum voltage seen by any switch of the SC is (M_{sc}–1)V_g, where M_{sc} is the conversion ratio of the SC converter, and the voltage stress of the inductive converter depends on the chosen topology. An appropriate selection of power division can ensure that the peak voltage stress in an HSO converter is always less than that in an inductive converter designed for the same power level.

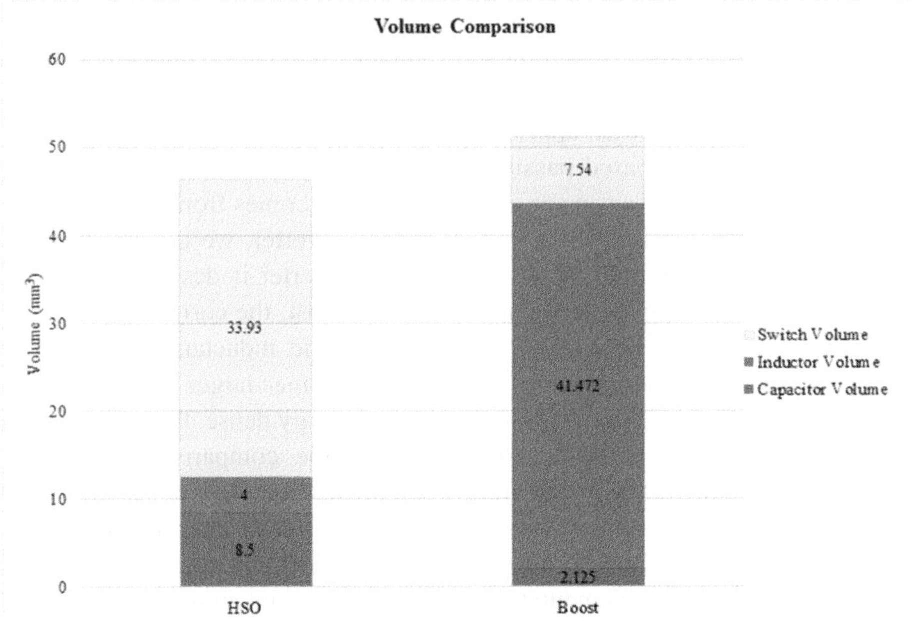

Figure 7.14. *Volume comparison between a discrete implementation of the SC-boost HSO and boost converters for the efficiencies shown above. It can be seen that although the capacitive volume has quadrupled compared with the boost, the inductor volume is significantly reduced. For the color version of this figure, see www.iste.co.uk/allard/systems.zip*

7.4. An on-chip integrated high-density power management solution for portable applications based on a multi-output switched-capacitor circuit

In modern battery-powered portable applications, such as cell phones and laptop or tablet computers, power management systems provide multiple voltage levels using many different DC/DC converters [HEN 09]. These voltages are supplied to various functional blocks, including digital processors, I/O interfaces and memory devices. A conventional power management architecture (PMA) is shown in Figure 7.15(a). In a two-stage DC/DC power conversion solution, a number of downstream SMPS and/or low-dropout (LDO) linear regulators are connected to a fixed intermediate bus voltage, and provide multiple output voltages that meet specific steady-state and dynamic voltage requirements [HEN 09, CLE 05, TEX 14]. One of the main drawbacks of conventional PMAs is their size. In numerous portable devices, PMAs are among the largest contributors to the overall size and weight of the entire device [KUM 12, KAI 04] and large consumers of printed circuit board (PCB) space.

This is primarily due to the bulky and costly reactive components of SMPS output filters, where the inductors are the largest and heaviest components of the filters [WAL 13]. SC voltage converters are proposed in numerous publications as an alternative solution to meet the strict volume requirement of modern portable devices [SEE 10, LE 10, SAN 11]. As these converters do not require bulky inductors, they can be implemented in a significantly smaller volume and can possibly be integrated on ICs [LE 10, SAN 11]. To provide a tight output voltage regulation under input voltage variations, as in [SUN 06, SEU 15], two-stage compact and power-efficient solutions are presented (Figure 7.15(b)). In these solutions and the ones shown in previous sections, a SC fixed-ratio front-end stage performs a large portion of voltage conversion at peak efficiency, and provides a bus voltage that is loosely regulated. However, for applications where multiple regulated output voltages are required, this concept still results in relatively large voltage swings for a number of downstream stages, since the bus voltage reduction is limited by the output of the downstream buck, producing the highest output voltage.

In this case, the concept of combined SC and inductor-based power processing is extended where the architectures of both front-end and downstream stages are completely redesigned. In the presented architecture namely the MSC-DB shown in Figure 7.15, the front-end converter is replaced with a multiple-output SC (MSC) stage, and instead of operating at the full bus voltage, the downstream converters are supplied by differential output taps, operating as differential-input buck (DB) converters. The MSC front-end stage, replacing the inductor-based converter of the conventional solution (Figure 7.15 (a)), increases the power density, as SC converters can be significantly more area-effective compared with their inductive counterparts [SEE 10].

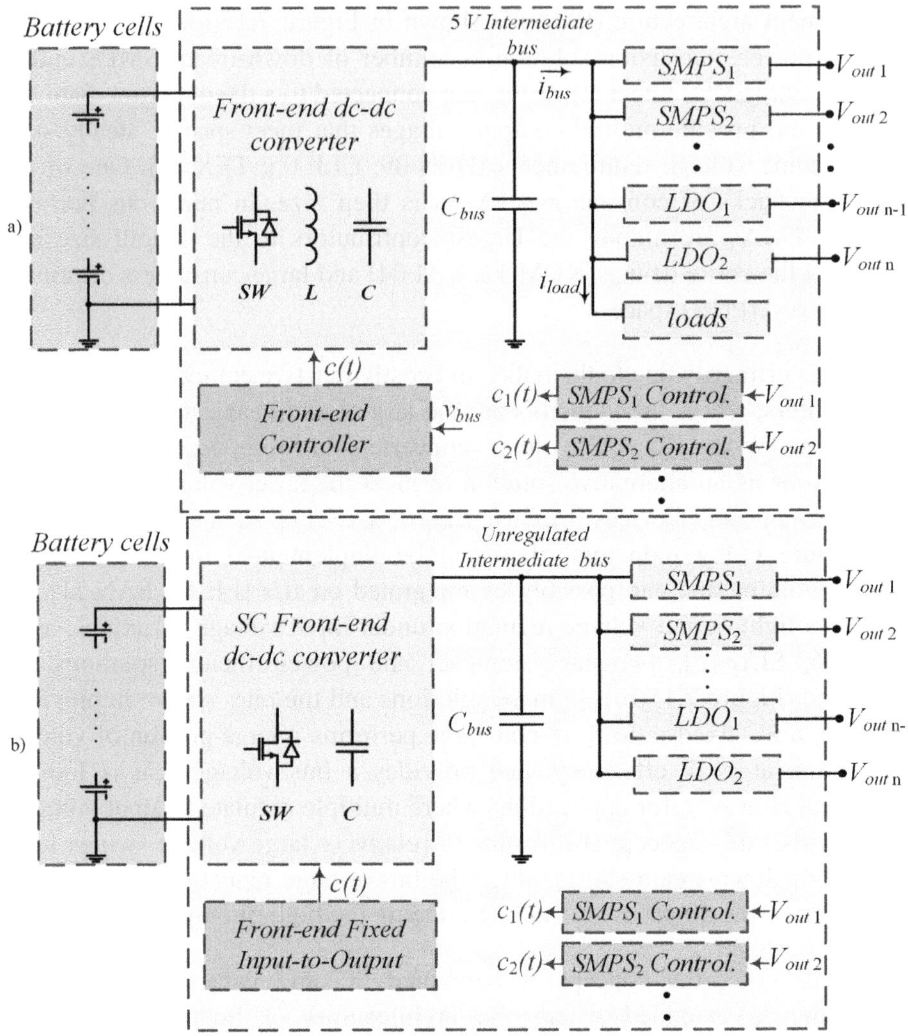

Figure 7.15. *a) Conventional two-stage PMA. b) Two-stage PMA with SC front end. For the color version of this figure, see www.iste.co.uk/allard/systems.zip*

Furthermore, the significantly lower energy density of the inductors makes them much more difficult to integrate than capacitors, due to the area and associated cost constraints of the IC processes. Each of these downstream buck converters is based on the dual input, i.e. differential-input buck converters, originally introduced in [SUN 06] for higher power AC/DC and DC/DC applications. In the presented architecture, the concept of the dual-input buck is extended to multiple regulated outputs, while maintaining all the advantages mentioned in [SUN 06].

In MSC-DB architecture, the inductors are drastically minimized through the reduction of their voltage swings. The voltage swing-based reduction principle and the size of reduction can be described by the following analysis, based on the constant current ripple criteria. The current ripple, Δ_{IL} and duty ratio, D equations for a general single inductor-based converter in continuous conduction mode are given in [7.9] and [7.10], where L is the inductance value, f_{sw} is the switching frequency, and $V_{L_{on}}$ and $V_{L_{off}}$ are the voltages across the inductor during on and off states of the main switch, respectively:

$$\Delta I_L = \frac{V_{L_{on}} \cdot D}{2 \cdot L \cdot f_{sw}} = \frac{V_{L_{off}} \cdot (1 - D)}{2 \cdot L \cdot f_{sw}} \qquad [7.9]$$

$$D = \frac{V_{L_{off}}}{V_{L_{on}} + V_{L_{off}}} \qquad [7.10]$$

In the conventional buck converter, $V_{L_{on}}$ = (V_{in}-V_{out}) and $V_{L_{off}}$= V_{out}, where V_{in} and V_{out} are the input and output voltages of the converter, respectively. The most common approach to minimizing the inductor value, while maintaining the same current ripple, is to increase the switching frequency [ERI 01]. However, this approach results in increased switching losses [AHS 11] and therefore poses a fundamental limit on the reduction of inductor size. In the modified buck converters shown in Figure 7.16, the inductors are reduced by minimizing $V_{L_{on}}$ and $V_{L_{off}}$ values of [7.9]. This is achieved by setting the two possible switching node (v_x of Figure 7.16) voltage values, V_{high} and V_{low}, to be slightly larger and slightly smaller than the desired converter output voltage (v_{out}), respectively. Following the analysis presented in [RAS 10], a simplified expression of the resulting reduction of the inductor values can be written as [7.11]:

$$\frac{L_{2-input}}{L_{buck}} = \frac{(1 - \frac{V_{out}}{V_{high}})}{(1 - \frac{V_{out}}{V_{in}})} \cdot \frac{(1 - \frac{V_{low}}{V_{out}})}{(1 - \frac{V_{low}}{V_{high}})} \qquad [7.11]$$

[7.11] proves that the differential buck has a smaller inductor, since $V_{high} < V_{in}$ and $V_{out} < V_{high}$. Considering both parts of [7.11], we obtain [7.12]:

$$\frac{(1-\frac{V_{out}}{V_{high}})}{(1-\frac{V_{out}}{V_{in}})} < 1 \cdots and \cdots \frac{(1-\frac{V_{low}}{V_{out}})}{(1-\frac{V_{low}}{V_{high}})} < 1 \qquad [7.12]$$

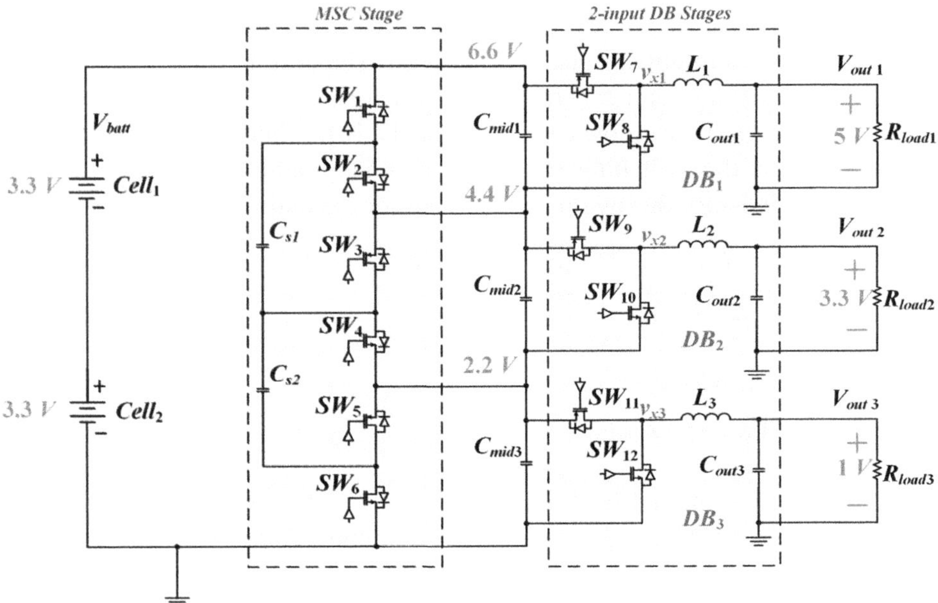

Figure 7.16. *MSC-DB PMA. For the color version of this figure, see www.iste.co.uk/allard/systems.zip*

In conventional converter architectures (Figure 7.15), voltage-mode controller is able to effectively compensate for load variation due to transients [ERI 01, PAT 03], and input voltage variation due to SoC of the battery, which is slow in nature [HUS 11]. However, in the MSC-DB architecture, it cannot achieve this without modification. Any high-frequency perturbation in the intermediate capacitors, i.e. in outputs of the MSC stage, would propagate to the outputs of the DB stages. This is mainly because the voltage-mode control is susceptible to high-frequency input voltage perturbations [ERI 01]. This drawback of the voltage-mode control could potentially create undesired cross-coupling between the output voltages. It can be seen that this problem exists due to the stacked-up intermediate capacitor configuration of the front-end SC stage. Since the two-input buck stages are connected across the intermediate capacitors, load transients at the output of one of them can create perturbation in stacked capacitors, affecting the outputs of the other downstream stages. To minimize this problem, the cross-regulation suppression module shown in Figure 7.17 is incorporated. Here, a feedforward-based approach [CAL 92, ARB 97] is used, where the analog-to-digital converter ADC2 is used to acquire information about the intermediate capacitor voltages. When there is a significant perturbation in intermediate capacitors, due to load transients, the cross-regulation suppression module adjusts the instantaneous duty ratio values, provided by PID compensators, such that the cross-regulation problem is minimized.

In a wide range of portable applications, using multiple battery cells, such as laptops, ultrabooks and tablet computers, cell balancing is a highly desirable feature. It extends the run-time of the battery, as well as the lifetime of the battery [WEN 14], by providing equal discharge of battery cells [KIM 14, GOO 15, BAU 07]. However, such a feature is rarely implemented in the targeted applications, mainly due to the need for dedicated power converters [LEE 05] that could significantly increase the volume, price and weight of portable devices. In the presented topology, the cell balancing can easily be incorporated, by adding a single small inductor, L_{bal} and a simple control circuit, as shown in Figure 7.17. Here, the balancing scheme is applied based on maintaining the same voltage across battery cells [MES 14, LI 12b, YUN 13]. Furthermore, the incorporated inductor improves the power processing efficiency of the MSC stage, by improving the balancing of flying capacitor voltages under large differences of output loads.

The effectiveness of the MSC-DB architecture in reducing the volume and improving efficiency is demonstrated here both through discrete and IC based implementation. The discrete prototype was built and tested. The discrete prototype implements the voltage-mode controlled architecture shown in Figure 7.17. Custom-made integrated power modules [AHS 15b] are used for on-chip implementation of the mixed-signal current programmed mode (CPM) architecture [PAT 03]. Furthermore, the IC-implemented prototype is co-packaged with on-chip inductors [WAN 08], demonstrating the ability of this architecture to drastically reduce the overall volume of power management systems in the targeted portable applications.

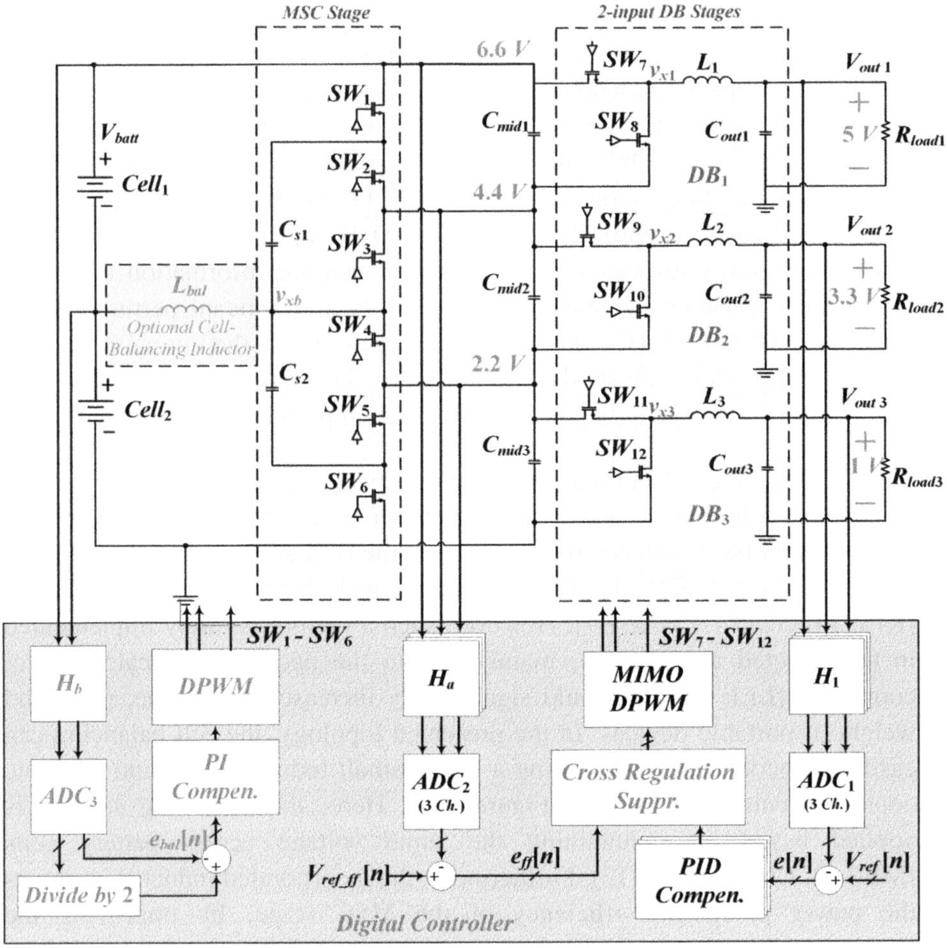

Figure 7.17. *Digital controller implementation with a cross-regulation suppression module and the cell balancing inductor, L_{bal}. For the color version of this figure, see www.iste.co.uk/allard/systems.zip*

The power stage of the system, shown in Figure 7.16, is developed with discrete transistors, gate drivers and passive components. For two 3.3 V Li-ion battery cells, these voltages are approximately 6.6 V, 4.4 V and 2.2 V. The MSC stage operates at a fixed switching frequency of 500 kHz while the downstream stages, providing 5 V/ 3 W, 3.3 V/ 9 W and 1 V/ 3 W, respectively, switch at 1 MHz. The cell balancing inductor, L_{bal}, whose value is 500 nH, provides equal discharge of both battery cells, while improving the efficiency of the MSC stage.

Figure 7.18 shows the comparison of inductor currents and switching node voltages of the discrete prototypes of the conventional and MSC-based differential buck downstream stages when providing 3.3 V at their outputs. As mentioned in section II, the conventional buck has a 5 V input and the input of the differential buck in this case is 2.2 V. For demonstration purposes, the inductors of both configurations are selected to be the same. The results demonstrate a lower switching node voltage swing and consequently about a 50% lower inductor current ripple for the 3.3 V 2-input buck of the MSC-DB architecture (Figure 7.17), allowing for the same percentage of inductor volume reduction while maintaining the same current ripple.

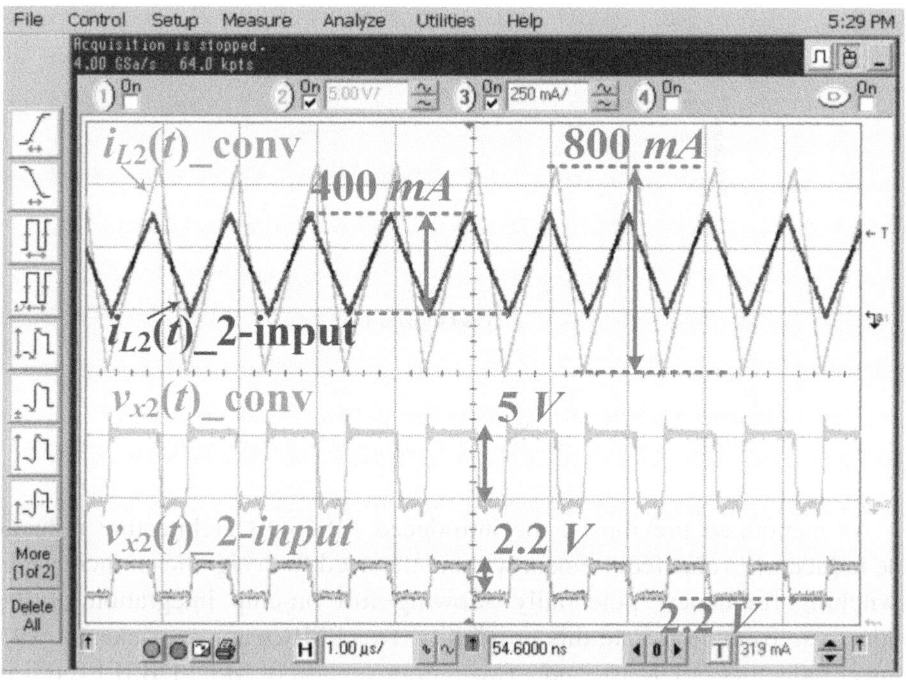

Figure 7.18. *Inductor currents and switching node voltages of the conventional and MSC-DB 2-input downstream stage for a 3.3 V output. Ch2: switching node, vx2 (5 V/div); Ch3: inductor current (250 mA/div). For the color version of this figure, see www.iste.co.uk/allard/systems.zip*

Figure 7.19 shows the comparison of measured efficiency between the conventional and 2-input downstream buck converters, providing 1 V output over a 250 mA to 3 A load variation. For the conventional case, the buck converter operates from a 5 V bus voltage, whereas in the introduced topology, it operates from the 2.2 V intermediate capacitor voltage. The results show up to 53% reduction in total losses, improving the overall efficiency by 12% at lighter loads, where the switching losses are dominant, as well as a significant improvement throughout the entire operating range. For portable applications, this improvement is quite important and extends the battery lifetime significantly, as portable devices spend a large portion of their on-time in sleep or idle mode [MUL 05].

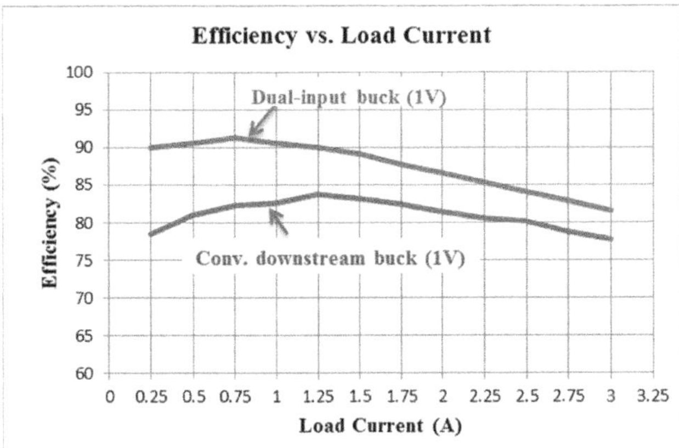

Figure 7.19. *Efficiency comparisons of MSC-DB 2-input buck architecture and the conventional downstream converter*

As mentioned previously, the introduced MSC-DB architecture reduces the inductance value requirement without the need for a significant increase in switching frequency, potentially allowing full on-chip integration of the inductors. To demonstrate this possibility, two PMICs are co-packaged with power inductors [VUK 15] and a two-output version of MSC-DB [LEE 05] is formed. In this architecture, the input voltage V_{in} is selected to be 3 V and the front-end SC stage provides $V_{in}/2$ =1.5 V across each of the two intermediate capacitors (C_{mid1} and C_{mid2}). Two differential buck converters provide output voltages, V_{out1} = 2.5 V, V_{out2} = 1 V. High-efficiency, low-profile, 100 nH on-chip inductors [WAN 08] are placed on top of the PMIC modules, as shown in Figure 7.20, and packaged together. It demonstrates that MSC-DB increases the level of on-chip implementation capability, creating a possibility of full on-chip integration of inductors with the power management integrated circuit (PMIC) in the near future.

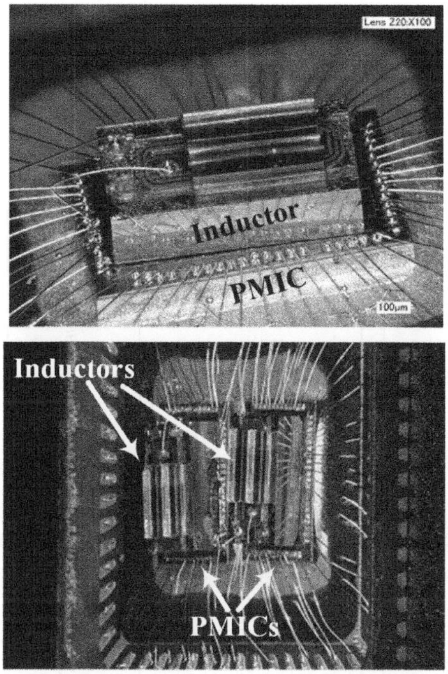

Figure 7.20. *Co-packaging of PMIC-BBs with on-chip power inductors. For the color version of this figure, see www.iste.co.uk/allard/systems.zip*

7.5. Multi-level and flying capacitor multi-level converters

In multi-level structures, the reduction of the voltage swing is achieved by the use of active capacitive divider structures. Figure 7.21 shows an active capacitor divider-based three-level buck converter [RAD 12] that can be derived by a source-load inversion of a three-level boost converter [MEY 95]. In this topology, the conventional SC stage is eliminated, and to balance the capacitive divider, the inductor of the buck converter is used. In this way, practically a single stage conversion is obtained and all switching components are shared between the divider and the buck stage.

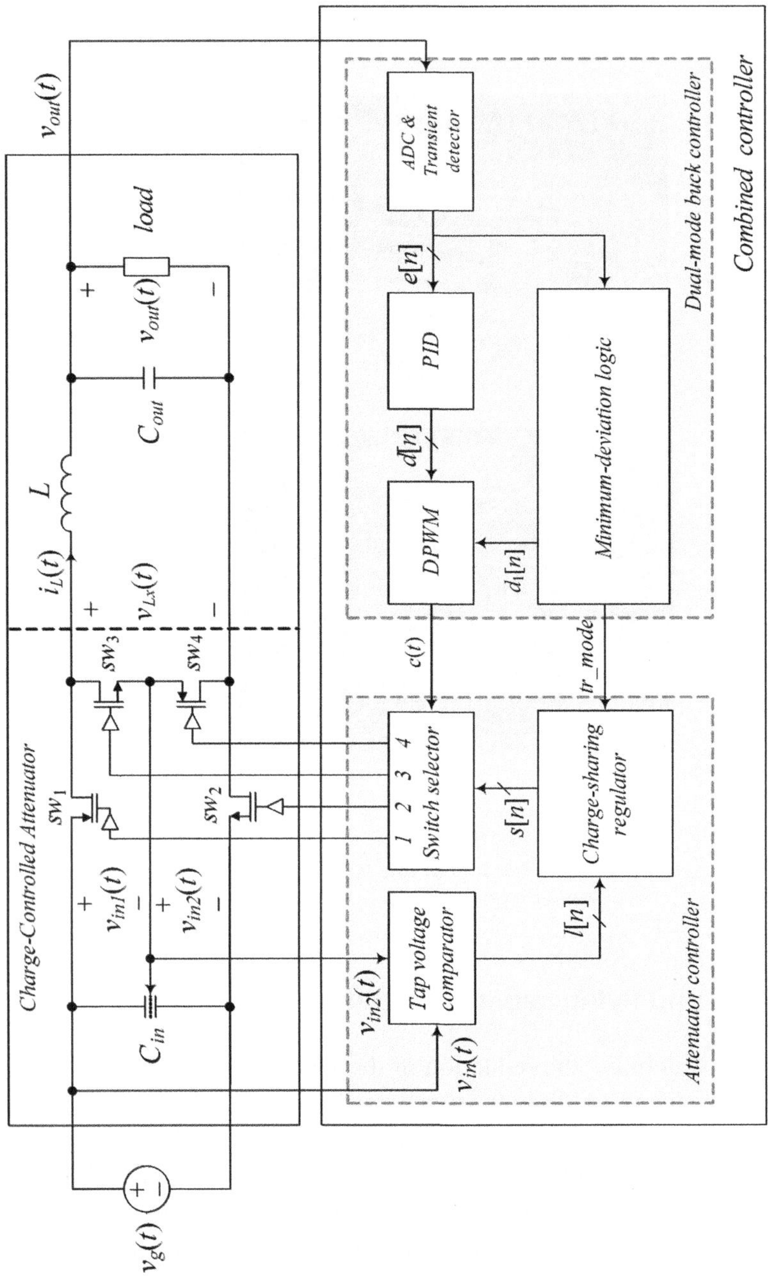

Figure 7.21. *Three-level capacitive divider-based buck converter*

In the topology, the capacitive divider guarantees a reduced voltage stress of switches and a lower inductor voltage swing, which in this case are reduced by half, compared with the conventional buck. Another interesting feature of this topology is that it replaces the input filter capacitor with a capacitive divider, theoretically, resulting in the same input filter capacitive energy storage requirement as the conventional buck stage. However, the main drawback of this topology for some of the applications is the non-existence of the common ground between the source and the output, which limits its use in some applications, where the common ground is preferred.

Flying capacitor-based multi-level converters, introduced by Meynard [MEY 95, MEY 92, MEY 02], have been widely adopted in high power applications. In these converters, the voltage swing reduction is obtained with a FC that behaves as a constant voltage source simultaneously reducing the voltage swing across the inductor and voltage stress of semiconductor components. For example, in the three-level buck version of this converter shown in Figure 7.6, the flying capacitor voltage is kept at $V_{in}/2$, and by shuffling it between the ground and Vin, as mentioned previously, advantages are obtained. Compared with the previously presented three-level buck, the three-level FC buck converter shown in Figure 7.22 provides the advantage of having the common ground between the battery and the load at the expense of an extra capacitor, i.e. flying capacitor, which, in the targeted applications, can be of a small size and often results in a favorable trade-off.

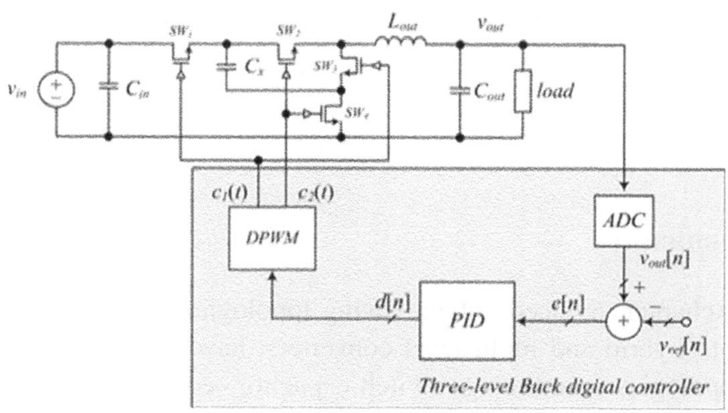

Figure 7.22. *Three-level FC-based buck converter*

A four-switch modification of a two-phase three-level FC double buck converter, introduced by Nishijima and Sato [NIS 05], is shown in Figure 7.23. This topology uses two inductors to balance the flying capacitor to provide a high step-down conversion ratio. To reduce the number of switching components, in this topology, trade-offs in terms of the conversion range and blocking voltage of a one transistor are made. This topology can only work for the input-to-output conversion ratios up to 1/4 and requires switch Q4 (Figure 7.22) to be rated for the full input voltage.

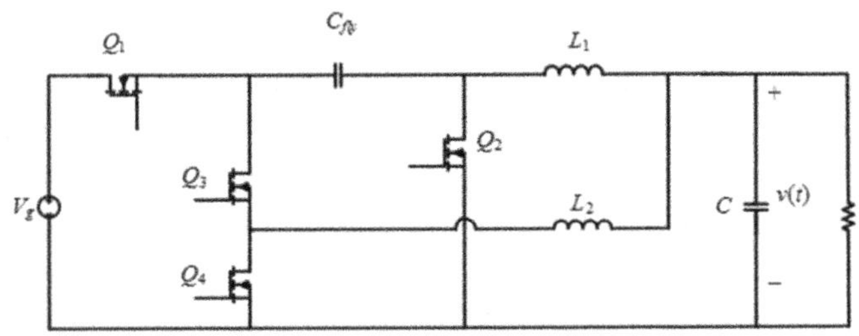

Figure 7.23. *Two-phase FC-based double buck converter*

In low power applications, FC multi-level converters still have not found a wider adoption. This is mostly due to more complex converter dynamics and challenges related to the implementation of cost-effective control. In recent publications [VUK 15, VUK 16], several of those challenges, such as adjustment of non-overlapping time for switches at zero inductor ripple operating point and flying capacitor balancing in the presence of non-negligible ripple, have been addressed and hardware-efficient solutions have been proposed.

7.6. Conclusion

In this chapter, reduced voltage swing topologies, which can be roughly divided into hybrid and multi-level converters, have been reviewed. Several hybrid topologies combining switch-capacitor converters and inductive converters suitable for on-chip implementation have been presented. Also, two main classes of multi-level topologies (capacitive divider based and flying capacitor based) suitable for on-chip implementation have been presented. In general, these topologies reduce the overall energy storage requirements of converters by minimizing inductor flux linkage while, at the same timer, are improving power processing efficiency. The reduction of inductor flux linkage also enables the use of lower voltage stress switches, which offsets the cost of a larger number of silicon components required.

The advantages of these architectures are experimentally verified by both discrete and IC-based implementations, where on-chip inductors are also co-packaged with silicon switches, to demonstrate that a significant reduction of the inductor volume allows for a higher level of integration, opening a possibility for complete on-chip system integration in the near future.

Bibliography

[ABE 07] ABEDINPOUR S., BAKKALOGLU B., KIAEI S., "A multistage interleaved synchronous buck converter with integrated output filter in 0.18 μm SiGe process", *IEEE Transactions on Power Electronics*, vol. 22, no. 6, pp. 2164–2175, 2007.

[AHN 94a] AHN C.H., KIM Y.J., ALLEN M.G., "A fully integrated planar toroidal inductor with a micromachined nickel-iron magnetic bar", *IEEE Transactions on Components, Packaging, and Manufacturing Technology: Part A*, vol. 17, no. 3, pp. 463–469, 1994.

[AHN 94b] AHN C.H., ALLEN M., "A new toroidal-meander type integrated inductor with a multilevel meander magnetic core", *IEEE Transactions on Magnetics*, vol. 30, no. 1, pp. 73–79, 1994.

[AHN 98] AHN C.H., ALLEN M., "Micromachined planar inductors on silicon wafers for MEMS applications", *IEEE Transactions on Industrial Electronics*, vol. 45, no. 6, pp. 866–876, 1998.

[AHS 11] AHSANUZZAMAN S.M., RADIĆ A., PRODIĆ A., "Adaptive switching frequency scaling digital controller for improving efficiency of battery powered DC–DC converters", *IEEE Applied Power Electronics Conference and Exposition (APEC)*, pp. 910–915, 2011.

[AHS 15a] AHSANUZZAMAN S.M., PRODIĆ A., JOHNS D.A., "A power management IC for distributed power supplies in low to medium power applications", *European Conference on Power Electronics and Applications (EPE)*, 2015.

[AHS 15b] AHSANUZZAMAN S.M., JOHNS D.A., PRODIĆ A., "A configurable power management IC for low-volume DC–DC converter applications with high frequency current programmed mode control", *IEEE Applied Power Electronics Conference and Exposition (APEC)*, 2015.

[ALI 09] ALIMADADI M., SHEIKHAEI S., LEMIEUX G. *et al.*, "A fully integrated 660 MHz low-swing energy-recycling DC–DC converter", *IEEE Transactions on Power Electronics*, vol. 24, no. 6, pp. 1475–1485, 2009.

[AME 09] AMELIFARD B., PEDRAM M., "Optimal design of the power-delivery network for multiple voltage-island system-on-chips", *IEEE Transactions on Computer-Aided Design of Integrated Circuits and Systems*, vol. 28, no. 6, pp. 888–900, 2009.

[AND 13a] ANDERSEN T., ZINGERLI C.M., KRISMER F. *et al.*, "Modeling and pareto optimization of microfabricated inductors for power supply on chip", *IEEE Transactions on Power Electronics*, vol. 28, no. 9, pp. 4422–4430, 2013.

[AND 13b] ANDERSEN T., KRISMER F., KOLAR J. *et al.*, "A 4.6W/mm^2 power density 86% efficiency on-chip switched capacitor DC–DC converter in 32 nm SOI CMOS", *Applied Power Electronics Conference and Exposition (APEC), 2013 Twenty-Eighth Annual IEEE*, pp. 692–699, March 2013.

[AND 14] ANDERSEN T., KRISMER F., KOLAR J. *et al.*, "A sub-ns response on-chip switched-capacitor DC–DC voltage regulator delivering 3.7W/mm^2 at 90% efficiency using deep-trench capacitors in 32 nm SOI CMOS", *2014 IEEE International on Solid-State Circuits Conference Digest of Technical Papers (ISSCC)*, pp. 90–91, February 2014.

[AND 16] ANDERSEN T., "A 10 W on-chip switched capacitor voltage regulator with feedforward regulation capability for granular microprocessor power delivery", *IEEE Transactions on Power Electronics*, 2016.

[ANS 15] ANS Y., ZHANG D.M., RODRIGUEZ M., "Monolithic very high frequency GaN switched-mode power converters", *IEEE Custom Integrated Circuits Conference, CICC*, pp. 1–4, 2015.

[APA 02] APARICIO R., HAJIMIRI A., "Capacity limits and matching properties of integrated capacitors", *IEEE Journal of Solid-State Circuits*, vol. 37, no. 3, pp. 384–393, 2002.

[ARB 97] ARBETTER B., MARKSIMOVIC D., "Feedforward pulse width modulators for switching power converters", *IEEE Transactions on Power Electronics*, vol. 12, no. 2, pp. 361–368, 1997.

[ARN 98] ARNTZEN B., MAKSIMOVIC D., "Switched-capacitor DC/DC converters with resonant gate drive", *IEEE Transactions on Power Electronics*, vol. 13, no. 5, pp. 892–902, 1998.

[BAE 05] BAE S., JUNG H.-M., MOON J.-S. *et al.*, "Embedded inductor using Ni-Zn ferrite film in polymer substrate", *INTERMAG Asia 2005, Digests of the IEEE International Magnetics Conference, 2005*, IEEE, pp. 1767–1768, 2005.

[BAK 08] BAK M., DUDEK M., DZIEDZIC A., "Chosen electrical and stability properties of surface and embedded planar PCB inductors", *2008 31st International Spring Seminar on Electronics Technology*, IEEE, pp. 545–549, 2008.

[BAL 08] BALIGA B.J., *Fundamentals of Power Semiconductor Devices*, Springer Science, 2008.

[BAN 00] BANERJEE S., ROY S., CHEN J. *et al.*, "Magnetic properties of oxide-coated iron nanoparticles synthesized by electrodeposition", *Journal of Magnetism and Magnetic Materials*, vol. 219, no. 1, pp. 45–52, 2000.

[BAT 12a] BATHILY M., ALLARD B., HASBANI F., "A 200 MHz integrated buck converter with resonant gate drivers for an RF power amplifier", *IEEE Transactions on Power Electronics*, vol. 27, no. 2, pp. 610–613, 2012.

[BAT 12b] BATHILY M., ALLARD B., HASBANI F. *et al.*, "Design flow for high switching frequency and large-bandwidth analog DC/DC step-down converters for a polar transmitter", *IEEE Transactions on Power Electronics*, vol. 27, no. 2, pp. 838–847, 2012.

[BAU 07] BAUGHMAN A.C., FERDOWSI M., "Double-tiered switched-capacitor battery charge equalization technique", *IEEE Transactions on Industrial Electronics*, vol. 55, no. 6, pp. 2277–2285, 2007.

[BAY 00] BAYER E., SCHMELLER H., "Charge pump with active cycle regulation-closing the gap between linear and skip modes", *IEEE Power Electronics Specialists Conference, PESC*, vol. 3, pp. 1497–1502, 2000.

[BEL 16] BEL POWER PRODUCT S.-F., Report, accessed 2016.

[BEN 09] BEN-YAAKOV S., EVZELMAN M., "Generic and unified model of switched capacitor converters", *IEEE Energy Conversion Congress and Exposition, ECCE*, pp. 3501–3508, 2009.

[BER 09] BERGVELD H., NOWAK K., KARADI R. *et al.*, "A 65-nm-CMOS 100-MHz 87%-efficient DC-DC down converter based on dual-die system-in-package integration", *Energy Conversion Congress and Exposition 2009, IEEE*, pp. 3698–3705, September 2009.

[BRA 00] BRANDON E.J., WESSELING E., WHITE V. *et al.*, "Microinductors for spacecraft power electronics", *Proceedings of International Symposium on Magnetic Materials, Processes and Devices, Electrochemical Society*, pp. pp. 559–567, October 2000.

[BRA 03] BRANDON E.J., WESSELING E., WHITE V. *et al.*, "Fabrication and characterization of microinductors for distributed power converters", *IEEE Transactions on Magnetics*, vol. 39, no. 4, pp. 2049–2056, 2003.

[BRE 09] BREUSSEGEM T.V. *et al.*, "A 82% efficiency 0.5% ripple 16-phase fully integrated capacitive voltage doubler", *Symposium on VLSI Circuits, 2009*, pp. 198–199, June 2009.

[BRU 02] BRUNET M., O'DONNELL T.C., BAUD L. *et al.*, "Electrical performance of microtransformers for DC–DC converter applications", *IEEE Transactions on Magnetics*, vol. 38, no. 5, pp. 3174–3176, September 2002.

[BRU 04] BRUNET M., CONNELL A., MCCLOSKEY P. *et al.*, "Plating of multi-layer structures", WO Patent App. PCT/IE2003/000,176, 8 July 2004.

[BUR 00] BURD T., BRODERSEN R., "Design issues for dynamic voltage scaling", *IEEE International Symposium on Low Power Electronics and Design, ISLPED*, pp. 9–14, 2000.

[BUR 14] BURTON E., SCHROM G., PAILLET F. *et al.*, "FIVR – fully integrated voltage regulators on 4th Generation Intel Core SoCs", *Applied Power Electronics Conference and Exposition (APEC), 2014 Twenty-Ninth Annual IEEE*, pp. 432–439, March 2014.

[CAL 92] CALDERONE L., PINOLA L., VAROLI V., "Optimal feed-forward compensation for PWM DC/DC converters with 'linear' and 'quadratic' conversion ratio", *IEEE Transactions on Power Electronics*, vol. 7, no. 2, pp. 349–355, 1992.

[CAS 07] CASTILLA M., GARCIA DE VICUNA L., GUERRERO J.M. *et al.*, "Simple low-cost hysteretic controller for single-phase synchronous buck converters", *IEEE Transactions on Power Electronics*, vol. 22, no. 4, pp. 1232–1241, 2007.

[CHA 10] CHANG L., MONTOYE R., JI B. *et al.*, "A fully-integrated switched-capacitor 2:1 voltage converter with regulation capability and 90% efficiency at 2.3A/mm^2", *IEEE Symposium on VLSI Circuits (VLSIC), 2010*, pp. 55–56, June 2010.

[CHE 05] CHEN R., CANALES F., YANG B. *et al.*, "Volumetric optimal design of passive integrated power electronics module (IPEM) for distributed power system (DPS) front-end DC/DC converter", *IEEE Transactions on Industrial Applications*, vol. 41, no. 1, pp. 9–17, 2005.

[CHE 14a] CHENG L., LIU Y., KI W.-H., "A 10/30 MHz fast reference-tracking buck converter with DDA-based type-III compensator", *IEEE Journal of Solid-State Circuits*, vol. 49, no. 12, pp. 2788–2799, December 2014.

[CHE 14b] CHENG P., VASIC M., GARCIA O. *et al.*, "Minimum time control for multiphase buck converter: analysis and application", *IEEE Transactions on Power Electronics*, vol. 29, no. 2, pp. 958–967, 2014.

[CHE 16] CHEN K., *Power Management Techniques for Integrated Circuit Design*, John Wiley and Sons, 2016.

[CHI 00] CHIESI L., KEJIK P., JANOSSY B. *et al.*, "CMOS planar 2D micro-fluxgate sensor", *Sensors and Actuators A: Physical*, vol. 82, no. 1–3, pp. 174–180, 2000.

[CLE 05] CLEVELAND T.L., "Bi-directional power system for laptop computers", *IEEE Applied Power Electronics Conference and Exposition (APEC)*, pp. 199–203, 2005.

[COO 00] COONLEY K.D., MEHAS G.J., SULLIVAN C.R. *et al.*, "Evaporatively deposited Co-MgF2 granular materials for thin-film inductors", *IEEE Transactions on Magnetics*, vol. 36, no. 5, pp. 3463–3465, September 2000.

[COR 09a] CORRADINI L., COSTABEBER A., MATTAVELLI P. *et al.*, "Parameter-independent time-optimal digital control for point-of-load converters", *IEEE Transactions on Power Electronics*, vol. 24, no. 10, pp. 2235–2248, 2009.

[COR 09b] CORRADINI L., ORIETTI E., MATTAVELLI P. *et al.*, "Digital hysteretic voltage-mode control for DC–DC converters based on asynchronous sampling", *IEEE Transactions on Electron Devices*, vol. 24, no. 201-211, 2009.

[COR 10] CORRADINI L., BABAZADEH A., BJELETIĆ A. *et al.*, "Current-limited time-optimal response in digitally controlled DC–DC converters", *IEEE Transactions on Power Electronics*, vol. 25, no. 11, pp. 2869–2880, 2010.

[COR 14a] CORTES J., SVIKOVIC V., ALOU P. *et al.*, "Comparison of the behavior of voltage mode, V^2 and V^2IC control of a buck converter for a very fast and robust dynamic response", *IEEE Applied Power Electronics Conference and Exposition (APEC)*, pp. 2888–2894, 2014.

[COR 14b] CORTES J., SVIKOVIC V., ALOU P. *et al.*, "An optimization algorithm to design fast and robust analog controls for Buck converters", *IEEE Workshop on Control and Modeling in Power Electronics*, pp. 1–10, 2014.

[COR 15a] CORTES J., A methodology to analyze, design and implement very fast and robust controls of Buck-type converters, PhD Thesis, Universita Politecnica de Madrid, 2015.

[COR 15b] CORTES J., SVIKOVIC V., ALOU P. *et al.*, "Improved transient response of controllers by synchronizing the modulator with the load step: application to V2 IC", *IEEE Transactions on Power Electronics*, vol. 30, no. 3, pp. 1577–1590, 2015.

[COR 15c] CORTES J., SVIKOVIC V., ALOU P. *et al.*, "Accurate analysis of subharmonic oscillations of V^2 and V^2IC controls applied to buck converter", *IEEE Transactions on Power Electronics*, vol. 30, no. 2, pp. 1005–1018, 2015.

[COR 15d] CORTES J., SVIKOVIC V., ALOU P. *et al.*, "v1 concept: designing a voltage-mode control as current mode with near time-optimal response for buck-type converters", *IEEE Transactions on Power Electronics*, vol. 30, no. 10, pp. 5829–5841, 2015.

[COS 08] COSTABEBER A., CORRADINI L., MATTAVELLI P. *et al.*, "Time optimal, parameters-insensitive digital controller for DC–DC buck converters", *IEEE Power Electronics Specialists Conference, PESC*, pp. 1243–1249, 2008.

[DAS 02] DAS D., ROY S., CHEN J.W. *et al.*, "Interface controlled electrical and magnetic properties in Fe–Fe_3O_4–silica gel nanocomposites", *Journal of Applied Physics*, vol. 91, no. 7, pp. 4573–4579, 2002.

[DEL 16] DELTA ELECTRONICS PRODUCT I., Report, accessed 2016.

[DEX 53] DEXTER D.L., HELLER W.R., "Quantum theory of the polarizability of an idealized crystal", *Physical Review*, vol. 91, p. 273, 1953.

[DHA 04] DHAGAT P., PRABHAKARAN S., SULLIVAN C.R., "Comparison of magnetic materials for V-groove inductors in optimized high-frequency DC–DC converters", *IEEE Transactions on Magnetics*, vol. 40, no. 4, pp. 2008–2010, 2004.

[DIB 10] DIBENE J.T., MORROW P., PARK C. *et al.*, "A 400 a fully integrated silicon voltage regulator with in-die magnetically coupled embedded inductors", *APEC–Special Presentation*, 2010.

[DIC 76] DICKSON J.F., "On-chip high-voltage generation in MNOS integrated circuits using an improved voltage multiplier technique", *IEEE Journal of Solid-State Circuits*, vol. 11, no. 3, pp. 374–378, 1976.

[DOG 05] DOGAN A., The reliability of the silicon nitride dielectric in capacitive MEMS switches, PhD Thesis, Pennsylvania State University, 2005.

[DOW 66] DOWELL P.L., "Effects of eddy currents in transformer windings", *Proceedings of the Institution of Electrical Engineers*, vol. 113, no. 8, pp. 1387–1394, 1966.

[EFF 08] EFFLER S., KELLY A., HALTON M. *et al.*, "Digital control law using a novel load current estimator principle for improved transient response", *IEEE Power Electronics Specialists Conference, PESC*, pp. 4585–4589, 2008.

[ELD 13] EL-DAMAK D., BANDYOPADHYAY S., CHANDRAKASAN A., "A 93% efficiency reconfigurable switched-capacitor DC–DC converter using on-chip ferroelectric capacitors", *Solid-State Circuits Conference Digest of Technical Papers (ISSCC)*, IEEE, pp. 374–375, February 2013.

[ENP 16a] ENPIRION PRODUCT, Report, accessed 2016.

[ENP 16b] ENPIRION PRODUCT, Report, accessed 2016.

[ERI 01] ERICKSON R.W., MAKSIMOVIĆ D., *Fundamentals of Power Electronics*, Springer Science+Business Media, 2001.

[EZH 00] EZHILVALAVAN S., TSENG T.-Y., "Progress in the developments of (Ba,Sr)TiO_3 (BST) thin films for Gigabit era DRAMs", *Materials Chemistry and Physics*, vol. 65, pp. 227–248, 2000.

[FAI 07] FAIRCHILD SEMICONDUCTORS, FDMF8700 data sheet, Datasheet, 2007.

[FEN 09] FENG S., WING-HUNG K., "Digitally assisted quasi-V2 hysteretic buck converter with fixed frequency and without using large-ESR capacitor", *2009 IEEE International Solid-State Circuits Conference – Digest of Technical Papers*, IEEE, pp. 446–447, February 2009.

[FLY 06] FLYNN D., DHARIWAL R., DESMULLIEZ M., "A design study of microscale magnetic components for operation in the MHz frequency range", *Journal of Micromechanics and Microengineering*, vol. 16, no. 9, p. 1811, 2006.

[FOL 10] FOLEY R., WALDRON F., SLOWEY J. *et al.*, "Technology roadmapping for Power Supply in Package (PSiP) and Power Supply on Chip (PwrSoC)", *Applied Power Electronics Conference and Exposition (APEC)*, pp. 525–532, February 2010.

[FUJ 06] FUJI ELECTRIC I., FB6831J, Datasheet, 2016.

[FUK 03] FUKUDA Y., INOUE T., MIZOGUCHI T. *et al.*, "Planar inductor with ferrite layers for DC–DC converter", *IEEE Transactions on Magnetics*, vol. 39, no. 4, pp. 2057–2061, July 2003.

[GAO 05] GAO X.-Y., ZHOU Y., DING W. *et al.*, "Fabrication of ultralow-profile micromachined inductor with magnetic core material", *IEEE Transactions on Magnetics*, vol. 41, no. 12, pp. 4397–4400, December 2005.

[GAR 08] GARDNER D.S., SCHROM G., HAZUCHA P. *et al.*, "Integrated on-chip inductors using magnetic material (invited)", *Journal of Applied Physics*, vol. 103, no. 7, pp. 1–7, 2008.

[GAR 09] GARDNER D., SCHROM G., PAILLET F. *et al.*, "Review of on-chip inductor structures with magnetic films", *IEEE Transactions on Magnetics*, vol. 45, no. 10, pp. 4760–4766, October 2009.

[GIL 02] GILDERSLEEVE M., FORGHANI-ZADEH H., RINCON-MORA G., "A comprehensive power analysis and a highly efficient, mode-hopping DC–DC converter", *Proceedings of the 2002 IEEE Asia-Pacific Conference on ASIC*, pp. 153–156, 2002.

[GJA 11] GJANCI J., CHOWDHURY M., "A hybrid scheme for on-chip voltage regulation in system-on-a-chip (SOC)", *IEEE Transactions on Very Large Scale Integration (VLSI) Systems*, vol. 19, no. 11, pp. 1949–1959, November 2011.

[GOD 11] GODSELL J.F., BALA T., RYAN K.M. *et al.*, "An ac susceptibility study in capped Ni/Ni(OH) 2 core–shell nanoassemblies: dual peak observations", *Journal of Physics D: Applied Physics*, vol. 44, no. 32, p. 325004, 2011.

[GOD 96] GODER D., PELLETIER W.R., "V^2 architecture provides ultra-fast transient response in switch mode power supplies", *Conference High Frequency Power Converters*, pp. 19–23, 1996.

[GOO 15] GOODARZI S., BEIRANVAND R., MOUSAVI S. *et al.*, "A new algorithm for increasing balancing speed of switched-capacitor lithium-ion battery cell equalizers", *IEEE Power Electronics, Drives Systems and Technologies Conference (PEDSTC)*, pp. 292–297, 2015.

[GRE 06] GREGOIRE B.R., "A compact switched-capacitor regulated charge pump power supply", *IEEE Journal of Solid-State Circuits*, vol. 41, no. 8, pp. 1944–1953, 2006.

[GUA 06] GUANG F., MEYER E., LIU Y.-F., "High performance digital control algorithms for DC–DC converters based on the principle of capacitor charge balance", *IEEE Power Electronics Specialists Conference, PESC*, pp. 1–7, 2006.

[GUA 07] GUANG F., MEYER E., LIU Y.-F., "A new digital control algorithm to achieve optimal dynamic performance in DC-to-DC converters", *IEEE Transactions on Power Electronics*, vol. 22, no. 4, pp. 1489–1498, 2007.

[GUO 09] GUO S., GAO Y., XU Y. *et al.*, "Digital PWM controller for high-frequency low-power DC–DC switching mode power supply", *Power Electronics and Motion Control Conference, 2009. IPEMC '09. IEEE 6th International*, pp. 1340–1346, May 2009.

[GUP 07] GUPTA M., OATLEY J., JOSEPH R. *et al.*, "Understanding voltage variations in chip multiprocessors using a distributed power-delivery network", *Design, Automation Test in Europe Conference Exhibition 2007*, pp. 1–6, April 2007.

[HAB 91] HABRAKEN F.H.P.M., ed., *LPCVD Silicon Nitride and Oxynitride Films, Material and Applications in Integrated Circuit Technology*, Springer-Verlag, 1991.

[HAY 03] HAYASHI Z., KATAYAMA Y., EDO M. *et al.*, "High-efficiency DC–DC converter chip size module with integrated soft ferrite", *IEEE Transactions on Magnetics*, vol. 39, no. 5, pp. 3068–3072, September 2003.

[HAZ 05] HAZUCHA P., SCHROM G., HAHN J. *et al.*, "A 233 MHz 80 %-87 % efficient four-phase DC–DC converter utilizing air-core inductors on package", *IEEE Journal of Solid-State Circuits*, vol. 40, no. 4, pp. 838–845, 2005.

[HEN 09] HENRY P., "New advances in portable electronics", *IEEE Applied Power Electronics Conference and Exposition (APEC)*, 2009.

[HER 90] HERZER G., "Grain size dependence of coercivity and permeability in nanocrystalline ferromagnets", *IEEE Transactions on Magnetics*, vol. 26, no. 5, pp. 1397–1402, September 1990.

[HER 93] HERZER G., "Nanocrystalline soft magnetic materials", *Physica Scripta*, vol. t49A, p. 307, 1993.

[HUA 01] HUANG W., "A new control for multi-phase buck converter with fast transient response", *IEEE Applied Power Electronics Conference and Exposition (APEC)*, pp. 273–279, 2001.

[HUA 07] HUA L., LUO S., "Design considerations for small signal modeling of DC–DC converters using inductor DCR current sensing under time constants mismatch conditions", *2007 IEEE Power Electronics Specialists Conference*, Orlando, IEEE, pp. 2182–2188, 2007.

[HUA 11] HUANG C.-S., WANG C.-Y., WANG J.-H. *et al.*, "A fast-transient quasi-V2 switching buck regulator using AOT control", *IEEE Asian Solid-State Circuits Conference 2011*, IEEE, pp. 53–56, November 2011.

[HUA 13] HUANG C., MOK P., "A 100 MHz 82.4% efficiency package-bondwire based four-phase fully-integrated buck converter with flying capacitor for area reduction", *IEEE Journal of Solid-State Circuits*, vol. 48, no. 12, pp. 2977–2988, December 2013.

[HUE 09a] HUERTA S., ALOU P., OLIVER J. *et al.*, "Design methodology of a non-invasive sensor to measure the current of the output capacitor for a very fast non-linear control", *IEEE Applied Power Electronics Conference and Exposition (APEC)*, pp. 806–811, 2009.

[HUE 09b] HUERTA S., ALOU P., OLIVER J. *et al.*, "A very fast control based on hysteresis of the C_{out} current with a frequency loop to operate at constant frequency", *2009 Twenty-Fourth Annual IEEE Applied Power Electronics Conference and Exposition*, IEEE, pp. 799–805, February 2009.

[HUE 13] HUERTA S., SOTO A., ALOU P. *et al.*, "Advanced control for very fast DC–DC converters based on hysteresis of the C_{out} current", *IEEE Transactions on Circuits and Systems I: Regular Papers*, vol. 60, no. 4, pp. 1052–1061, April 2013.

[HUS 11] HUSSEIN H.-H., BATARSEH I., "State-of-charge estimation for a single lithium battery cell using extended Kalman Filter", *IEEE Power and Energy Society General Meeting*, pp. 1–5, 2011.

[INO 98] INOUE T., NISHIJIMA K., YATABE S. *et al.*, "The effect of magnetic film structure on the inductance of a planar inductor", *IEEE Transactions on Magnetics*, vol. 34, no. 4, pp. 1372–1374, July 1998.

[INT 08] INTEL, Voltage regulator module (VRM) and enterprise voltage regulator-down (EVRD) 11.0, Specifications, 2008.

[INT 14] INTERSIL, ISL8201M, Datasheet, 2014.

[INT 15] INTERSIL, ISL6545, Datasheet, 2015.

[IRF 16] IRF PRODUCT O., Report, accessed 2016.

[ISH 10] ISHIDA K., TAKEMURA K., BABA K. *et al.*, "3D stacked buck converter with 15 um thick spiral inductor on silicon interposer for fine-grain power-supply voltage control in SiPs", *3D Systems Integration Conference (3DIC), 2010 IEEE International*, pp. 1–4, November 2010.

[JAC 13] JACQUELINE S., DOMENGÉS B., VOIRON F. *et al.*, "Conduction mechanisms in 2D and 3D SIS capacitors", *Semiconductor Science Technologies*, vol. 28, 2013.

[JAI 14] JAIN R., GEUSKENS B., KIM S. *et al.*, "A 0.45–1 V fully-integrated distributed switched capacitor DC–DC converter with high density MIM capacitor in 22 nm Tri-Gate CMOS", *IEEE Journal of Solid-State Circuits*, vol. 49, no. 4, pp. 917–927, April 2014.

[JAM 10] JAMIESON B., O'DONNELL T., KULKARNI S. *et al.*, "Shape-independent permeability model for uniaxially-anisotropic ferromagnetic thin films", *Applied Physics Letters*, vol. 96, p. 202509, 2010.

[JED 04] JEDEC, "Foundry process qualification guidelines, JP001.01", available at www.jedec.org/standards-documents/docs/jp-0010/, May 2004.

[JEO 12] JEON H., KIM Y.-B., "A fully integrated switched-capacitor DC–DC converter with dual output for low power application", *Proceedings of the Great Lakes Symposium on VLSI*, pp. 83–86, 2012.

[JIA 08] LU J., HONGWEI, J. ARIAS A. *et al.*, "On-chip bondwire magnetics with ferrite-epoxy glob coating for power systems on chip", *International Journal of Power Management Electronics*, vol. 2008, p. 9, 2008.

[JIA 10] JIA L., WANG D., FU J. *et al.*, "A novel parameter-independent digital optimal control algorithm for DC–DC buck converters based on parabolic curve fitting", *IEEE Energy Conversion Congress and Exposition, ECCE*, pp. 500–507, 2010.

[KAI 04] KAIWEI Y., High-frequency and high-performance VRM design for the next generations of processors, PhD Thesis, Virginia Polytechnic Institute and State University, 2004.

[KAT 00] KATAYAMA Y., SUGAHARA S., NAKAZAWA H. *et al.*, "High-power-density MHz-switching monolithic DC–DC converter with thin-film inductor", *Power Electronics Specialists Conference, IEEE 31st Annual*, vol. 3, pp. 1485–1490, 2000.

[KES 14] KESARWANI K., SANGWAN R., STAUTH J., "Resonant switched-capacitor converters for chip-scale power delivery: design amp; implementation", *IEEE Transactions on Power Electronics*, vol. PP, no. 99, pp. 1–1, 2014.

[KIM 02] KIM K.H., KIM J., KIM H.-J. *et al.*, "A megahertz switching DC/DC converter using FeBN thin film inductor", *IEEE Transactions on Magnetics*, vol. 38, no. 5, pp. 3162–3164, September 2002.

[KIM 05a] KIMBALL J.W., KREIN P.T., "Analysis and design of switched capacitor converters", *Twentieth Annual IEEE Applied Power Electronics Conference and Exposition, APEC*, vol. 3, pp. 1473–1477, 2005.

[KIM 05b] KIMBALL J.W., KREIN P.T., CAHILL K.R., "Modeling of capacitor impedance in switching converters", *IEEE Transactions on Power Electronics*, vol. 3, no. 4, pp. 136–140, 2005.

[KIM 14] KIM M.Y., KIM C.H., KIM J.H. *et al.*, "A chain structure of switched capacitor for improved cell balancing speed of lithium-ion batteries", *IEEE Transactions on Industrial Electronics*, vol. 61, no. 8, pp. 3989–3999, 2014.

[KOS 12] KOSE S., FRIEDMAN E., "Distributed on-chip power delivery", *IEEE Journal on Emerging and Selected Topics in Circuits and Systems*, vol. 2, no. 4, pp. 704–713, December 2012.

[KOW 05] KOWASE I., SATO T., YAMASAWA K. *et al.*, "A planar inductor using Mn-Zn ferrite/polyimide composite thick film for low-voltage and large-current DC–DC converter", *IEEE Transactions on Magnetics*, vol. 41, no. 10, pp. 3991–3993, October 2005.

[KUD 11] KUDVA S., HARJANI R., "Fully-integrated on-chip DC-DC converter with a 450X output range", *IEEE Journal of Solid-State Circuits*, vol. 46, no. 8, pp. 1940–1951, August 2011.

[KUM 12] KUMAR P., PROEFROCK W., "Novel switched capacitor based Triple Output Fixed Ratio Converter (TOFRC)", *IEEE Applied Power Electronics Conference and Exposition (APEC)*, pp. 2352–2356, 2012.

[KUR 15] KURD N., CHOWDHURY M., BURTON E. *et al.*, "Haswell: A family of IA 22 nm processors", *IEEE Journal of Solid-State Circuits*, vol. 50, no. 1, pp. 49–58, January 2015.

[LAB 13] LABBE B., A contribution to synchronization of the sliding-mode control-based integrated step-down DC/DC converter, PhD Thesis, INSA-Lyon, France, 2013.

[LAL 13] LALLEMAND F., VOIRON F., "Silicon interposers with integrated passive devices, an excellent alternative to discrete components", *Microelectronics Packaging Conference (EMPC), 2013 European*, pp. 1–6, September 2013.

[LE 10] LE H.-P., SEEMAN M., SANDERS S. *et al.*, "A 32nm fully integrated reconfigurable switched-capacitor DC–DC converter delivering 0.55W/mm^2 at 81% efficiency", *Solid-State Circuits Conference Digest of Technical Papers (ISSCC), 2010 IEEE International*, pp. 210–211, February 2010.

[LE 11] LE H.-P., SANDERS S.R., ALON E., "Design techniques for fully integrated switched-capacitor DC–DC converters", *IEEE Journal of Solid-State Circuits*, vol. 46, no. 9, pp. 2120–2131, 2011.

[LE 13] LE H.-P., CROSSLEY J., SANDERS S. *et al.*, "A sub-ns response fully integrated battery-connected switched-capacitor voltage regulator delivering 0.19W/mm^2 at 73% efficiency", *Solid-State Circuits Conference Digest of Technical Papers (ISSCC), 2013 IEEE International*, pp. 372–373, February 2013.

[LEE 05] LEE Y.S., CHENG M.W., "Intelligent control battery equalization for series connected lithium-ion battery strings", *IEEE Transactions on Industrial Electronics*, vol. 52, no. 5, pp. 1297–1307, 2005.

[LEE 06] LEE J.-Y., KIM S.-E., SONG S.-J. *et al.*, "A regulated charge pump with small ripple voltage and fast start-up", *IEEE Journal of Solid-State Circuits*, vol. 41, no. 2, pp. 425–432, February 2006.

[LEE 08] LEE D.W., HWANG K.P., WANG S.X., "Fabrication and analysis of high-performance integrated solenoid inductor with magnetic core", *IEEE Transactions on Magnetics*, vol. 44, no. 11, pp. 4089–4095, November 2008.

[LEE 09] LEE F., "A hysteretic control method for multiphase voltage regulator", *IEEE Transactions on Power Electronics*, vol. 24, no. 12, pp. 2726–2734, December 2009.

[LEU 05] LEUNG K.K.-S., CHUNG H.S.-H., "Dynamic hysteresis band control of the buck converter with fast transient response", *IEEE Transactions on Circuits and Systems II*, vol. 52, pp. 398–402, 2005.

[LI 05] LI W., SUN Y., SULLIVAN C.R., "High-frequency resistivity of soft magnetic granular films", *IEEE Transactions on Magnetics*, vol. 41, no. 10, pp. 3283–3285, October 2005.

[LI 09] LI Q., LIM M., SUN J. *et al.*, "Technology roadmap for high frequency integrated DC–DC converter", *Power Electronics and Motion Control Conference. IEEE 6th International*, pp. 1–8, May 2009.

[LI 12a] LI Q., "A fully-integrated buck converter design and implementation for on-chip power supplies", *Journal of Computers*, vol. 7, no. 5, pp. 1270–1277, May 2012.

[LI 12b] LI S., MI C., ZHANG M., "A high efficiency low cost direct battery balancing circuit using a multi-winding transformer with reduced switch count", *IEEE Applied Power Electronics Conference and Exposition (APEC)*, pp. 2128–2133, 2012.

[LIM 15] LIM S., RANSON J., OTTEN D.M. *et al.*, "Two-stage power conversion architecture suitable for wide range input voltage", *IEEE Transactions on Power Electronics*, vol. 30, no. 2, pp. 805–816, 2015.

[LIN 07] LINEAR TECHNOLOGY CORP., LTM4608, Datasheet, 2007.

[LIN 08] LINEAR TECHNOLOGY CORP., LTC3851, Datasheet, 2008.

[LIN 16] LINEAGE POWER PRODUCT P., Report, 2016.

[LOT 98] LOTFI A., DOVER R.B.V., SCHNEEMEYER L. *et al.*, "Micro-transformer devices using thin-film electroplated deposition", *29th Annual IEEE Power Electronics Specialists Conference, 1998*, vol. 2, pp. 1511–1515, May 1998.

[LU 07] LU S., SUN Y., GOLDBECK M., "30 MHz power inductor using nano-granular magnetic material", *Power Electronics Specialists Conference, 2007*, pp. 1773–1776, June 2007.

[LU 10] LU J., JIA H., WANG X. *et al.*, "Modeling, design, and characterization of multiturn bondwire inductors with ferrite epoxy glob cores for power supply system-on-chip or system-in-package applications", *IEEE Transactions on Power Electronics*, vol. 25, no. 8, pp. 2010–2017, 2010.

[LU 14] LU D., QIAN Y., HONG Z., "An 87%-peak-efficiency DVS-capable single-inductor 4-output DC–DC buck converter with ripple-based adaptive off-time control", *IEEE International Solid-State Circuits Conference Digest of Technical Papers (ISSCC)*, pp. 82–83, February 2014.

[LU 15] LU Y., JIANG J., KI W.-H. *et al.*, "A 123-phase DC–DC converter-ring with fast-DVS for microprocessors", *IEEE International Solid-State Circuits Conference – (ISSCC)*, pp. 1–3, February 2015.

[LUD 03] LUDWIG M., DUFFY M., O'DONNELL T. *et al.*, "PCB integrated inductors for low power DC/DC converter", *IEEE Transactions on Power Electronics*, vol. 18, no. 4, pp. 937–945, 2003.

[LUK 07] LUKIC Z., ZHAO Z., PRODIC A. *et al.*, "Digital controller for multi-phase DC–DC converters with logarithmic current sharing", *IEEE Power Electronics Specialists Conference, PESC*, pp. 119–123, 2007.

[MAH 12] MAHDAVIKHAH B., JAIN P., PRODIC A., "Digitally controlled multi-phase buck-converter with merged capacitive attenuator", *IEEE Applied Power Electronics Conference and Exposition (APEC)*, pp. 1083–1087, 2012.

[MAK 95] MAKOWSKI M.S., MAKSIMOVIC D., "Performance limits of switched-capacitor DC–DC converters", *26th Annual IEEE Power Electronics Specialists Conference, PESC'95*, vol. 2, pp. 1215–1221, 1995.

[MAK 99] MAKSIMOVIC D., DHAR S., "Switched-capacitor DC–DC converters for low-power on-chip applications", *30th Annual IEEE Power Electronics Specialists Conference, 1999*, vol. 1, pp. 54–59, 1999.

[MAN 05] MANO Y., BAE S., MOON J. *et al.*, "Planar inductor with ferrite layers for DC–DC converter", *The 13th International Conference on Solid-State Sensors, Actuators and Microsystems, 2005*, vol. 1, pp. 891–894, 2005.

[MAN 12] MANIKTALA S., *Switching Power Supplies A-Z*, Newnes, 2012.

[MAS 16] MASSACHUSETTS INSTITUTE OF TECHNOLOGY, Principles of optimal control, course by J.P. How, accessed 2016.

[MAT 12] MATHUNA C., WANG N., KULKARNI S. *et al.*, "Review of integrated magnetics for power supply on chip (PwrSoC)", *IEEE Transactions on Power Electronics*, vol. 27, no. 11, pp. 4799–4816, 2012.

[MAT 13] MATHUNA C.O., WANG N., KULKARNI S. *et al.*, "PwrSoC (integration of micro-magnetic inductors/transformers with active semiconductors) for more than Moore technologies", *The European Physical Journal Applied Physics*, vol. 63, no. 1, p. 14408, 2013.

[MAT 14] MATHUNA C.O., WANG N., KULKARNI S. *et al.*, "Power supply with integrated passivEs – the EU FP7 powerswipe project", *8th International Conference on Integrated Power Systems (CIPS)*, pp. 1–7, February 2014.

[MAX 14] MAXIM INTEGRATED PRODUCTS CORP., MAX15046, Datasheet, 2016.

[MCR 15] MCRAE T., PRODIC A., "Hybrid serial-output converter topology for volume and weight restricted LED lighting applications", *IEEE Energy Conversion Congress and Exposition, ECCE-Asia*, 2015.

[MCR 16] MCRAE T., PRODIC A., LISI G. *et al.*, "Hybrid serial-output converter for integrated LED lighting applications", *IEEE Applied Power Electronics Conference and Exposition (APEC)*, 2016.

[MEE 09a] MEERE R., O'DONNELL T., BERGVELD H. *et al.*, "Analysis of microinductor performance in a 20–100 MHz DC/DC converter", *IEEE Transactions on Power Electronics*, vol. 24, no. 9, pp. 2212–2218, 2009.

[MEE 09b] MEERE R., O'DONNELL T., WANG N. *et al.*, "Size and performance tradeoffs in micro-inductors for high frequency DC–DC conversion", *IEEE Transactions on Magnetics*, vol. 45, no. 10, pp. 4234–4237, 2009.

[MEE 11] MEERE R., WANG N., O'DONNELL T. *et al.*, "Magnetic-core and air-core inductors on silicon: a performance comparison up to 100 MHz", *IEEE Transactions on Magnetics*, vol. 47, no. 10, pp. 4429–4432, 2011.

[MEN 05] MENSI L., COLALONGO L., RICHELLI A. *et al.*, "A new integrated charge pump architecture using dynamic biasing of pass transistors", *Proceedings of the 31st European Solid-State Circuits Conference, 2005. ESSCIRC 2005.*, pp. 85–88, September 2005.

[MES 14] MESTRALLET F., KERACHEV L., CREBIER J. *et al.*, "Multiphase interleaved converter for lithium battery active balancing", *IEEE Transactions on Power Electronics*, vol. 29, no. 6, pp. 2874–2881, 2014.

[MEY 92] MEYNARD T., FOCH H., "Multi-level conversion: high voltage choppers and voltage-source inverters", *IEEE Power Electronics Specialists Conference, PESC*, pp. 397–403, 1992.

[MEY 95] MEYNARD T., FOCH H., "Multilevel converters and derived topologies for high power conversion", *IEEE Internation Conference Industrial Electronics, Control, and Instrumentation*, pp. 21–26, 1995.

[MEY 02] MEYNARD T., FOCH H., THOMAS P. *et al.*, "Multicell converters: basic concepts and industry applications", *IEEE Transactions on Industrial Electronics*, vol. 49, no. 5, pp. 955–964, 2002.

[MEY 10] MEYER C.D., BEDAIR S.S., MORGAN B.C. *et al.*, "High-inductance-density, air-core, power inductors, and transformers designed for operation at 100–500 MHz", *IEEE Transactions on Magnetics*, vol. 46, no. 6, pp. 2236–2239, 2010.

[MEY 11] MEYVAERT H., VAN BREUSSEGEM T., STEYAERT M., "A 1.65W fully integrated 90 nm Bulk CMOS intrinsic charge recycling capacitive DC–DC converter: design amp; techniques for high power density", *Energy Conversion Congress and Exposition (ECCE), 2011 IEEE*, pp. 3234–3241, September 2011.

[MEY 12] MEYER E., ZHANG Z., LIU Y.-F., "Digital charge balance controller to improve the loading/unloading transient response of buck converters", *IEEE Transactions on Power Electronics*, vol. 27, no. 3, pp. 1314–1326, 2012.

[MIC 08] MICREL INCORPORATED, MIC2130, Datasheet, 2008.

[MIC 14] MICHAL V., "Peak-efficiency detection and peak-efficiency tracking algorithm for switched-mode DC–DC power converters", *IEEE Transactions on Power Electronics*, vol. 29, no. 12, pp. 6555–6568, 2014.

[MIC 15] MICREL INCORPORATED, MIC3385, Datasheet, 2015.

[MIN 96a] MINO M., TSUKAMOTO K., YANAGISAWA K. *et al.*, "A compact buck-converter using a thin-film inductor", *Eleventh Annual Applied Power Electronics Conference and Exposition. Conference Proceedings 1996.*, vol. 1, pp. 422–426, 1996.

[MIN 96b] MINO M., YACHI T., TAGO A. *et al.*, "Planar microtransformer with monolithically-integrated rectifier diodes for micro-switching converters", *IEEE Transactions on Magnetics*, vol. 32, no. 2, pp. 291–296, 1996.

[MIY 14] MIYAZAKI T., OGAWA T., "Constant on-time DC–DC converter using ripple injection filter with inherent adaptive voltage positioning", *Applied Power Electronics Conference and Exposition (APEC), 2014 Twenty-Ninth Annual IEEE*, pp. 460–463, March 2014.

[MOR 91] MORI S.M. *et al.*, "ONO inter-poly dielectric scaling for nonvolatile memory applications", *IEEE Transactions on Electron Devices*, vol. 38, pp. 386–391, 1991.

[MOR 11] MORROW P., PARK C.-M., KOERTZEN H. *et al.*, "Design and fabrication of on-chip coupled inductors integrated with magnetic material for voltage regulators", *IEEE Transactions on Magnetics*, vol. 47, no. 6, pp. 1678–1686, 2011.

[MUL 05] MULLIGAN M.D., BROACH B., LEE T.H., "A constant-frequency method for improving light-load efficiency", *IEEE Power Electronics Letters*, vol. 3, no. 1, pp. 24–29, 2005.

[MUR 16] MURATA PRODUCT M., Report, accessed 2016.

[MUR 06] MURRAY F.M. *et al.*, "Silicon based system-in-package: breakthroughs in miniaturization and "nano"-integration supported by very high quality passives and system level design tools", *MRS Fall Meeting*, 2006.

[MUS 05] MUSUNURI S., CHAPMAN P.L., "Design of low power monolithic DC–DC buck converter with integrated inductor", *2005 IEEE 36th Power Electronics Specialists Conference*, pp. 1773–1779, June 2005.

[NAK 00] NAKAZAWA H., EDO M., KATAYAMA Y. *et al.*, "Micro-DC/DC converter that integrates planar inductor on power IC", *IEEE Transactions on Magnetics*, vol. 36, no. 5, pp. 3518–3520, 2000.

[NAT 05] NATIONAL SEMICCONDUCTOR PRODUCTS, LMS1585A, Report, 2005.

[NAT 09] NATIONAL SEMICONDUCTOR, LM3218, Datasheet, 2009.

[NEV 14] NEVEU F., MARTIN C., ALLARD B., "Review of high frequency, highly integrated inductive DC–DC converters", *8th International Conference on Integrated Power Systems (CIPS)*, pp. 1–7, February 2014.

[NEV 16] NEVEU F., ALLARD B., MARTIN C., "A review of state-of-the-art and proposal for high frequency inductive step-down DC–DC converter in advanced CMOS", *Journal Analog Integrated Circuits and Signal Processing*, vol. 87, no. 2, pp. 201–211, 2016.

[NG 09] NG V.W., SEEMAN M.D., SANDERS S.R., "Minimum PCB footprint point-of-load DC–DC converter realized with switched-capacitor architecture", *IEEE Energy Conversion Congress and Exposition, ECCE*, pp. 1575–1581, 2009.

[NI 09a] NI J., HONG Z., LIU B.Y., "Improved on-chip components for integrated DC–DC converters in 0.13μm CMOS", *Proceedings of ESSCIRC, 2009*, pp. 448–451, September 2009.

[NI 09b] NI Y., XU J., "A fixed-frequency sliding mode controller for buck converter in continuous conduction mode", *Twenty-Fourth Annual IEEE Applied Power Electronics Conference and Exposition*, IEEE, pp. 794–798, February 2009.

[NIS 05] NISHIJIMA K., HARADA K., NAKANO T. *et al.*, "Analysis of double step-down two-phase buck converter for VRM", *International Telecommunication Conference, INTELEC*, pp. 497–502, 2005.

[NXP 07a] NXP SEMICONDUCTORS, PIP212-12M, Datasheet, 2007.

[NXP 07b] NXP SEMICONDUCTORS, UM10117 PIP212-12M demonstration board, 2007.

[OHN 95] OHNUMA S., FUJIMORI H., FURUKAWA S. *et al.*, "Proceedings of the International Conference of Nitromagnetics Co (N, O)-based granular thin films and their soft magnetic properties", *Journal of Alloys and Compounds*, vol. 222, no. 1, pp. 167–172, 1995.

[ON 09] ON SEMICONDUCTOR CORP., Theory of operation of V^2 controllers, application note AND8276, 2009.

[ON 11] ON SEMICONDUCTOR CORP., NCP3030, Datasheet, 2011.

[ONI 06] ONIZUKA K., KAWAGUCHI H., TAKAMIYA M. *et al.*, "Stacked-chip implementation of on-chip buck converter for power-aware distributed power supply systems", *Solid-State Circuits Conference, 2006, IEEE Asian*, pp. 127–130, November 2006.

[ONI 07] ONIZUKA K., INAGAKI K., KAWAGUCHI H. *et al.*, "Stacked-chip implementation of on-chip buck converter for distributed power supply system in SiPs", *IEEE Journal of Solid-State Circuits*, vol. 42, no. 11, pp. 2404–2410, November 2007.

[ORL 06] ORLANDO B., HIDA R., CUCHET R. *et al.*, "Low-resistance integrated toroidal inductor for power management", *IEEE Transactions on Magnetics*, vol. 42, no. 10, pp. 3374–3376, 2006.

[OSA 98] OSAKA T., TAKAI M., HAYASHI K. *et al.*, "A soft magnetic CoNiFe film with high saturation magnetic flux density and low coercivity", *Nature*, vol. 392, p. 796, 1998.

[OSA 00] OSAKA T., "Electrodeposition of highly functional thin films for magnetic recording devices of the next century", *Electrochimica Acta*, vol. 45, no. 20, pp. 3311–3321, 2000.

[OSA 02] OSAKA T., Method of making a Co–Fe–Ni soft magnetic thin film, US Patent 6,337,007, 8 January 2002.

[OST 14] OSTMAN K., JARVENHAARA J., BROUSSEV S. *et al.*, "A 3.6-to-1.8 V cascode buck converter with a stacked LC filter in 65-nm CMOS", *IEEE Transactions on Circuits and Systems II: Express Briefs*, vol. PP, no. 99, pp. 1–5, 2014.

[PAL 12] PALA V., PENG H., WRIGHT P. *et al.*, "Integrated high-frequency power converters based on GaAs pHEMT: technology characterization and design examples", *IEEE Transactions on Power Electronics*, vol. 27, no. 5, pp. 2644–2656, 2012.

[PAN 01] PANOV Y., JOVANOVIĆ M., "Design considerations for 12-V/1.5-V, 50-A voltage regulator modules", *IEEE Transactions on Power Electronics*, vol. 16, pp. 776–783, 2001.

[PAN 05] PAN T., BALDI A., DAVIES-VENN E. *et al.*, "Fabrication and modeling of silicon-embedded high-Q inductors", *Journal of Micromechanics and Microengineering*, vol. 15, no. 4, p. 849, 2005.

[PAR 98a] PARK J.Y., ALLEN M.G., "Packaging-compatible microinductors and microtransformers with screen-printed ferrite using low temperature processes", *IEEE Transactions on Magnetics*, vol. 34, no. 4, pp. 1366–1368, 1998.

[PAR 98b] PARK J.Y., ALLEN M.G., "Development of magnetic materials and processing techniques applicable to integrated micromagnetic devices", *Journal of Micromechanics and Microengineering*, vol. 8, no. 4, p. 307, 1998.

[PAR 99] PARK J.Y., HAN S., ALLEN M.G., "Batch-fabricated microinductors with electroplated magnetically anisotropic and laminated alloy cores", *IEEE Transactions on Magnetics*, vol. 35, no. 5, pp. 4291–4300, 1999.

[PAR 02] PARK J.-W., CROS F., ALLEN M.G., "A sacrificial layer approach to highly laminated magnetic cores", *The Fifteenth IEEE International Conference on Micro Electro Mechanical Systems*, pp. 380–383, January 2002.

[PAR 03a] PARK J.Y., BU J.U., "Packaging compatible microtransformers on a silicon substrate", *IEEE Transactions on Advanced Packaging*, vol. 26, no. 2, pp. 160–164, 2003.

[PAR 03b] PARK J.-W., ALLEN M., "Ultralow-profile micromachined power inductors with highly laminated Ni/Fe cores: application to low-megahertz DC–DC converters", *IEEE Transactions on Magnetics*, vol. 39, no. 5, pp. 3184–3186, 2003.

[PAR 04] PARK J.-W., CROS F., ALLEN M., "Planar spiral inductors with multilayer micrometer-scale laminated cores for compact-packaging power converter applications", *IEEE Transactions on Magnetics*, vol. 40, no. 4, pp. 2020–2022, 2004.

[PAS 15] PASCAL Y., PILLONNET G., "Efficiency comparison of inductor-, capacitor-, and resonant-based converters fully integrated in CMOS technology", *IEEE Journal on Emerging and Selected Topics in Circuits and Systems*, vol. PP, no. 99, pp. 1–9, 2015.

[PAT 03] PATELLA B.J., PRODIC A., ZIRGER A. *et al.*, "High-frequency digital PWM controller IC for DC–DC converters", *IEEE Transactions on Power Electronics*, vol. 18, no. 1, pp. 438–446, 2003.

[PAT 04] PATOUNAKIS G., LI Y.W., SHEPARD K.L., "A fully integrated on-chip DC–DC conversion and power management system", *IEEE Journal of Solid-State Circuits*, vol. 39, no. 3, pp. 443–451, 2004.

[PEL 03] PELLICONI R., IEZZI D., BARONI A. *et al.*, "Power efficient charge pump in deep submicron standard CMOS technology", *IEEE Journal of Solid-State Circuits*, vol. 38, no. 6, pp. 1068–1071, 2003.

[PEN 13] PENG L., WU R., FANG X. *et al.*, "A novel 3D TSV transformer technology for digital isolator gate driver applications", *25th International Symposium on Power Semiconductor Devices IC's (ISPSD)*, pp. 69–72, May 2013.

[PIL 08] PILAWA-PODGURSKI R., GIULIANO D., PERREAULT D., "Merged two-stage power converter architecture with soft charging switched-capacitor energy transfer", *IEEE Power Electronics Specialists Conference, PESC*, pp. 4008–4015, 2008.

[PIL 15] PILLONNET G., ANDRIEU A., ALON E., "Dual-input switched capacitor converter suitable for wide voltage gain range", *IEEE Journal on Emerging and Selected Topics in Circuits and Systems*, vol. 5, no. 3, pp. 413–420, 2015.

[PIL 16] PILLONNET G., "Modeling and efficiency analysis of multiphase resonant-switched capacitive converters", *IEEE Transactions on Power Electronics*, vol. 31, pp. 11–14, 2016.

[PRA 03] PRABHAKARAN S., SULLIVAN C.R., VENKATACHALAM K., "Measured electrical performance of V-groove inductors for microprocessor power delivery", *IEEE Transactions on Magnetics*, vol. 39, no. 5, pp. 3190–3192, 2003.

[PRA 04] PRABHAKARAN S., SULLIVAN C.R., O'DONNELL T. *et al.*, "Microfabricated coupled inductors for DC–DC converters for microprocessor power delivery", *Power Electronics Specialists Conference*, vol. 6, pp. 4467–4472, June 2004.

[PRA 05] PRABHAKARAN S., SUN Y., DHAGAT P. *et al.*, "Microfabricated V-groove power inductors for high-current low-voltage fast-transient DC–DC converters", *Power Electronics Specialists Conference*, pp. 1513–1519, June 2005.

[PRO 02] PRODIĆ A., MAKSIMOVIĆ D., "Design of a digital PID regulator based on look-up tables for control of high-frequency DC–DC converters", *IEEE Computers in Power Electronics Conference*, pp. 18–22, 2002.

[RAD 10] RADIĆ A., LUKIĆ Z., PRODIĆ A. *et al.*, "Minimum deviation digital controller IC for single and two phase DC–DC switch-mode power supplies", *IEEE Applied Power Electronics Conference and Exposition (APEC)*, pp. 1–6, 2010.

[RAD 11] RADIĆ A., PRODIĆ A., "Buck converter with merged active charge-controlled capacitive attenuation", *IEEE Transactions on Power Electronics*, vol. 27, pp. 1049–1054, 2011.

[RAD 12] RADIĆ A., PRODIĆ A., "Buck converter with merged active charge-controlled capacitive attenuation", *IEEE Transactions on Power Electronics*, vol. 27, no. 3, pp. 1049–1054, 2012.

[RAD 13] RADIĆ A., LUKIĆ Z., PRODIĆ A. *et al.*, "Minimum-deviation digital controller IC for DC–DC switch-mode power supplies", *IEEE Transactions on Power Electronics*, vol. 28, no. 9, pp. 4281–4298, 2013.

[RAD 14] RADIĆ A., STRAKA A., PRODIĆ A., "Low-volume stackable flyback converter with near minimum deviation controller", *IEEE Applied Power Electronics Conference and Exposition (APEC)*, pp. 1948–1953, 2014.

[RAH 06] RAHMAN N., PARAYANDEH A., WANG K. *et al.*, "Multimode digital SMPS controller IC for low-power management", *IEEE International Symposium on Circuits and Systems, ISCAS*, pp. 5327–5330, 2006.

[RAM 10a] RAMADASS Y., FAYED A., HAROUN B. *et al.*, "A 0.16 mm^2 completely on-chip switched-capacitor DC–DC converter using digital capacitance modulation for LDO replacement in 45nm CMOS", *2010 IEEE International Solid-State Circuits Conference (ISSCC)*, pp. 208–209, February 2010.

[RAM 10b] RAMADASS Y.K., FAYED A.A., CHANDRAKASAN A.P., "A fully-integrated switched-capacitor step-down DC–DC converter with digital capacitance modulation in 45 nm CMOS", *IEEE Journal of Solid-State Circuits*, vol. 45, no. 12, pp. 2557-2565, 2010.

[RAS 10] RASHID M.H., *Power Electronics Handbook*, 3rd edition, Elsevier, 2010.

[RED 09] REDL R., "Ripple-based control of switching regulators—an overview", *IEEE Transactions on Power Electronics*, vol. 24, no. 12, pp. 2669–2680, 2009.

[REN 06] REN Y., XU M., ZHOU J. *et al.*, "Analytical loss model of power MOSFET", *IEEE Transactions on Power Electronics*, vol. 21, no. 2, pp. 310–319, 2006.

[RFP 16] RF POWER AMPLIFIER LM 3218 N. S., Report, accessed 2016.

[RIC 04] RICHARD J.F., SAVARIA Y., "High voltage charge pump using standard CMOS technology", *The 2nd Annual IEEE Northeast Workshop on Circuits and Systems*, pp. 317–320, June 2004.

[RID 03] RIDEAU D., SCHEER P., ROY D. *et al.*, "Series resistance estimation and C (V) measurements on ultra thin oxide MOS capacitors", *International Conference on Microelectronic Test Structures*, pp. 191–196, 2003.

[ROB 11] ROBERTSON J., FORNARI R., "New high-K materials for CMOS applications", *Comprehensive Semiconductor Science and Technology*, vol. 4, pp. 132–176, 2011.

[ROD 98] RODRIGUEZ E., GARCIA O., COBOS J. *et al.*, "A single-stage rectifier with PFC and fast regulation of the output voltage", *IEEE Power Electronics Specialists Conference, PESC*, pp. 1642–1648, 1998.

[ROH 16] ROHM PRODUCT B., Report.

[ROM 70] ROMANKIW L., CROLL I., HATZAKIS M., "Batch-fabricated thin-film magnetic recording heads", *IEEE Transactions on Magnetics*, vol. 6, no. 3, pp. 597–601, 1970.

[ROS 08] ROSA E.B., "The self and mutual inductances of linear conductors", *Bulletin of the Bureau of Standards*, vol. 4, no. 2, pp. 301–343, 1908.

[SAG 04] SAGGINI S., GHIONI M., GERACI A., "An innovative digital control architecture for low-voltage, high-current DC–DC converters with tight voltage regulation", *IEEE Transactions on Power Electronics*, vol. 19, pp. 210–218, 2004.

[SAN 07] SANO I., USUI Y., SEKI T., "A 2nd generation micro DC–DC converter", *Fuji Electric Review*, vol. 53, pp. 89–92, 2007.

[SAN 11] SANTA T., AUER M., SANDNER C. *et al.*, "Switched capacitor DC–DC converter in 65nm CMOS technology with a peak efficiency of 97%", *IEEE International Symposium on Circuits and Systems, ISCAS*, pp. 1351–1354, 2011.

[SAN 13] SANDERS S., ALON E., LE H.-P. *et al.*, "The road to fully integrated DC:DC conversion via the switched-capacitor approach", *IEEE Transactions on Power Electronics*, vol. 28, no. 9, pp. 4146–4155, 2013.

[SAS 97] SASAKI Y., MORITA S., HATANAI T. *et al.*, "Viewpoint set of selected papers from a Special Symposium on 4th International Union of Materials Research Societies (IUMRS) International Conference High frequency soft magnetic properties of nanocrystalline Fe–(Co)–Hf–O films with high electrical resistivity and their application to micro DC–DC converter", *Nanostructured Materials*, vol. 8, no. 8, pp. 1025–1032, 1997.

[SAS 00] SASAKI Y., KOMAI E., HATANAI T. *et al.*, "Thin film inductor for microswitching converters using a high resistive Fe–Zr–O film", *Proceedings of the International Power Electronics Conference, IPEC'00, Tokyo*, pp. 315–319, 2000.

[SAT 96] SATO T., INOUE T., TOMITA H. *et al.*, "5 MHz switching micro DC–DC converter using planar inductor", *18th International Telecommunications Energy Conference*, pp. 485–490, October 1996.

[SAT 97] SATO T., KOMAI E., YAMASAWA K. *et al.*, "Application of nanocrystalline Fe(or Co–Fe)–Hf–O magnetic films with high electrical resistivity to micro DC–DC converters", *IEEE Transactions on Magnetics*, vol. 33, no. 5, pp. 3310–3312, 1997.

[SAT 00] SATO T., YAMASAWA K., TOMITA H. *et al.*, "Planar power inductor using FeCoBN magnetic film with high saturation magnetization and high electrical resistivity", *Proceedings of the International Power Electronics Conference*, pp. 303–308, 2000.

[SAT 04] SATO F., ONO T., WAKO N. *et al.*, "All-in-one package ultracompact micropower module using thin-film inductor", *IEEE Transactions on Magnetics*, vol. 40, no. 4, pp. 2029–2031, 2004.

[SCH 04] SCHROM G., HAZUCHA P., HAHN J.-H. *et al.*, "Feasibility of monolithic and 3D-stacked DC–DC converters for microprocessors in 90nm technology generation", *Proceedings of the 2004 International Symposium on Low Power Electronics and Design*, pp. 263–268, 2004.

[SCH 07] SCHROM G., HAZUCHA P., PAILLET F. *et al.*, "A 100MHz eight-phase buck converter delivering 12A in 25 mm^2 using air-core inductors", *Applied Power Electronics Conference, APEC 2007*, pp. 727–730, February 2007.

[SCH 15] SCHAEF C., STAUTH J.T., "A 3-phase resonant switched capacitor converter delivering 7.7 W at 85% efficiency using 1.1 nH PCB trace inductors", *IEEE Journal of Solid-State Circuits*, vol. 50, no. 12, pp. 2861–2869, 2015.

[SEE 08a] SEEMAN M.D., SANDERS S.R., RABAEY J.M., "An ultra-low-power power management IC for energy-scavenged wireless sensor nodes", *IEEE Power Electronics Specialists Conference, PESC*, pp. 925–931, 2008.

[SEE 08b] SEEMAN M.D., SANDERS S.R., "Analysis and optimization of switched-capacitor DC–DC converters", *IEEE Transactions on Power Electronics*, vol. 23, no. 2, pp. 841–851, 2008.

[SEE 09] SEEMAN M.D., A design methodology for switched-capacitor DC–DC converters, Report, DTIC document, BWRC, UC Berkeley, 2009.

[SEE 10] SEEMAN M.D., NG V.W., LE H.-P. *et al.*, "A comparative analysis of switched-capacitor and inductor-based DC–DC conversion technologies", *12th IEEE Workshop on Control and Modeling for Power Electronics (COMPEL)*, pp. 1–7, 2010.

[SER 05] SERNEELS B., PIESSENS T., STEYAERT M. *et al.*, "A high-voltage output driver in a 2.5-V 0.25- mu;m CMOS technology", *IEEE Journal of Solid-State Circuits*, vol. 40, no. 3, pp. 576–583, 2005.

[SER 07] SERNEELS B., STEYAERT M., DEHAENE W., "A 237mW aDSL2+ CO line driver in standard 1.2V 0.13μm CMOS", *2007 IEEE International Solid-State Circuits Conference, Digest of Technical Papers*, pp. 524–619, February 2007.

[SER 12] SERNEELS B., GEUKENS E., MUER B.D. *et al.*, "A 1.5W 10V-output Class-D amplifier using a boosted supply from a single 3.3V input in standard 1.8V/3.3V 0.18μm CMOS", *2012 IEEE International Solid-State Circuits Conference*, pp. 94–96, February 2012.

[SEU 15] SEUNGBUM L., RANSON J., OTTEN D. *et al.*, "Two-stage power conversion architecture suitable for wide range input voltage", *IEEE Transactions on Power Electronics*, vol. 30, no. 2, pp. 805–816, 2015.

[SHA 08] SHARAFI N., NEKOUI M.A., "Analyzing optimization of fluxgate sensor", *International Conference on Electronic Design*, pp. 1–4, December 2008.

[SHE 07] SHEN Z.J., LU J., CHENG X. *et al.*, "On-chip bondwire inductor with ferrite-epoxy coating: a cost-effective approach to realize power systems on chip", *2007 IEEE Power Electronics Specialists Conference*, pp. 1599–1604, June 2007.

[SHO 14] SHOUSHA M., MCRAE T., PRODIĆ A. *et al.*, "Assisting converter based integrated battery management system for low power applications", *IEEE Applied Power Electronics Conference and Exposition (APEC)*, pp. 1579–1583, 2014.

[SIM 71] SIMMONS G., "Conduction in thin dielectric films", *Journal of Physics D: Applied Physics*, vol. 4, pp. 613–657, 1971.

[SOM 06] SOMASUNDARAM M.N., MA D., "Integrated low-ripple-voltage fast-response switched-capacitor power converter with interleaving regulation scheme", *IEEE International Symposium on Circuits and Systems, ISCAS*, 2006.

[SOM 09] SOMASEKHAR D., SRINIVASAN B., PANDYA G. *et al.*, "Multi-phase 1GHz voltage doubler charge-pump in 32nm logic process", *Symposium on VLSI Circuits*, pp. 196–197, June 2009.

[SON 14] SONG M.K., DEHGHANPOUR M., SANKMAN J. *et al.*, "A VHF-level fully integrated multi-phase switching converter using bond-wire inductors, on-chip decoupling capacitors and DLL phase synchronization", *Twenty-Ninth Annual IEEE Applied Power Electronics Conference and Exposition (APEC)*, pp. 1422–1425, March 2014.

[SOT 05] SOTO A., ALOU P., COBOS J., "Design methodology for dynamic voltage scaling in the buck converter", *IEEE Applied Power Electronics Conference and Exposition (APEC)*, 2005.

[SOT 07] SOTO A., CASTRO A.D., ALOU P. *et al.*, "Analysis of the buck converter for scaling the supply voltage of digital circuits", *IEEE Transactions on Power Electronics*, vol. 22, no. 6, pp. 2432–2443, 2007.

[STA 01] STARZYK J.A., JAN Y.-W., QIU F., "A DC–DC charge pump design based on voltage doublers", *IEEE Transactions on Circuits and Systems I: Fundamental Theory and Applications*, vol. 48, no. 3, pp. 350–359, 2001.

[STA 13] STAUTH J., SEEMAN M., KESARWANI K., "Resonant switched-capacitor converters for sub-module distributed photovoltaic power management", *IEEE Transactions on Power Electronics*, vol. 28, no. 3, pp. 1189–1198, 2013.

[STM 13] STM PRODUCTS LD1580XX L. K., Report, 2013.

[STR 99] STRATAKOS A., High-efficiency, low-voltage DC–DC conversion for portable applications, PhD Thesis, University of California, Berkeley, 1999.

[STU 13] STURCKEN N., O'SULLIVAN E.J., WANG N. *et al.*, "A 2.5D integrated voltage regulator using coupled-magnetic-core inductors on silicon interposer", *IEEE Journal of Solid-State Circuits*, vol. 48, no. 1, pp. 244–254, 2013.

[SU 08a] SU F., KI W.-H., TSUI C.-Y., "Ultra fast fixed-frequency hysteretic buck converter with maximum charging current control and adaptive delay compensation for DVS applications", *IEEE Journal of Solid-State Circuits*, vol. 43, pp. 815–822, 2008.

[SU 08b] SU L., MA D., BROKAW A.P., "A monolithic step-down SC power converter with frequency-programmable subthreshold z-domain DPWM control for ultra-low power microsystems", *Solid-State Circuits Conference, 2008*, pp. 58–61, September 2008.

[SUL 93] SULLIVAN C.R., SANDERS S.R.,"Microfabrication of transformers and inductors for high frequency power conversion", *IEEE Power Electronics Specialists Conference*, pp. 33–33, 1993.

[SUL 96] SULLIVAN C.R., SANDERS S.R., "Design of microfabricated transformers and inductors for high-frequency power conversion", *IEEE Transactions on Power Electronics*, vol. 11, no. 2, pp. 228–238, 1996.

[SUL 09] SULLIVAN C.R., "Integrating magnetics for on-chip power: Challenges and opportunities", *2009 IEEE Custom Integrated Circuits Conference*, pp. 291–298, September 2009.

[SUL 13] SULLIVAN C.R., HARBURG D., QIU J. *et al.*, "Integrating magnetics for on-chip power: A perspective", *IEEE Transactions on Power Electronics*, vol. 28, no. 9, pp. 4342–4353, 2013.

[SUN 06] SUN J., MING X., YUCHENG Y. *et al.*, "High power density, high efficiency system two-stage power architecture for laptop computers", *IEEE Power Electronics Specialists Conference, PESC*, pp. 1–7, 2006.

[SUN 07a] SUN J., XU M., LEE F.C., "Transient analysis of the novel voltage divider", *IEEE Applied Power Electronics Conference and Exposition (APEC)*, pp. 550–556, 2007.

[SUN 07b] SUN J., LU J.-Q., GIULIANO D. *et al.*, "3D power delivery for microprocessors and high-performance ASICs", *Applied Power Electronics Conference, APEC 2007*, pp. 127–133, 2007.

[SVI 14] SVIKOVIC V., CORTES J., ALOU P. *et al.*, "Optimization and analysis of PwrSoC Buck converter with integrated passives for automotive application", *Energy Conversion Congress and Exposition (ECCE)*, pp. 2164–2171, September 2014.

[SYE 04] SYED X., AHMED E., MAKSIMOVIC D. *et al.*, "Digital pulse width modulator architectures", *IEEE Power Electronics Specialists Conference, PESC*, pp. 4689–4695, 2004.

[TAN 05] TAN S.-C., LAI Y., CHEUNG M. *et al.*, "On the practical design of a sliding mode voltage controlled buck converter", *IEEE Transactions on Power Electronics*, vol. 20, no. 2, pp. 425–437, 2005.

[TEX 07] TEXAS INSTRUMENTS CORP., TPS40180, Datasheet, 2007.

[TEX 11] TEXAS INSTRUMENTS CORP., D-CAP Mode With All-Ceramic Output Capacitor Application, Application Report SLVA453, 2011.

[TEX 14] TEXAS INSTRUMENTS CORP., Texas Instruments Multimedia Tablet Reference Design, 2014.

[TEX 16] TEXAS INSTRUMENTS CORP., TPS82671, Datasheet, 2016.

[TIP 16] TI PRODUCTS T., Report, accessed 2016.

[TIP 05] TIPEK A., O'DONNELL T., RIPKA P. *et al.*, "Excitation and temperature stability of PCB fluxgate sensor", *IEEE Sensors Journal*, vol. 5, no. 6, pp. 1264–1269, 2005.

[TOM 96] TOMITA H., INOUE T., MIZOGUCHI T.,"Hetero-amorphous Fe-Co-B-C soft magnetic thin films with uniaxial magnetic anisotropy and large magnetostriction", *IEEE Transactions on Magnetics*, vol. 32, no. 5, pp. 4529–4531, 1996.

[TOP 14] TOPRAK-DENIZ Z., SPERLING M., BULZACCHELLI J. *et al.*, "5.2 Distributed system of digitally controlled microregulators enabling per-core DVFS for the POWER8TM microprocessor", *Solid-State Circuits Conference Digest of Technical Papers*, pp. 98–99, February 2014.

[TRO 05] TROCHUT S., Contribution à l'étude de stabilité des convertisseurs à découpage monolithiques. Application à la téléphonie mobile, PhD Thesis, INSA Lyon, 2005.

[ULR 00] ULRICH R.U. *et al.*, "Comparison of paraelectric and ferroelectric materials for applications as dielectrics in thin film integrated capacitors", *The International Journal of Microcircuits and Electronic Packaging*, vol. 23, no. 2, 2000.

[UTK 09] UTKIN V., GULDNER J., SHI J. *Sliding Mode Control in Electro-Mechanical Systems*, CRC Press, 2009.

[UZU 14] UZUN O.A., KOSE S., "Converter-gating: a power efficient and secure on-chip power delivery system", *IEEE Journal on Emerging and Selected Topics in Circuits and Systems*, vol. 4, no. 2, pp. 169–179, 2014.

[VAI 13] VAISBAND I., FRIEDMAN E.G., "Heterogeneous methodology for energy efficient distribution of on-chip power supplies", *IEEE Transactions on Power Electronics*, vol. 28, no. 9, pp. 4267–4280, 2013.

[VAL 69] VALSTYN E., KOSY D., "The write field of a magnetic-film recording head", *IEEE Transactions on Magnetics*, vol. 5, no. 3, pp. 442–445, 1969.

[VAN 10] VAN BREUSSEGEM T., STEYAERT M., "A fully integrated gearbox capacitive DC/DC-converter in 90 nm CMOS: Optimization, control and measurements", *Control and Modeling for Power Electronics (COMPEL), 2010 IEEE 12th Workshop on*, pp. 1–5, June 2010.

[VIE 11] DEL VIEJO M., ALOU P., OLIVER J. *et al.*, "V2IC control: A novel control technique with very fast response under load and voltage steps", *Applied Power Electronics Conference and Exposition*, pp. 231–237, March 2011.

[VIL 03] VILLAR-PIQUE G., ALARCON E., GUINJOAN F. *et al.*, "Optimized design of MOS capacitors in standard CMOS technology and evaluation of their equivalent series resistance for power applications", *IEEE International Symposium on Circuits and Systems, ISCAS*, vol. III, p. 451, 2003.

[VIL 08] VILLAR G., ALARCON E., "Monolithic integration of a 3-level DCM-operated low-floating-capacitor buck converter for DC–DC step-down donversion in standard CMOS", *Power Electronics Specialists Conference, IEEE*, pp. 4229–4235, June 2008.

[VIL 12] VILLAR-PIQUE G., "A 41-phase switched-capacitor power converter with 3.8 mV output ripple and 81% efficiency in baseline 90 nm CMOS", *Solid-State Circuits Conference Digest of Technical Papers*, pp. 98–100, February 2012.

[VIL 13] VILLAR-PIQUE G., BERGVELD H., ALARCON E.,"Survey and benchmark of fully integrated switching power converters: Switched-capacitor versus inductive approach", *IEEE Transactions on Power Electronics*, vol. 28, no. 9, pp. 4156–4167, 2013.

[VIS 05] VISHAY, FX5545G108, Datasheet, 2005.

[VOI 14] VOIGTLANDER P., "Capacitive current-mode control of a DC/DC converter", U.S. Patent US8 736 244 B1, May 2014.

[VOI 15] VOIRON F., FOURNAUD L., "Silicon high-density capacitors for power decoupling applications", *IEEE International Workshop on Integrated Power Packaging*, pp. 48–51, 2015.

[VUK 15] VUKADIOVIC N., PRODIC A., MIWA B. *et al.*, "Skip-duty control method for minimizing switching stress in low-power multi-level DC–DC converters", *IEEE Computers in Power Electronics Conference (COMPEL)*, 2015.

[VUK 16] VUKADIOVIC N., PRODIC A., MIWA B. *et al.*, "Extended wide-load range model for multi-level DC–DC converters and a practical dual-mode digital controller", *IEEE Applied Power Electronics Conference and Exposition (APEC)*, 2016.

[WAL 13] WALDRON F., FOLEY R., SLOWEY J. *et al.*, "Technology roadmapping for power supply in package (PSiP) and power supply on chip (PwrSoC)", *IEEE Transactions on Power Electronics*, vol. 28, no. 9, pp. 4137–4145, 2013.

[WAN 05] WANG N., O'DONNELL T., ROY S. *et al.*,"High-frequency micro-machined power inductors", *Journal of Magnetism and Magnetic Materials*, vol. 290–291, Part 2, pp. 1347–1350, 2005.

[WAN 07a] WANG N., O'DONNELL T., ROY S. *et al.*, "Thin film microtransformer integrated on silicon for signal isolation", *IEEE Transactions on Magnetics*, vol. 43, no. 6, pp. 2719–2721, 2007.

[WAN 07b] WANG N., O'DONNELL T., ROY S. *et al.*, "Micro-inductors integrated on silicon for power supply on chip", *Journal of Magnetism and Magnetic Materials*, vol. 316, no. 2, pp. 233–237, 2007.

[WAN 08] WANG N., O'DONNELL T., MEERE R. *et al.*,"Thin-film-integrated power inductor on Si and its performance in an 8 MHz buck converter", *IEEE Transactions on Magnetics*, vol. 44, no. 11, pp. 4096–4099, 2008.

[WAN 10a] WANG M., LI J., NGO K.D. *et al.*,"A novel integrated power inductor in silicon substrate for ultra-compact power supplies", *Applied Power Electronics Conference and Exposition (APEC)*, IEEE, pp. 2036–2041, 2010.

[WAN 10b] WANG N., HANNON J., FOLEY R. *et al.*, "Integrated magnetics on silicon for power supply in package (PSiP) and power supply on chip (PwrSoC)", *Electronic System-Integration Technology Conference (ESTC)*, pp. 1–6, September 2010.

[WAN 13] WANG N., BARRY J., HANNON J. *et al.*, "High frequency DC–DC converter with co-packaged planar inductor and power IC", *IEEE 63rd Electronic Components and Technology Conference (ECTC)*, pp. 1946–1952, 2013.

[WAN 15] WANG N., MIFTAKHUTDINOV R., KULKARNI S. *et al.*, "High efficiency on Si-integrated microtransformers for isolated power conversion applications", *IEEE Transactions on Power Electronics*, vol. 30, no. 10, pp. 5746–5754, 2015.

[WEN 08] WENS M., STEYAERT M., "A fully-integrated 0.18 m CMOS DC–DC step-down converter, using a bondwire spiral inductor", *Custom Integrated Circuits Conference, 2008. IEEE*, pp. 17–20, 2008.

[WEN 11] WENS M., STEYAERT M., "A fully integrated CMOS 800-mW four-phase semiconstant ON/OFF-time step-down converter", *IEEE Transactions on Power Electronics*, vol. 26, no. 2, pp. 326–333, 2011.

[WEN 14] WEN S., "Cell balancing buys extra run time and battery life", *Texas Instruments Analog Applications Journal*, 2014.

[WES 00] WESSELING E., BRANDON E., LIENEWEG U., "AC and DC current dependence of on-chip inductors", *Proceedings of International Symposium on Magnetic Materials, Processes and Devices, Electrochemical Society*, pp. 547–558., October 2000.

[WIB 08] WIBBEN J., HARJANI R., "A high-efficiency DC–DC converter using 2 nH integrated inductors", *IEEE Journal of Solid-State Circuits*, vol. 43, no. 4, pp. 844–854, 2008.

[WU 01] WU A.M., SANDERS S.R., "An active clamp circuit for voltage regulation module (VRM) applications", *IEEE Transactions on Power Electronics*, vol. 16, no. 5, pp. 623–634, 2001.

[WU 11] WU R., SIN J.K., "A novel silicon-embedded coreless inductor for high-frequency power management applications", *IEEE Electron Device Letters*, vol. 32, no. 1, pp. 60–62, 2011.

[WU 13] WU H., PhD Thesis, Arizona State University, 2013.

[XIA 04] XIAO J., PETERCHEV A., ZHANG J. *et al.*, "A 4-μA quiescent-current dual-mode digitally controlled buck converter IC for cellular phone applications", *IEEE Journal of Solid-State Circuits*, vol. 39, pp. 2342–2348, 2004.

[XIN 13] XING X., SUN N.X., CHEN B.,"High-bandwidth low-insertion loss solenoid transformers using FeCoB multilayers", *IEEE Transactions on Power Electronics*, vol. 28, no. 9, pp. 4395–4401, 2013.

[YAM 93] YAMAGUCHI K., OHNUMA S., IMAGAWA T. *et al.*,"Characteristics of a thin film microtransformer with circular spiral coils", *IEEE Transactions on Magnetics*, vol. 29, no. 5, pp. 2232–2237, 1993.

[YAN 13] YAN Y., LEE F.C., MATTAVELLI P. *et al.*, "Small signal analysis of V^2 control using current mode equivalent circuit model", *IEEE Applied Power Electronics Conference and Exposition (APEC)*, pp. 1709–1716, 2013.

[YAO 11] YAO D., LEVEY C.G., SULLIVAN C.R.,"Microfabricated V-groove power inductors using multilayer Co-Zr-O thin films for very-high-frequency DC–DC converters", *2011 IEEE Energy Conversion Congress and Exposition*, pp. 1845–1852, September 2011.

[YEA 14] YEATES K., YEHUI H., "Quasi-parallel switched-capacitor and regulating PWM DC–DC converter", *IEEE Applied Power Electronics Conference and Exposition (APEC)*, pp. 2015–2110, 2014.

[YEU 04] YEUNG Y.P.B., CHENG K.W.E., HO S.L. *et al.*, "Unified analysis of switched-capacitor resonant converters", *IEEE Transactions on Industrial Electronics*, vol. 51, no. 4, pp. 864–873, 2004.

[YOU 08] YOUSEFZADEH V., BABAZADEH A., RAMACHANDRAN B. *et al.*,"Proximate time-optimal digital control for synchronous buck DC–DC converters", *IEEE Transactions on Power Electronics*, vol. 23, pp. 2018–2026, 2008.

[YUN 13] YUN J., YEO T., PARK J., "High efficiency active cell balancing circuit with soft-switching technique for series-connected battery string", *IEEE Applied Power Electronics Conference and Exposition (APEC)*, pp. 3301–3304, 2013.

[ZEN 10] ZENG Z., YE X., FENG Z. *et al.*,"Tradeoff analysis and optimization of power delivery networks with on-chip voltage regulation", *Design Automation Conference (DAC)*, pp. 831–836, June 2010.

[ZHA 99] ZHANG Y.E., SANDERS S.R., "In-board magnetics processes", *Power Electronics Specialists Conference*, vol. 1, pp. 561–567, 1999.

[ZHA 04] ZHANG Y.E., SANDERS S., "Design, manufacture and application of in-board magnetic devices", *Power Electronics and Motion Control Conference*, vol. 3, pp. 1791–1798, 2004.

[ZHA 08] ZHAO Z., PRODIĆ A., "Continuous-time digital controller for high-frequency DC–DC converters", *IEEE Transactions on Power Electronics*, vol. 23, no. 564–573, 2008.

[ZHA 13] ZHAO J., YEATES K., HAN Y., "Analysis of high efficiency DC/DC converter processing partial input/output power", *IEEE Computers in Power Electronics Conference (COMPEL)*, pp. 1–8, 2013.

[ZHU 96] ZHU G., IOINOVICI A., "Switched-capacitor power supplies: DC voltage ratio, efficiency, ripple, regulation", *IEEE International Symposium on Circuits and Systems, ISCAS*, vol. 1, pp. 553–556, 1996.

List of Acronyms

AVP	Adaptive Voltage Positioning
AW	Active Well
BJT	Bipolar Junction Transistor
CAD	Computer-Aided Design
CCM	Continuous Conduction Mode
CMC	Current-Mode Control
CMOS	Complementary Metal-Oxide-Semiconductor
DCM	Discontinuous Conduction Mode
DRC	Design Rule Check
DSP	Digital Signal Processor
DVS	Dynamic Voltage Scaling
ECU	Engine Control Unit
EEF	Efficiency Enhancement Factor
EMI	ElectroMagnetic Interference
ESD	Electro-Static Discharge
ESL	Equivalent Series Inductance
ESR	Equivalent Series Resistance
FC	Flying Capacitor
FFB	Fast FeedBack path
FIVR	Fully Integrated Voltage Regulator

FoM	Figure of Merit
FPGA	Field Programmable Gate Array
IC	Integrated Circuit
I/O	Inputs/Outputs
IoT	Internet of Things
JFET	Junction Field-Effect Transistor
LDMOSFET	Laterally-Diffused MOSFET
LDO	Low Drop-Out
LRF	Losses Reduction Factor
MIM	Metal-Insulator-Metal
MLCC	Multi-Layer Ceramic Capacitor
MOM	Metal-Oxide-Metal
MOS	Metal-Oxide-Semiconductor
MOSFET	Metal-Oxide-Semiconductor Field-Effect Transistor
PCB	Printed Circuit Board
PCM	Peak Current-mode Control
PDN	Power Delivery Network
PFM	Pulse Frequency Modulation
PICS	Passive Integrated Connecting Substrate
PLS	Post Layout Simulation
PMIC	Power Management Integrated Circuit (or Power Module IC)
PMU	Power Management Unit
PowerSiP	Power Supply in Package
PowerSoC	Power System on Chip
PWM	Pulse Width Modulation
RMS	Root Mean Square
RPM	Ramp Pulse Modulation
SC	Switched-Capacitor

SCR	Space-Charge Region
SFB	Slow FeedBack path
SiP	System in Package
SMD	Surface Mounting Device
SMPS	Switched-Mode Power Supply
SMU	Source-Measurement Unit
SoC	System on Chip
SPICE	Simulation Program With Integrated Circuit Emphasis
SRF	Self-Resonant Frequency
VMC	Voltage Mode Control
VNA	Vector Network Analyzer
VRM	Voltage Regulation Module
ZCS	Zero Current Switching
ZVS	Zero Voltage Switching

List of Authors

Sheikh Mohammad AHSANUZZAMAN
University of Toronto
Toronto
Canada

Bruno ALLARD
INSA Lyon
Lyon
France

Pedro ALOU
CEI
Madrid
Spain

José A. COBOS
CEI
Madrid
Spain

Santosh KULKARNI
Tyndall National Institute
Cork
Ireland

Benoît LABBE
INSA Lyon
Lyon
France

Behzad MAHDAVIKHAH
University of Toronto
Toronto
Canada

Christian MARTIN
University of Lyon
Lyon
France

Timothy MCRAE
University of Toronto
Toronto
Canada

Florian NEVEU
University of Lyon
Lyon
France

Jesus A. OLIVER
CEI
Madrid
Spain

Cian OMATHUNA
Tyndall National Institute
Cork
Ireland

Gael PILLONNET
CEA-LETI
Grenoble
France

Aleksander PRODIC
University of Toronto
Toronto
Canada

Aleksandar RADIC
University of Toronto
Toronto
Canada

Thomas SOUVIGNET
INSA Lyon
Lyon
France

Frédéric VOIRON
IPDiA
Caen
France

Index

S, T, V

Other titles from

iSTE

in

Energy

2016

ANDRÉ Michel, SAMARAS Zissis
Energy and Environment
(Research for Innovative Transports Set - Volume 1)

2015

CROGUENNEC Laurence, MONCONDUIT Laure, DEDRYVÈRE Rémi
Electrodes for Li-ion Batteries
(Energy Storage – Batteries, Supercapacitors Set – Volume 2)

LEPRINCE-WANG Yamin
Piezoelectric ZnO Nanostructure for Energy Harvesting
(Nanotechnologies for Energy Recovery Set – Volume 1)

ROBYNS Benoît, FRANÇOIS Bruno, DELILLE Gauthier, SAUDEMONT Christophe
Energy Storage in Electric Power Grids

ROSSI Carole
Al-based Energetic Nanomaterials
(Nanotechnologies for Energy Recovery Set – Volume 2)

TARASCON Jean-Marie, SIMON Patrice
Electrochemical Energy Storage
(Energy Storage – Batteries, Supercapacitors Set – Volume 1)

2013

LALOUI Lyesse and DI DONNA Alice
Energy Geostructures: Innovation in Underground Engineering

2012

BECKERS Benoit
Solar Energy at Urban Scale

ROBYNS Benoît, DAVIGNY Arnaud, FRANÇOIS Bruno, HENNETON Antoine, SPROOTEN Jonathan
Electricity Production from Renewable Energies

2011

GAO Fei, BLUNIER Benjamin, MIRAOUI Abdellatif
Proton Exchange Membrane Fuel Cell Modeling

MULTON Bernard
Marine Renewable Energy Handbook

2010

BRUNET Yves
Energy Storage

2009

SABONNADIERE Jean-Claude
Low Emission Power Generation Technologies and Energy Management

SABONNADIERE Jean-Claude
Renewable Energy Technologies

Printed and bound by CPI Group (UK) Ltd, Croydon, CR0 4YY

15/09/2026

14962356-0003